# Calculus the Maple® Way

**Second Edition**

ROBERT B. ISRAEL

DEPARTMENT OF MATHEMATICS
UNIVERSITY OF BRITISH COLUMBIA



Addison-Wesley

An imprint of Addison Wesley Longman Ltd.

Don Mills, Ontario • Reading, Massachusetts • Harlow, England
Melbourne, Australia • Amsterdam, The Netherlands • Bonn, Germany

Publisher: Ron Doleman
Managing Editor: Linda Scott
Production and Manufacturing Coordinator: Alexandra Odulak
Cover Design: Anthony Leung
Printing and Binding: Maracle Press



**Canadian Cataloguing in Publication Data**

Israel, Robert B.
Calculus: the Maple way

2nd ed.
ISBN 0-201-61383-2
Includes index.

1. Maple (Computer file). 2. Calculus—Data processing. I. Title.

QA303.5.D37I87 1999 515'.0285'5369 C99-931354-1



ISBN 0-201-61383-2 ✓

Printed and bound in Canada.

A B C D E -MP- 03 02 01 00 99

# Preface

*Calculus: The Maple Way* is intended to be used in integrating the computer algebra system Maple into a calculus course. This powerful tool has a number of potential benefits. Its graphics capabilities, including animation, can be especially helpful in enhancing the understanding of the concepts of calculus. The active involvement of the student in producing the Maple graphics, rather than just looking at a diagram printed in a textbook or one sketched on a blackboard, is beneficial for learning. Moreover, there is the opportunity for exploration, to try out different functions or "zoom in" on an area of interest, or in three dimensions to look at an object from different directions. Maple's capability for symbolic computation can reduce the amount of time spent in routine algebraic manipulation. And it greatly increases the range of problems that can be tackled, especially with its facilities for numerical computation to handle problems that have no "closed-form" solution.

To obtain these benefits requires a significant investment of time and effort in learning to use Maple. A one-hour lab session every second week is the bare minimum, with students also having access to Maple on their own time (either in an open lab or on their own personal computers). Two hours per week would be better. This book is organized into 30 labs, covering the material of a four-semester calculus course. As a general rule, each lab is one week's work, but there are exceptions. In particular, Lab 1 has a lot of material, covering the basic operation of the Maple system. It need not be done all at once; the student can be referred back to its later sections as needed. However, Labs 1 and 2 are very important, and you should resist the temptation to jump directly into limits and derivatives without them. Many sections or whole labs can be skipped, depending on what is appropriate for a particular course. The order of the labs can be changed as well. Appendix IV shows the correspondence between the labs and sections of several current calculus texts. It is important for classroom instruction on the theoretical aspects and concepts involved to precede the lab work. However, it is probably impossible to coordinate the classroom and the lab precisely, and at times the lab will be dealing with material that was seen in the classroom several weeks before. The exercises at the end of each lab are at various levels of difficulty, some completely routine and some that will challenge the best students. It is not intended that every student should do every exercise in the lab. Some might be turned into more extended "projects".

The following sections of the book contain what might be called "core" material, which is especially important to cover:

- Labs 1 and 2
- Labs 4 and 5
- "Linear Approximation" in Lab 7
- Lab 9
- Lab 14
- "Sums in Maple" in Lab 15
- "Taylor Series" in Lab 16
- Labs 17 through 19
- Labs 23 and 24
- Labs 26 through 28.

Everything else can be considered as optional, at the instructor's discretion. In particular, Example 5 in Lab 6 and the sections "Error Bounds for Newton's Method" in Lab 7, "Proving Inequalities" in Lab 10, "Upper and Lower Sums" in Lab 11, "Romberg Integration" in Lab 13, and "Approximating the Sum" in Lab 15 might be reserved for an "honours" class. The material of Lab 8 is not covered in most "traditional" calculus courses, but I would hope that a place could be found for it — this provides a very nice application of the Chain Rule, and a connection to areas of active research with practical applications. Lab 12 can be omitted if the instructor believes that integration techniques beyond simple substitution should be entirely left up to the computer (or, on the other hand, that the student should integrate entirely by hand). In Lab 25 the instructor should probably either make a choice between "The Skier" and "Curvature, Torsion, and the Frenet Frame", or spend more than one lab session on these topics. I doubt that many classes will do Lab 30 as a single lab: it is included for those courses that incorporate material on differential equations, and its sections can be done at different times when this material is introduced.

The student is not required to have prior experience with computers. The instructor should provide some basic information on logging in (if necessary), printing, editing, and transferring files, which will differ from one installation to another. Knowledge of computer programming is not necessary (nor will it be found in *Calculus: The Maple Way*). I have kept the use of control structures to an absolute minimum — a few "`for`" loops and one "`if`". Fortunately, a lot can be accomplished in Maple by stringing a few commands together. There are several cases in which I have programmed some useful commands, and these can be read into Maple from some small files that should be made available to the student. The files (only about 5 kilobytes in all) can be obtained at the URL

`http://www.geocities.com/CapeCanaveral/Campus/7705`

The instructor need not be an expert on Maple, but a fair degree of familiarity with it is advisable in order to be able to answer the students' questions. Besides the many books on the subject, the Usenet newsgroup `sci.math.symbolic` and the Maple Users' Group are good sources of information and answers to questions. Among the many web sites containing information about Maple, I might mention my site, the Maple Advisor Database at

`http://www.math.ubc.ca/~israel/advisor`

which contains answers to many common questions about Maple, explanations of error messages, bug fixes, and some useful Maple procedures.

This book covers Maple V, Releases 4, 5, and 5.1, both Student and Professional editions. I would recommend that you use the most recent release available. *Calculus: The Maple Way* can be used with Maple on any platform that supports good-quality graphics.

In nearly all cases, I end a Maple command with "`;`", which causes Maple to put the result of the command on the screen, rather than "`:`" which does not show its result. This includes many times when I don't bother to reproduce that result in the book. It is a good idea to use "`;`" as much as possible, even when you don't think you need to see the result, because it is helpful in catching mistakes: that result may not be what you expected. The main exception is in assigning graphics to a variable, where "`;`" would result in several screens full of uninteresting output.

The figures in *Calculus: The Maple Way* were produced by Maple as PostScript files, with a few exceptions produced using MG Graphics software (developed by my colleague Robert A. Adams and myself). Some have been edited slightly to improve positioning of labels. When a Maple command produces one of the figures, I indicate that fact, e.g.,

```
> plot(x^2 - 2 * x + 2, x=-2..2); (Figure 1.1)
```

## To the Student

*Calculus: The Maple Way* is intended to be read beside the computer. When you see some Maple commands,

```
> in this style of type,
```

you should enter the commands and observe the results. Mathematics is not a spectator sport: you learn by doing it. The worked examples in the text show how to use the various Maple commands, and the strategies employed in them can often be used in the exercises at the end of the lab.

Maple is a powerful tool, but it is not a substitute for thought. You should think about what we are trying to do, and the possible strategies for achieving it. My method is often not the only one that works. Please feel free to experiment and try various alternative approaches. After all, this is not a chemistry lab—nothing will explode if you try to do things differently!

Appendix III contains answers to most of the odd-numbered exercises. However, you should remember that these are not complete solutions, and that sometimes answers may vary slightly (especially in the last one or two decimal places of a numerical answer) depending on the version of Maple and the details of how the answer was obtained. The method of obtaining the answer is usually more important than the answer itself.

This "dangerous curve" sign indicates particularly tricky aspects of Maple, where you should pay close attention to avoid mistakes.

You will certainly find that you make mistakes in using Maple. Like most computer systems, it is very finicky about commas and parentheses being in the right places. Fortunately, most mistakes are easy to discover and correct. You may find Appendix II helpful: it lists a variety of the most common errors that people make with Maple.

You may also find that Maple itself makes mistakes. "Bugs" occur in almost every complicated software package, and Maple certainly has its share of them. You should not put too much blind faith in Maple's results, any more than you would in a calculation by a human (who can also make mistakes). Your first line of defence against mistakes, whether by humans or by Maple, is to ask yourself, "is this result reasonable?"

If you find what appears to be a bug in Maple, you should first try to check that it really is a bug rather than a mistake on your part. Read the Maple help pages on the commands that you use, to make sure that you are using them correctly. Then show the bug to your instructor. Bugs or technical problems can be reported to the Maple developers by e-mail to `support@maplesoft.on.ca`.

## Acknowledgements

I am grateful to my very patient publisher Ron Doleman and managing editor Linda Scott, to Bob Adams for his frequent advice and encouragement, to Waterloo Maple Software for providing pre-release copies of the software, and to the members of the Maple Users' Group and the `sci.math.symbolic` newsgroup for questions and answers that have greatly increased my knowledge of Maple. In preparing this second edition, I have benefited from the helpful comments and bug reports of Rafał Abłamowicz, John Doner, Soren Eilers, Bob Jantzen, Harald Pleym, and Doug Sly. Finally, I'd like to thank my wife and children for all they have put up with while I was writing this.

I would be happy to receive your comments and suggestions at `israel@math.ubc.ca`.

Robert B. Israel
Vancouver, Canada
April 27, 1999

# Contents

CHAPTER 1

# Preliminaries

## LAB 1 GETTING ACQUAINTED WITH MAPLE

**Goals:** *To learn some of the basics of the Maple system, including how to perform simple calculations, solve equations, plot graphs and animations, use the help system, save your work, and write a report.*

Maple is a very powerful tool, combining symbolic, numerical, and graphical methods in mathematics. It is widely used by professional mathematicians, scientists, and engineers, yet can also be very helpful to the beginning calculus student. Essentially the same Maple program runs on Macintoshes, PC's, and supercomputers.

There are some differences between different releases of Maple, as the program is revised and improved from time to time. There are also differences from one platform (1) to another, mainly in the user interface.(2) However, most of these differences are rather minor: in recent releases, Maple has tried to achieve as much uniformity as possible between platforms. It would be impossible to describe in detail all the different platforms and releases here, so I'll concentrate on the Windows(3) versions of Maple V, Releases 4 to 5.1. Nearly all the Windows-specific material in this book is in Lab 1.

If Maple is not yet installed on your computer, you can find out how to do so in the documentation that came with your copy of Maple (perhaps in a booklet called *Installation Guide*). That booklet also contains useful information about the user interface for your platform. There is one important point to emphasize for the Windows versions: the default setup will set the working directory to the one that contains Maple's executable files (something like "`C:\Program Files\Maple V Release 5.1\Bin.wnt`"). This is a very bad idea, because there is a danger of accidentally deleting or overwriting important files. You should make a new folder (e.g., "`C:\working`"), and change the setup to use it as a working directory. In Windows 95, find the shortcut you will use to start Maple (typically something like "`C:\Windows\Start Menu\Programs\Maple V Release 5.1\Maple V Release 5.1`" for the entry in the Start menu), click the right mouse button on it, and choose Properties. In the window that pops up, choose Shortcut, change the entry for "`Start in:`" to your new folder, and click on the OK button. While you are here, it may also be convenient to copy the shortcut onto the desktop.

Start the Maple program and try it out. In XWindows, this involves entering the command "`xmaple`" in a terminal window, while in Windows 95 or 98 the most convenient way is to double-click on a shortcut in the Start menu or on the desktop. Maple provides you with a "worksheet". You enter commands (input), and Maple responds with output, which may be either text or graphics. You should see a prompt, which should look something ">". This tells you that Maple is waiting for your next command. Type "`2 + 3;`" and press the `Enter` key. The result should be as follows:

---

(1) The **platform** refers to the type of computer and operating system. The program you run must be made for the platform you are using.
(2) The term **user interface** describes the way that you, the user, interact with the computer program: what the screen looks like, what keys or mouse buttons you press, etc.
(3) "Windows" includes Windows 3.1, NT, 95 and 98.

```
> 2+3;
```

$$5$$

```
>
```

The prompt indicates that Maple is ready for another command. The semicolon "`;`", followed by `Enter` [4], is necessary to end your command (in cases where you don't want to see Maple's response to your command, use a colon "`:`" instead). If you press `Enter` without the semicolon or colon, Maple will give you another prompt, together with a warning that the command is incomplete, and wait for more input. This allows long commands to stretch over more than one line. If you press `Shift` + `Enter` at the end of a line, you go to a new line with no warning.

On Macintosh computers, the `Enter` and `Return` keys are different: `Enter` (at the far right of the keyboard) is needed at the end of your input, while `Return` (next to the alphabetic keys) puts the cursor on a new line and waits for more input.

## The Maple Window

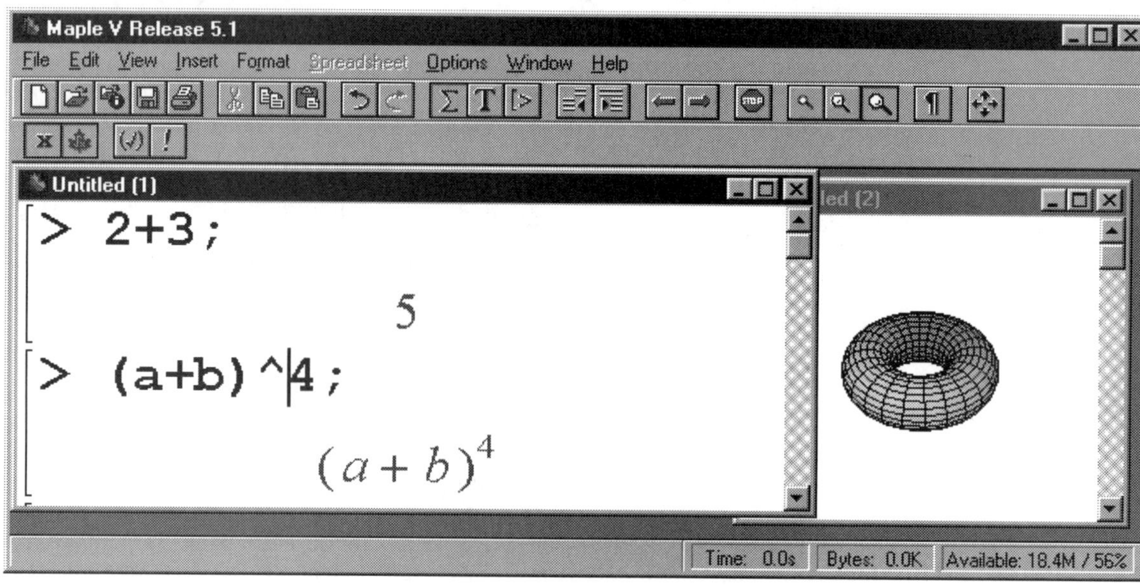


Figure 1.1. Release 5.1 Maple window

Figure 1.1 shows the Maple window of Release 5.1 under Windows 95. This has all the standard features of most application windows. On its outside edge is a border, which can be dragged with the mouse to change the window's size. The window can be moved around on the screen by dragging the title bar (which reads "Maple V Release 5.1"). At the right are the (Minimize), (Maximize), and (Close) buttons: these reduce the window to a button on the taskbar, expand it to fill the screen, and close it respectively. Below the title bar is the menu bar. Clicking on any of its items brings down a menu.

(4) Or pressing the `!` button with the mouse.

Below the menu bar is the tool bar containing a number of buttons[5]. If you turn Balloon Help on (so that it has a check mark) in the Help menu, a cartoon-style balloon shows the function of each button when the mouse cursor is over it. Particularly useful are the (Stop) button, which interrupts a computation, and the button, which inserts a new region and prompt below the current one. Below the tool bar is the context bar. Then is the main work area, which in this case contains a worksheet window ("Untitled(1)") and behind it a plot window. Help windows also go in this area. All these windows have their own borders, title bars, , , and buttons (maximizing the window makes it fill the work area). We will usually have only one worksheet window, in which case we might as well maximize it, but if necessary we can look at several windows with different worksheets, help pages, and plots at the same time. Because our worksheet extends beyond the boundaries of its window, it has a scroll bar on its right edge that can be used to move through the worksheet.

The worksheet is organized into **groups**, two of which are shown here. They are delimited by tall "`[`" brackets at the left edge of the window. Each group here contains a prompt with math input (an **input region**) and the output from that input (an **output region**).

## Quitting and Interrupting

Just as in driving a car, it's a good idea to learn how to stop before you start. You can quit Maple with the "`quit`" command. In the Windows version, you can also use one of the standard Windows methods to close the Maple window: choose "Exit" from the File menu or "Close" from the Control menu[6], click , or press `Alt` + `F4`.

Sometimes Maple takes a very long time to execute your command, and after a while you may decide that you have changed your mind and want to try something else instead. You can click on in the tool bar or press `Alt` + `F4`, which will usually interrupt the calculation (perhaps after a delay of a second or two) and give you a new input prompt. However, sometimes Maple refuses to be interrupted. If all else fails you can press `Alt` + `Ctrl` + `Del`, which will let you terminate Maple, but doesn't let you save your work.

## Numerical Calculations

Let's try out some more arithmetic.

```
> 2+3*4;
```

$$14$$

The multiplication sign in Maple is the asterisk "*"[7]. Note that Maple uses algebraic precedence rules: the multiplication is done before the addition, as is normal in algebra.

```
> 2^1000;
```

107150860718626732094842504906000181056140481170553360744375038837\
035105112493612249319837881569585812759467291755314682518714 5\
285692314043598457757469857480393456777482423098542107460506 2\
371141877954182153046474983581941267398767559165543946077062 9\
14571196477686542167660429831652624386837205668069376

[5] The tool bar, the context bar and the status line may or may not be displayed. Items in the View menu control this.

[6] The Control menu for the overall Maple window, not the one for the worksheet window.

[7] In the Courier font this book uses for inputs to Maple, this has five points, but on most keyboards it has six (*).

For powers, you can use either "^" or "**". Maple can deal with some very large numbers (the only limitation is the computer's memory capacity). The "\" at the end of a line indicates that the number continues onto the next line.

```
> 37/97;
```

$$\frac{37}{97}$$

Here's our first hint that Maple is more than a fancy calculator. Maple uses a fraction, rather than approximating $37/97$ by a decimal representation like $0.3814433$. If you do want to see a decimal result, you can use the "`evalf`" command:

```
> evalf(37/97);
```

$$.3814432990$$

Maple normally calculates decimal results using 10 digits. This can be changed, however, using a variable called "`Digits`".

```
> Digits:= 25:
> evalf(37/97);
```

$$.3814432989690721649484536$$

Some points to note here:

(1) Maple is case-sensitive. "`DIGITS`" or "`digits`" would be considered as completely different variables, which would not influence the number of digits Maple uses.

(2) "`:=`" is the assignment sign. The value on the right is assigned to the variable on the left. If you leave out the "`:`", a common mistake, you have an equation rather than an assignment.

(3) I didn't need the result of the "`Digits:= 25`" command (which would have been "$Digits := 25$"), so I used "`:`" instead of "`;`"[8].

(4) Once "`Digits`" has been assigned, it affects all subsequent decimal calculations until the next "`Digits:=...`" or the end of the session. You can also specify a number of digits within the "`evalf`" command, affecting only that calculation.

```
> evalf(37/97, 20);
```

$$.38144329896907216495$$

In student editions of Maple, `Digits` is restricted to a maximum of 100. The professional edition has no restriction (up to the memory capacity of the machine).

## Symbolic Calculations

```
> x^2 - 3 * x - 4;
```

$$x^2 - 3x - 4$$

Now here's something really different from a calculator. The "$x$" here is a variable, and the result is an algebraic expression. Note: don't forget the "*" between `3` and `x`, which is necessary in the input even though you don't see it in the output. Maple's output looks like ordinary typeset mathematics. All the usual algebraic manipulations can be done on expressions.

`> "/(x-4);` (Release 4), or

`> %/(x-4);` (Release 5)

$$\frac{x^2 - 3x - 4}{x - 4}$$

[8] Generally you should use a semicolon unless you are really sure that you won't want to look at the response. Even if you don't need it now, it may be useful later in locating errors. I will almost always use semicolons in this book, although I will not always include Maple's responses.

The very useful symbol `"` (in Release 4) or `%` (in Release 5) stands for "the last output". You can also use `""` or `%%` for the second-last output and `"""` or `%%%` for the third-last (but `""""` or `%%%%` doesn't work). Hmmm, it should be possible to simplify this latest fraction.

> `simplify(");` or > `simplify(%);`

$$x+1$$

The "`simplify`" command will often make a complicated expression simpler by finding cancellations. Its ideas of what is simpler may not always agree with yours, however. Some other commands that you may find useful are "`expand`" and "`collect`". The "`expand`" command tries to make an expression into a sum of simple terms. The "`collect`" function writes an expression as a sum of coefficients times powers of a specified variable.

```
> expand((x+1)*(x-1));
```

$$x^2-1$$

```
> collect(4 * x^2 + 2 * x + y * x + 1, x);
```

$$4x^2+(2+y)x+1$$

## Assigning Values to Variables

If you might want to use something later in your Maple session, you can assign a name to it using "`:=`".

```
> p:= x^2 - a * x * y - 4;
```

$$p := x^2 - axy - 4$$

We have just assigned the variable "p" the value $x^2 - axy - 4$. In Maple a variable's value can be a name or expression, as well as a number. Before anything has been assigned to it, a variable's value is its own name.

```
> (p+1)^2;
```

$$(x^2-axy-3)^2$$

After we assign a value to "p", this value will be used whenever "p" appears in an expression. This includes definitions using "p" that have occurred before. For example, the definition of "p" uses the variable "a", which does not have a value yet. Let's now assign it a value.

```
> a:= 3;
```

This new value for "a" affects the value of "p":

```
> p;
```

$$x^2-3xy-4$$

## Solving Equations

Maple can solve equations.

```
> solve(p = 3, x);
```

$$\frac{3}{2}y+\frac{1}{2}\sqrt{9y^2+28},\ \frac{3}{2}y-\frac{1}{2}\sqrt{9y^2+28}$$

Note that "`p = 3`" is an equation, not an assignment (we are not changing the value of `p`), so we use "`=`", not "`:=`". Since the equation "`p = 3`", i.e., $x^2 - 3xy - 4 = 3$, contains two variables $x$ and $y$, we had to specify the variable to solve for as well as the equation. With only one variable, only the equation would have been needed. In some cases, the solutions that "`solve`" looks for may be very complicated. In other cases, it may be impossible to express the solutions in terms of the usual operations of algebra.

```
> solve(x^5 - 3*x + 1 = 0, x);
```

$$\text{RootOf}\left(\_Z^5-3\_Z+1\right)$$

To find a numerical solution, you can use "`fsolve`".

```
> fsolve(x^5 - 3*x + 1 = 0);
```

$$-1.388791984, .3347341419, 1.214648043$$

For a polynomial equation, this finds all the real roots. In other cases, it tries to find one root. If Maple can't find a real solution, it either doesn't return anything or returns the "`fsolve`" command unevaluated. The first equation below has no real solutions (can you see why?). The second has two, but "`fsolve`" just finds one[9].

```
> fsolve(x^4 + 2/(x^2 + 1) = 0);
> fsolve(x^4 + 2/(x^2 + 1) = 3);
```

$$-1.217043263$$

## Graphs

A picture being worth a thousand words, one of the most useful commands in Maple is "`plot`", which makes a graph. Maple can make many different types of graph. The simplest type of plot command plots a function of one variable. You specify the interval to use for that variable (the "$x$ axis"). Maple automatically chooses the interval for the $y$ axis so the curve fits nicely in the graph. Here is the graph of $y = x^2 - 2x + 2$:

```
> plot(x^2 - 2 * x + 2, x=-2..2);
```

(Figure 1.2)

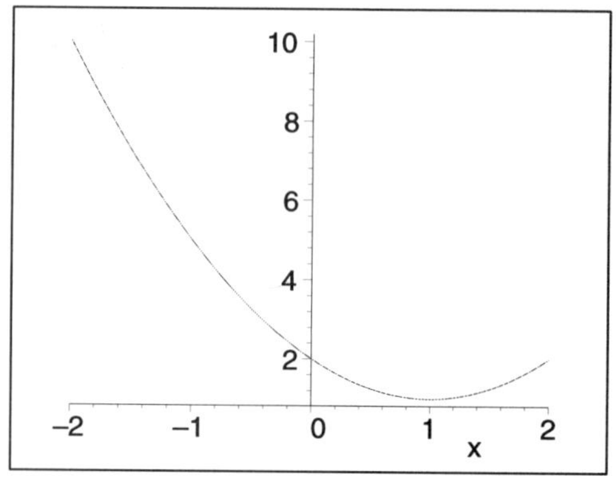


Figure 1.2. $y = x^2 - 2x + 2$

Note the two dots for "to" in "`x=-2..2`". Three dots would also work[10]. If you wish, you can specify an interval for $y$ as a third input to the "`plot`" function. Note that "`y=`" is not needed in this one.

```
> plot(x^2 - 2 * x + 2, x=-2 .. 2, 0 .. 4);
```

Supplying an interval for $y$ is especially useful in cases where $y$ would be very large for some values of $x$ in the given interval.

The plot normally appears as an "inline" plot in the worksheet itself. The menu bar and context bar change when you click the cursor in the plot area, and a frame appears around the plot. You can use the mouse to drag the little black boxes on the edges and corners of the frame to change the size and shape of the plot. You can also choose to have plots appear in their own windows by setting "Plot Display" to "Window" in the Options menu, or with the command

```
> interface(plotdevice=window);
```

To go back to inline plots, set "Plot Display" to "Inline", or use

```
> interface(plotdevice=inline);
```

A plot window has its own title bar with [minimize], [maximize], and [close] buttons. When you are finished looking at the graph, you can close the window by clicking its [close] button. If you may want to look at the same plot again, you can keep it (perhaps making it an icon with [minimize]), and return to the worksheet window.

---

[9] By symmetry, in this case the other root is 1.217043263. In Lab 5 we will see how to find other roots more generally.

[10] Note that "`0...1`" means 0 to 1 rather than 0 to .1. To avoid any confusion, I prefer to leave spaces around "`..`", and also to write .1 as "`0.1`".

## Getting Help

Maple has thousands of functions and commands. Fortunately, you are very unlikely ever to need most of them. If you need a command you are unfamiliar with, the on-line help facility is very useful. For example, to get help on the "`expand`" command, you just enter

```
> ?expand
```

Instead of using "?", another way to get help is to press `Ctrl` + `F1` when the cursor is located in (or directly before) the word you want to look up. This is especially useful when you are in the middle of typing a complicated command.

Sometimes the help page can be rather technical, and discusses all sorts of options you don't need. It almost always ends with some examples, however, and these can be useful even when the text is hard to understand.

The help page comes up in a separate window which is similar to a worksheet window. You should close this (just as you closed plot windows in the last section) when you don't need it any more. Underlined words in the help page (usually in the "See Also" section) are **hyperlinks**: clicking on one brings up the help page for that word (in Release 4, a new help window appears, while in Release 5, the new page replaces the old one in the same window). You can use the ← and → buttons on the tool bar to go back and forth between help pages you have seen in this session.

There are several features you can use if you don't know the name of the command you need. "Topic Search" on the Help menu finds help topics starting with a given string of characters, so you can use it if you can guess the first few letters of the command. Even more useful is "Full Text Search", which finds all help pages that mention one or more given words. Thus if you forget the name of the command that can be used to plot an implicit function, you can try entering the words "`plot implicit function`" into "Full Text Search"; one of the matching topics listed should be "`plots,implicitplot`", which is the help page you need. The only problem is that a lot of pages may contain these words.

In Release 5, the help browser appears in the top part of each help window. At the far left is a list of top-level topics. If you click on a topic that has a "...", its subtopics appear in the next section to the right. Topics (or subtopics) with no "..." correspond to individual help pages, which are displayed when you click on them. In Release 4, "Contents" on the Help menu brings up the top level of a hierarchical system of help pages.

For help on the worksheet interface, including menu items and buttons, try "`?worksheet`" and follow the hyperlinks.

## Editing

Maple has some features that will save you a lot of typing. It often happens that you want to repeat earlier commands, perhaps with small variations. Instead of typing the whole thing out again, you can reuse what you typed before. In both Windows and Unix, the standard copy-and-paste method is to drag the mouse cursor over the text to be copied, press `Ctrl` + `C`, then position the cursor where you want to put the material you copied and press `Ctrl` + `V`. On the Macintosh you use `Command` + `C` and `Command` + `V`.

You can copy Maple output and paste it to use as input. Although the output is "pretty-printed" to look like standard mathematics, when pasted to an input region it will appear in Maple notation suitable for input. You should be careful not to include the blank space between the region delimiter (the long "`[`") and the Maple output, or what you copy will be pasted as output.

You can also go to a previously entered input region (by clicking the mouse on it), change it, and press `Enter` to get a new result. If "Replace Output" has a check mark in the Options menu, when you press `Enter` the new output will replace the old output. If it is not checked, both outputs will be seen with the new one above the old one.

Editing can produce very strange-looking results. For example:

```
> a:= 1:
> a + 1;
```

$$3$$

No, Maple doesn't think that $1 + 1 = 3$. The first line originally read "`a:= 2:`". Then after entering the line "`a + 1;`" I went back and sneakily changed the 2 to 1. Until I press `Enter` on the line "`a + 1;`" again, the 3 will remain there. Another way to produce similar unsettling effects would be to have the value of "a" changed somewhere else in the worksheet. After making this change, you go back to the line "`a + 1;`" and press `Enter`. The result reflects the new value of "a", not the value shown on the previous line. The moral of the story is that a Maple worksheet that has been edited may not be an accurate representation of the sequence of calculations that have occurred.

You should generally have the group ranges (brackets) displayed, by having "Show Group Ranges" checked in the View menu. Pressing `Enter` on any line in a group executes all the commands in that group, which may include several input prompts. The group ranges let you see which commands are in the same group.

## Saving Your Work

In any computer application, it is often important to save your work. Even the best computer system can "crash" from time to time, or you may make an error that is hard to recover from. Therefore you should get into the habit of saving your work every 15 minutes or so. It takes a few seconds, but can prevent a lot of aggravation and wasted time recovering from a crash. Moreover, instead of doing an entire lab in one session, you may wish to take a break and later resume where you left off.

To save your Maple worksheet, use "Save" or "Save As..." in the File menu. Initially (when the worksheet is "`Untitled`") these are the same, but after a file name has been chosen "Save" will use that name again, while "Save As..." lets you choose a new name. The button on the tool bar is equivalent to "Save". The worksheet file normally has extension "`.mws`".

To re-load a saved worksheet, use "Open" in the File menu, or choose one of the recently used file names listed near the bottom of the File menu. Or (in Windows) you can click on the icon for your "`.mws`" file to start Maple with that file. Note that what is saved in the "`.mws`" file is only a visual representation of the inputs and outputs of the Maple session, not the internal results of the commands. Therefore loading it does not actually perform the commands contained in the worksheet. For example, variables that are given values in the worksheet will not not have those values after you load the worksheet. After loading, you can execute the worksheet with "Execute", "Worksheet" in the Edit menu; this will assign the variables their values.

If some of the commands in your worksheet took a long time, they will presumably take a long time to execute after loading the worksheet again. Therefore it may be worthwhile to save the results of these commands. This is done using the "`save`" command. To save the values of one or more variables, e.g., `name1`, `name2`, `name3`, in a file "`myfile.txt`", you enter a command such as

```
> save name1, name2, name3, `myfile.txt`;
```

You can save all names that have been assigned values in your session if you don't include any variable names, just the file name (which in this case must have extension `.m`):

```
> save  `myfile.m`;
```

To read the file later:

```
> read `myfile.txt`;
```

Thus instead of executing a worksheet, if you have saved the values that you would need to continue working with the worksheet, you only need to enter that one "`read`" command.

The "`save`" and "`read`" commands are also useful in other contexts. Suppose, for example, you have defined a very useful, but rather complicated, function, let's say "`myfunc`". It would be inconvenient to have to type in the definition every time you wanted to use that function. Instead, you can "`save`" the definition to a file with the command

```
> save myfunc, `myfunc.def`;
```

and then in later sessions you will be able to "`read`" it whenever you wish. In this book, we will use "`read`" to read in the definitions of a number of useful functions.

Some points to note about the "`save`" and "`read`" commands:

(1) These are two of the very few commands where we do not put the inputs in parentheses.

(2) If you use the name of an already-existing file in "`save`', that file will be destroyed and replaced by whatever you are saving now.

(3) The file name is enclosed in "back quotes", not ordinary single quotes (apostrophes). The [~ `] key is at the top left of a standard 101-key IBM-compatible keyboard, but its location may vary.

(4) If you need to specify a directory, use forward slashes (/) instead of the back slashes (\) that are usual in DOS and Windows, e.g.,

```
> read `c:/maple/myfile.txt`;
```

(5) If you use a file name with the extension "`.m`", the file will be saved in Maple's internal format, otherwise the file will be a text file. In general, the "`.m`" file format changes in new releases of Maple, so a "`.m`" file saved using one release of Maple won't be readable with another. About the only advantage of using "`.m`", besides not having to specify particular names, is that you can put the file in your Maple library directory and use "`readlib(...)`" to retrieve the definition.

## Restarting Maple

During a long Maple session, many definitions will be stored in your computer's memory. After you finish working on a problem, the definitions used in that problem may not be useful any more. You can clear Maple's memory with the "`restart`" command. This will reset everything, (almost) as if you had quit Maple and then started it again. Of course, before you restart you should be careful to save anything that you might need later on.

```
> restart;
```

I wrote "(almost)" above because there may be less memory available to Maple after a "`restart`" than when Maple is started from scratch. This is especially important on personal computers, where memory is often a significant limitation for Maple. As the amount of available memory decreases, Maple will take more and more time to perform even simple commands (because it will spend a lot of time accessing its "virtual memory" on disk). Eventually you may get the dreaded message: "Maple has run out of resources, and will shut down". To combat this, you should quit Maple and start a new session if you are at a convenient stopping place and notice that available memory has decreased significantly, or that Maple's speed has deteriorated. Before quitting, save the worksheet and whatever variables and functions you have defined and still need. It will probably take less time and memory to read the values of the variables and definitions of the functions from the file that you saved than it would to recalculate them.

## Writing a Report

You may want to make a record of your Maple calculations, perhaps to hand in as a homework assignment. There are a number of ways of doing this. Your instructor may tell you what format to use.

One possibility is to save your worksheet as indicated above, and submit it electronically. A second possibility is to print the worksheet. Use "Print..." from the File menu. Maple does **not** have the capability to print only part of a worksheet. Instead, you should copy the part that you want to print, paste it into a new worksheet, and print that worksheet.

In Windows, there is also the possibility of copying Maple text and graphics into a document made on a Windows word processor such as Microsoft Word. By default, Maple output pastes into word processors in the "Maple notation" style, which is usually not what you want. However, it is possible [11] to paste output as graphics. Use "Paste As" rather than "Paste" in your word processor, and choose "picture" or "graphics". [12]. This graphics object can be moved or resized, but can't be edited (except by a graphics application such as Paintbrush). You can change the "Output Display" to "Character Notation" in the Options menu so that Maple's output will be editable text. However, this looks much cruder, e.g., the fraction $1/(x^2+4)$ becomes

```
          1
       ------
         2
        x  + 4
```

You need to use a monospaced font such as Courier in your word processor to make sure this will line up properly.

Whichever method you use to make your report, be sure to include more than just the Maple input and output. You must explain what you are doing and how you interpret the results. This is useful even if you are the only person who will ever read the document, because tomorrow you may not remember the reasoning you used today.

You can put these explanations and interpretations into "text regions" in your worksheet. To start a text region, either press `F5` or click **T** on the tool bar. The text region can start on its own line, or after math input. When you are finished typing text and want a new input prompt, click **[>** on the tool bar. To start a new paragraph on a new line, choose "Paragraph" and "Before" or "After" from the Insert menu. Also useful is the **[>** button on the tool bar, which produces a new group with input prompt after the cursor. Text regions can contain in-line math, which you start by clicking **Σ** on the tool bar, and stop by clicking **T**.

## Graphs of Functions and Equations

To plot the graph of a parabola in its standard form, use the "`plot`" function:

`> plot(x^2/4, x=-10..10);` (Figure 1.3)

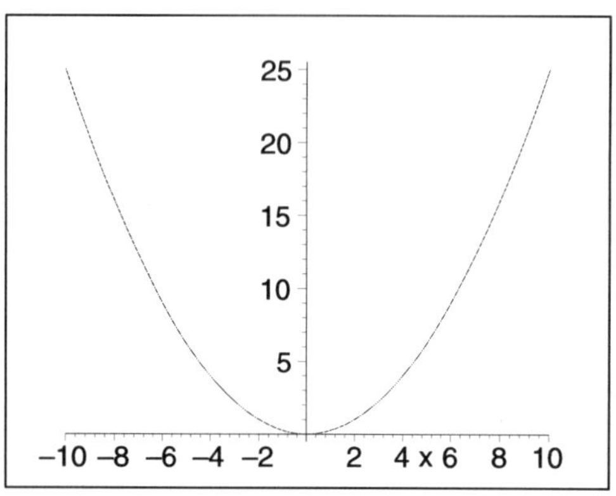


Figure 1.3. $y = x^2/4$

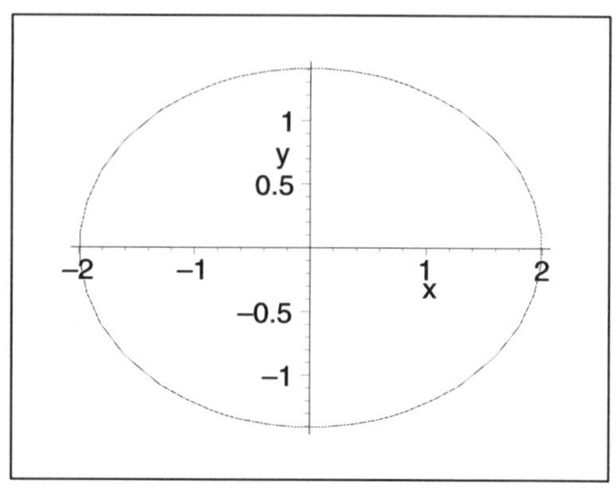


Figure 1.4. $x^2 + 2y^2 = 4$

For ellipses and hyperbolas, however, instead of $y$ being given as an expression involving $x$, we have an equation involving both $x$ and $y$. For this we can use the "`implicitplot`" function. This is part of the "`plots`" package, so we first load that package.

---

(11) In Release 4.00b and later.

(12) Note that even if you just select part of a typeset output, the entire output will be pasted.

```
> with(plots):
```

The package only needs to be loaded once during a Maple session, after which you can use all the commands and functions in the package for the rest of the session. If you use a semicolon rather than a colon after the "`with`" command, Maple will print out a list of all the commands contained in the package. Now here is the actual `implicitplot` command.

```
> implicitplot(x^2 + 2 * y^2 = 4, x = -3..3, y = -3..3);
```
**(Figure 1.4)**

Even though we specified the bounds for $x$ and $y$ as $-3$ to 3, Maple only shows a rectangle within these bounds that contains the curve.

One of a number of options that applies to the plot commands is "`scaling`". Ordinarily, scaling is "`unconstrained`". This means that the scales on the $x$ and $y$ axes are chosen automatically so that the desired $x$ and $y$ intervals fit neatly in the plotting window. That may not be the best choice when you want to see the shape of a graph, as the scales in the $x$ and $y$ directions may be different. In order to have the same scale in both directions, i.e., one unit of $x$ the same length as one unit of $y$, you can choose to have the scaling "`constrained`". You can do this in the "`plot`" command, e.g.,

```
> plot(x^2/4, x=-10..10, scaling=constrained);
```

or you can change it using the menu bar or context bar. Click on the plot to make the menu bar and context bar change to the ones for plots. Now choose "Constrained" or "Unconstrained" in the Projection menu, or click the 1:1 button ("Constrained" corresponds to this being depressed). Changing the option in the menu or context bar immediately changes the plot. Of course, when you use "constrained" scaling you may get a very long and narrow graph (depending on what is being plotted). You can experiment to see what is the best choice in any particular situation.

## Several Curves in One Graph

Maple can do several plots or several implicit plots on the same set of axes. Instead of one expression or equation, you use several, separated by commas and enclosed by braces `{}`.

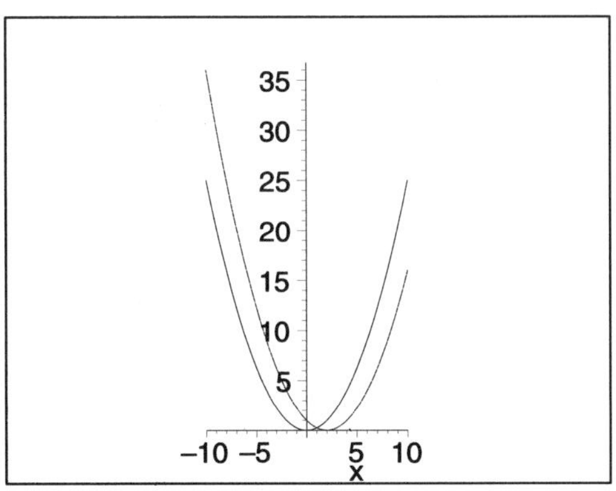


Figure 1.5. Two parabolas

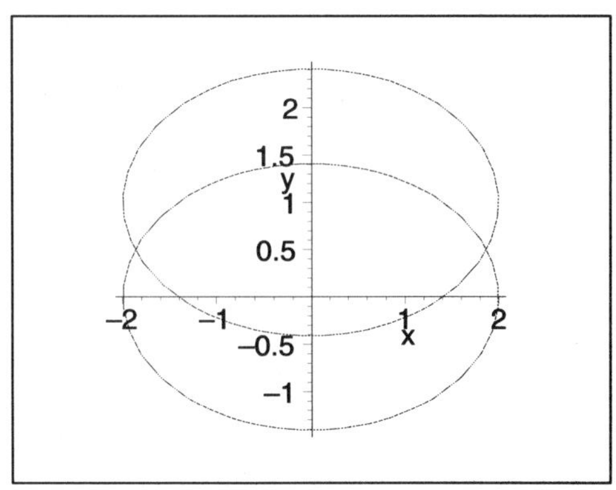


Figure 1.6. Two ellipses

```
> plot({ x^2/4, (x-2)^2/4 },
>       x=-10 .. 10, scaling=constrained);
```
**(Figure 1.5)**

```
> implicitplot({ x^2 + 2*y^2 = 4, x^2 + 2*(y-1)^2 = 4 },
>              x=-3 .. 3, y=-3 .. 3, scaling=constrained);
```
**(Figure 1.6)**

On a colour monitor, "`plot`" shows the different curves in different colours.

For a family of curves depending on a parameter, you can either enter the expression or equation several times for different values of the parameter, e.g.,

```
> plot({ 1 + x - 2 * (x^2 - 1), 1 + x - 1 * (x^2 - 1),
>        1 + x + 0 * (x^2 - 1), 1 + x + 1 * (x^2 - 1),
>        1 + x + 2 * (x^2 - 1) }, x=-3 .. 3);
```

or you can have these generated automatically using the "seq" operator.

```
> plot({ seq(1 + x + k * (x^2 - 1), k = -2 .. 2) },
>        x=-3 .. 3);
```

In general, "seq(*expr*, *v* = *a* .. *b*)" generates a sequence of values of *expr* for each value of the loop variable $v$ from $a$ to $b$ (increasing by 1 each time). Thus

```
> seq(k^3, k=-2 .. 2);
```

$$-8,\ -1,\ 0,\ 1,\ 8$$

## Animation

Instead of seeing all the curves at once, it is possible to make a "movie" that will show you one curve after another. This can be done for "plot" curves using "animate". Note that "animate", like "implicitplot", is part of the "plots" package, so you will need to use the command "with(plots):" in your Maple session before you can use "animate".

```
> animate(1 + x + k * (x^2 - 1), x=-3 .. 3, k = -2 .. 2);
```

An animation has a set of "VCR controls" on the context bar.

 starts the action and  stops it.

 advances one frame at a time.

 slows the action down while  speeds it up.

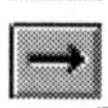 , if depressed, means the movie runs forward (from lower to higher values of the parameter k). 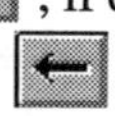 means it runs backwards. Click on the button to switch between these.

 if depressed, means the movie will stop when it reaches the end.  means it will start again and repeat until you stop it. Click on the button to switch between these modes.

While "animate" will not produce implicit plots, you can use "display" with the option "insequence=true" to animate any sequence of plots. Here I will use "seq" to produce a sequence of implicit plots, and then display them. Note that square brackets [ ] rather than braces { } must be used[13]. The "display" command is part of the "plots" package.

```
> display([seq(implicitplot(x^2 - y^2 = k*y, x=-5 .. 5,
>               y=-5 .. 5), k=-3 .. 3)], insequence=true);
```

One difference between using "animate" and "display([seq(..." is in the number of frames produced. By default, "animate" produces 16 frames, dividing the parameter range into equal intervals; the number can be changed by inserting an extra input "frames=*number*". On the other hand, with "seq" the parameter always increases by 1 from frame to frame. Of course, you could change the equation, e.g., to make 16 frames in the last example you could say

```
> display([seq(implicitplot(x^2 - y^2 = (-3 + 6/15 * k) * y,
>     x=-5 .. 5, y=-5 .. 5), k=0 .. 15)], insequence=true);
```

Don't do this if you're in a hurry and your computer is slow! Constructing a large number of implicit plots can take a lot of time, and also use a large amount of memory.

Animations can be printed as part of worksheets (what is printed is the currently shown frame). In Release 5, where worksheets can be be made into web pages using "Export As", "HTML..." in the File menu, animations in these worksheets are exported as animated GIF files.

---

(13) This is because we want to have the plots displayed in the correct order. Maple uses braces to enclose a set, where the order of the elements does not matter, and square brackets to enclose a list, where order does matter.

## EXERCISES

1. Enter $\dfrac{(2x^2-1)^4-(4x^2-1)^2}{2(x^2-1)^2-1}$ into Maple, and simplify it. *Hint:* don't forget to use `*` for multiplication.
2. **(a)** Graph the polynomials $(x-2)^{12}$ and $(x+12)^2$ from $x=0$ to 3. Which curve is which?

   **(b)** Graph these polynomials on an interval of length 0.1 containing a point where the two curves intersect.
3. Let $P$ be the polynomial $x^3-4x^2+3x+2$.

   **(a)** Find all the roots of $P$.

   **(b)** Graph $P$ on an interval containing all the roots.
4. What are the questions that the following Maple commands answer? The help facility will be useful here.

   **(a)** `sum(i^2, i= 1..10);`

   **(b)** `ifactor(1995);`

   **(c)** `solve({x + 2*y=5, x^2 + y^2 = 10}, {x,y});`
5. Use Maple to find out:

   **(a)** Which is larger, $\pi^3$ or $3^\pi$? *Hint:* the constant $\pi$ is "`Pi`", not "`pi`".

   **(b)** What is the 50th digit after the decimal point in $\sqrt{2}$? *Hint:* you can use either "`sqrt(2)`" or "`2^(1/2)`" here.

   **(c)** Is $3^{1000}-1$ divisible by 10 000?

   **(d)** Is $x^{1000}-1$ divisible by $x^4-x^3+x^2-x+1$?

   **(e)** What is the smallest positive integer $n$ for which $n!$ ends in exactly 10 zeros?
6. The command below will produce a strange-looking result. Explain. What should be used instead to get a nice-looking graph of $y=1/x$?

   ```
   > plot(1/x, x=-1 .. 1);
   ```
7. A locomotive is on a straight section of track, 100 metres from a level crossing and moving toward it at 30 metres per second. A car is 70 metres from the crossing on a straight road at right angles to the track, and moving toward the crossing at 25 metres per second. Suppose that both vehicles continue moving at constant speed[14].

   **(a)** Using Maple, plot the distance between the locomotive and the car as a function of the time $t$.

   **(b)** Estimate from the graph the closest distance between the locomotive and the car, and the time at which this occurs. Redo the graph using a shorter interval of time to improve your estimate.

   **(c)** Try some different speeds for the car. Find a speed so that the closest distance between the locomotive and the car is between 5 and 6 metres (with the car crossing before the locomotive).
8. Begin with the graph of the parabola $y=1-x^2$ for $x$ from $-2$ to 2. Using either multiple plots on one screen or animation, examine what happens when you

   **(a)** scale the graph vertically, horizontally, or both (by the same factors).

   **(b)** shift the graph vertically, horizontally, or both.

   **(c)** Repeat (a) and (b) for the hyperbola $x^2-y^2=1$.

---

(14) Don't try this in real life!

**9.** Examine the one-parameter family of curves $y = 1 + x + k(x^2 - 1)$ for $k$ from -2 to 2.

**(a)** What do they all have in common?

**(b)** Which value of $k$ is particularly "special" in terms of the shape of the curve?

**(c)** Write the equation of a one-parameter family of parabolas with vertical axes, all of which pass through the points $(a, b)$ and $(c, d)$. Try this out in the case $(a, b) = (0, 1)$, $(c, d) = (2, 3)$.

**10.** How are the one-parameter family of parabolas $y = (x - k)^2 + k^2$ related to the curve $y = x^2/2$? Find and plot a family of downward-pointing parabolas with a similar relation to this curve.

**11.** *(A Maple puzzle)* Try the following sequence of Maple commands. Then interchange the second and third lines and try again. Why is the final result different the second time? *Hint:* work through the commands, step by step.

```
> p:= 'p';
> q:= 1 + p;
> p:= 2;
> q;
> p:= 1;
> q;
```

# LAB 2 FUNCTIONS IN MAPLE

**Goals:** *To learn how to define and manipulate functions in Maple, including piecewise-defined functions and trigonometric functions.*

It is possible to define functions in Maple. The simplest way is by a mapping rule such as

```
> f := x -> x^2 + 1;
```

$$f := x \to x^2 + 1$$

You can then evaluate the function at a certain value of the independent variable, putting the independent variable in parentheses.

```
> f(3);
```

$$10$$

```
> f(p);
```

$$p^2 + 1$$

Note the difference between the function $f$ and the expression which you would get by

```
> g:= x^2 + 1;
```

$$g := x^2 + 1$$

The expression is tied to the particular name $x$ for the independent variable. If you give $x$ a certain value, the value of $g$ changes to match it. On the other hand, the $x$ in the definition of $f$ is just a "dummy variable", a name that stands for "the independent variable of the function $f$". The definition of $f$ is not affected by any values you assign to this name.

```
> x:= 5:
> f(u);
```

$$u^2 + 1$$

It's often more convenient to use functions rather than expressions, especially when you want to evaluate the function at many different values of the independent variable. For most purposes, however, you could get along with only expressions and not functions. To evaluate an expression at a particular value of the independent variable, you can use the "`subs`" command. To show you an example, I'll first undo the assignment of 5 to $x$.

```
> x:= 'x':
> subs(x=3, g);
```

$$10$$

An idiosyncracy of "`subs`" is that it performs only the most basic simplifications after doing the substitution.[15] It wasn't needed in this case, but sometimes you will want to follow "`subs`" by "`eval`":

```
> eval(subs(x=3, g));
```

In Release 5, you can use a form of "`eval`" instead of "`subs`". This time, the equation comes after the expression rather than before, and no extra "`eval`" is needed.

```
> eval(g, x=3);
```

You can also make an expression into a function, using the "`unapply`" command. If $x$ is a variable and *expr* an expression, "`unapply(`*expr*`,`$x$`)`" is the function taking $x$ to *expr*.

```
> h:= unapply(g,x);
```

$$h := x \to x^2 + 1$$

## Domains and Ranges

Maple does not deal directly with domains and ranges of functions. If $f$ is defined by a formula, it generally assumes the domain of $f$ includes all numbers $x$ for which the formula $f(x)$ makes sense (e.g., doesn't involve division by 0). But since Maple deals with complex numbers[16] rather than just real numbers, it has no qualms about allowing square roots of negative numbers.

One of the best ways to determine the range of a function is by graphing it.

■ **EXAMPLE 1** Find the range of the function

$$f(x) = \frac{x^2 + 3x}{x^4 + x + 1}$$

to within four significant figures.

**SOLUTION** My first decision: what interval for $x$ should I use to plot it? Surprisingly enough, I can simply say

---

(15) More technically, it doesn't evaluate the result of the substitution.

(16) Complex numbers are of the form $a+bi$, where $i = \sqrt{-1}$. They are widely used in more advanced mathematics. We won't use them here, but you will sometimes encounter them in Maple output. In Maple, $i$ is written as $I$.

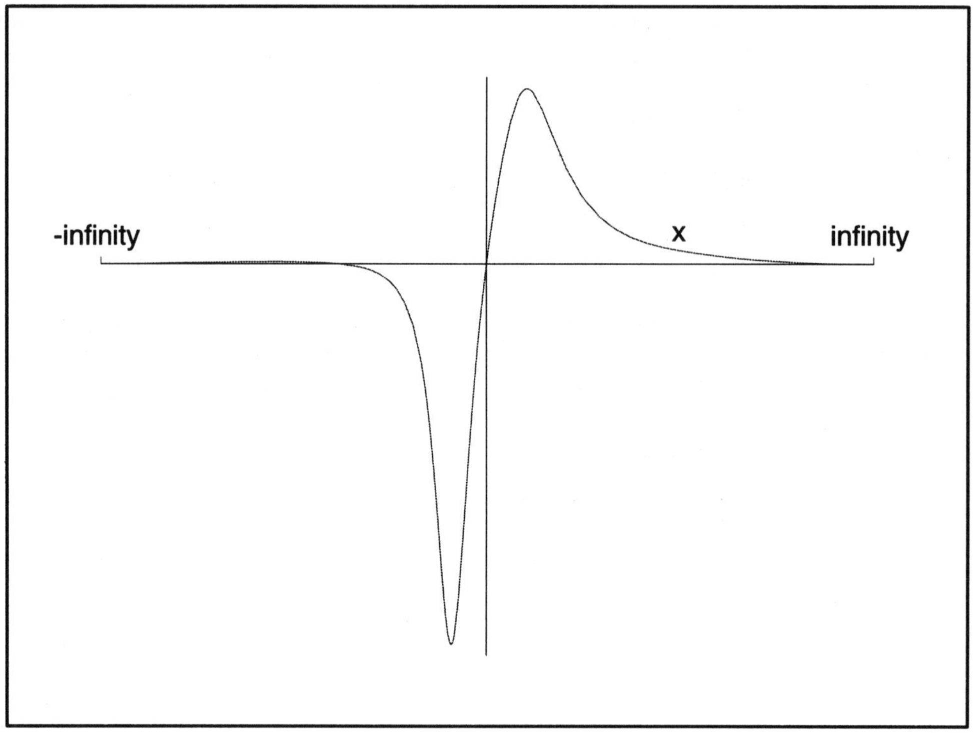


Figure 1.7. An "infinity plot"

```
> f:= x -> (x^2 + 3*x)/(x^4 + x + 1):
> plot(f(x), x=-infinity..infinity); (Figure 1.7)
```

Here we have an infinite interval "squeezed" onto a finite graph. From this I can see that there is a deep "valley" at some negative $x$ and a "peak" at some positive $x$; the range of the function will be the interval from the depth of the valley to the height of the peak.

To find the actual range, I must locate the valley and the peak more precisely. I don't know much about the location of the valley or the peak, because Maple won't put a scale on a graph with an infinite interval. But I'll take a guess and try `x=-5..5`. If I see a valley or a peak, these are the ones I want. If not, I will need a larger interval. In fact, both the valley and the peak appear. The valley bottom appears to be around $x = -1, y = -3$ and the peak around $x = 1, y = 1.5$.

```
> plot(f(x), x = -5 .. 5);
```

Here the worksheet interface has a very nice feature: if you click the left mouse button on a plot, the context bar will indicate the X and Y coordinates of the tip of the arrow cursor. According to this, the valley bottom seems to be at $x = -0.7229, y = -3.02$ and the peak at $x = 0.8162, y = 1.345$. I wouldn't believe more than about one decimal place of any of these numbers, because both the steadiness of my hand and the resolution of the screen are limited. Without this feature, however, we would be limited to much rougher estimates by eye.

The next step is to "zoom in", plotting a smaller interval for $x$ which appears to contain the valley or peak, until we have located it with sufficient accuracy. Thus to find the valley bottom, having determined with the mouse that it is in the interval $x = -0.8$ to $-0.7$, I would plot the function on that interval. This shows me a close-up of the valley bottom and lets me locate it more precisely at about $x = -0.7255$, certainly between $-0.727$ and $-0.723$. Plotting the function on that interval produces a graph (Figure 1.8) on which it is clear that the depth of the valley, to four significant figures, is $-2.992$. ■

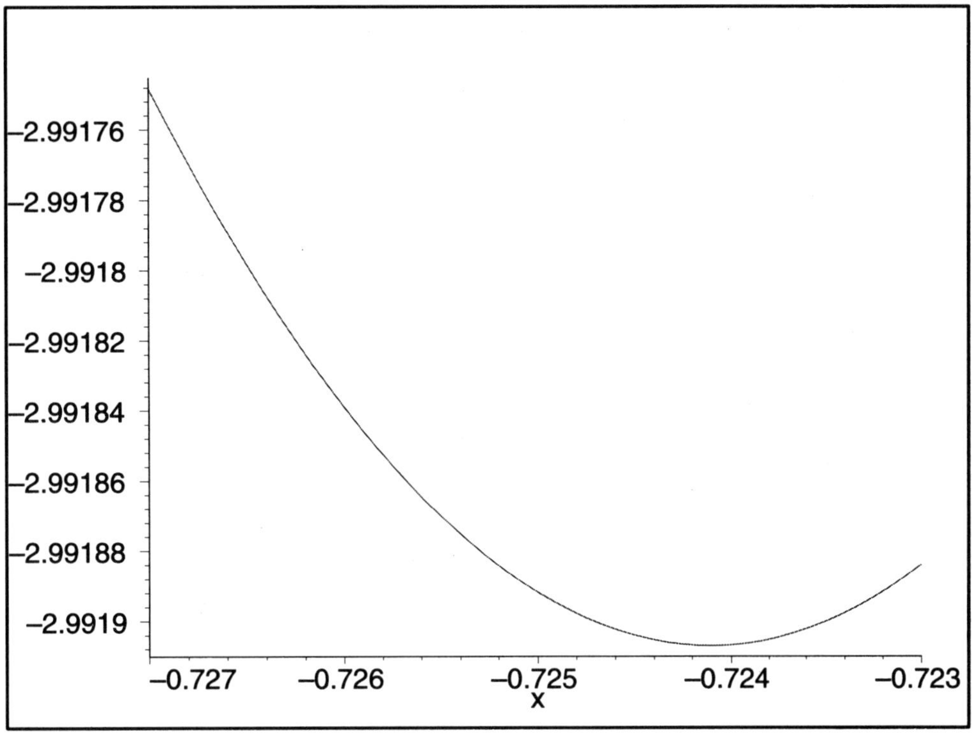


Figure 1.8. Close-up of the valley

## Constructing Functions

Composition of functions can be done in Maple. There is no "∘" on the keyboard, so Maple uses "`@`" instead. There is another important change from standard mathematical notation: when evaluating a composite function, you should put it in parentheses: "`(f@g)(x)`", not "`f@g(x)`".

```
> f:= x -> sqrt(x);
```

$$f := \mathrm{sqrt}$$

Note: "`sqrt`" is the square root function. You could also use "`x^(1/2)`".

```
> g:= x -> x + 1:
```

$$g := x \to x + 1$$

```
> (f@g)(x);
```

$$\sqrt{x+1}$$

There are a number of useful functions that are defined "piecewise", with one formula in one part of the domain and a different formula in another. The most important is the **absolute value** function:

$$|x| = \begin{cases} x & \text{if } x \geq 0, \\ -x & \text{if } x < 0 \end{cases}$$

In Maple this function is called "`abs`". Thus in Maple input we write `abs(x)` rather than the standard mathematical notation $|x|$. However, Maple uses the $|x|$ notation in output.

While the absolute value function, you might say, hides the distinction between $x$ and $-x$, the **signum** function reveals it:

$$\mathrm{sgn}(x) = \begin{cases} 1 & \text{if } x > 0, \\ -1 & \text{if } x < 0 \end{cases}$$

You may notice that I haven't said anything about the case $x = 0$. In fact, mathematicians are not all in agreement about what sgn(0) should be: some say 1, some 0, and some leave it undefined. Maple, whose version of this function is called "`signum`"[17], is also undecided: `signum(0)` is 0 by default, but can be changed to follow whichever convention you prefer[18].

It is quite easy to define functions "piecewise", using the "`piecewise`" command.

**EXAMPLE 2** Define in Maple the functions

$$f(x) = \begin{cases} (x+1)^2 & \text{for } x < -1 \\ -x & \text{otherwise} \end{cases}$$

$$g(x) = \begin{cases} (x+1)^2 & \text{for } x < -1 \\ -x & \text{for } -1 \le x < 1 \\ \sqrt{x-1} & \text{otherwise} \end{cases}$$

**SOLUTION** You can simply write

```
> f:= x -> piecewise(x < -1, (x+1)^2, -x);
```

$$f := x \to \text{piecewise}\left(x < -1, (x+1)^2, -x\right)$$

The first input to "`piecewise`" is the condition for the first "piece" of the function, and the second is the formula in that piece. The last input is the formula to be used when the condition is false (the "otherwise" piece). If you evaluate "`f`" at a symbolic variable, the result looks like ordinary mathematical notation.

```
> f(x);
```

$$\begin{cases} (x+1)^2 & x < -1 \\ -x & \mathit{otherwise} \end{cases}$$

For $g$ we must use two conditions.

```
> g:= x -> piecewise(x < -1, (x+1)^2,
>                    x >= -1 and x < 1, -x,
>                    sqrt(x-1));
```

We have, in order:

- the first condition `x < -1`
- the formula `(x+1)^2` to be used when the first condition is true
- the second condition `x >= -1 and x < 1`
- the formula `-x` to be used when the second condition is true
- the formula `sqrt(x-1)` to be used when none of the conditions are true (the "otherwise" piece).

Actually, you could leave out the "`x >= -1`" from the second condition because "`piecewise`" allows overlapping intervals: if the first condition is true it uses the first formula, otherwise if the second condition is true it uses the second formula, etc. I included the "`x >= -1`" because standard mathematical notation doesn't allow overlapping intervals. Notice also that you can't write a "double inequality" `-1 <= x < 1`; you must write it as two inequalities connected with `and`. ■

## Plotting With Jumps

There is a problem with plotting functions that have "jumps", such as $g$ of Example 2:

`> plot(g(x), x=-2..2);` (Figure 1.9)

Maple joins each point $(x, y)$ that it computes to the next one, so the graph looks like it has almost-vertical segments around the jumps. The solution is to include an option "`discont=true`" in the "`plot`" command:

---

(17) There is also another one called "`csgn`": for our purposes these are the same, as the distinction between them arises only for complex numbers.

(18) You assign the value you want for `signum(0)` to the variable "`_Envsignum0`".

```
> plot(g(x), x=-2..2, discont=true);
```
(Figure 1.10)

There are some situations (not necessarily involving the `piecewise` function, and more common in Release 4 than Release 5) in which "`discont=true`" doesn't work. In such a case a solution would be to combine several different "`plot`" commands with the "`display`" command[19]. Each "`plot`" command will deal with the function on one interval. When the interval doesn't contain its left or right endpoint, we move the endpoint in the "`plot`" command a tiny bit to the right or left respectively, e.g., $-1.001$ instead of $-1$, so that the value at $x = -1$ (which belongs to the second interval) is not plotted with the first interval.

```
> with(plots):
> display({ plot(g(x),x=-2..-1.001),
>              plot(g(x),x=-1..0.999),
>              plot(g(x),x=1..2)});
```

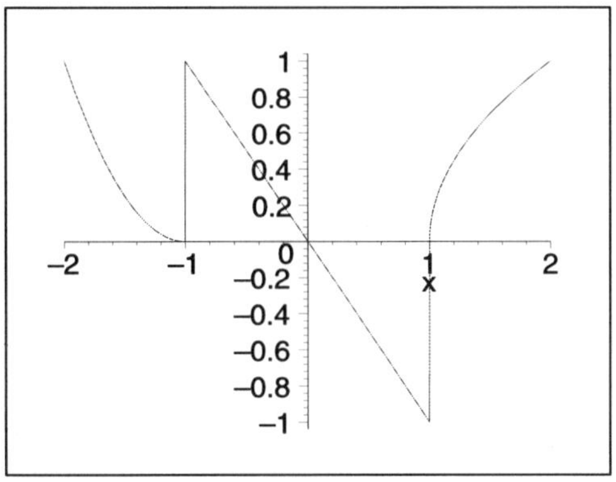


Figure 1.9. (Incorrect)

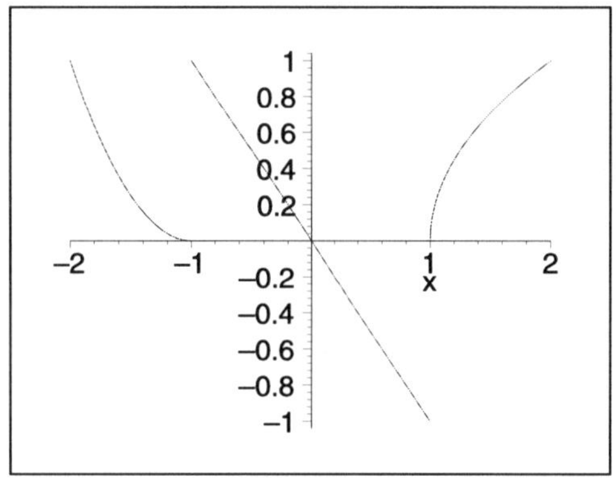


Figure 1.10. (Correct)

## Trigonometric Functions

Maple has all the usual trigonometric functions: `sin`, `cos`, `tan`, `cot`, `sec`, and `csc`. There are some notational points to watch out for: while in ordinary mathematics we can write $\cos x$ or $\cos^2 x$, in Maple these must be `cos(x)` and `cos(x)^2` respectively[20]. The "`simplify`" command can recognize most trigonometric identities. In order to prove an identity $a = b$, the trick is to simplify $a - b$. If the identity is true, this should return 0.

```
> simplify(cos(x)^4 - sin(x)^4 - cos(2*x));
```

$$0$$

The "`simplify`" command usually expresses all other trigonometric functions in terms of `sin` and `cos`. The "`normal`" command does some basic simplification without using trigonometric identities (and thus without expressing other functions in terms of `sin` and `cos`). The "`expand`" command will apply the addition formulas for the trigonometric functions, thus changing an expression involving a trigonometric function of a sum of terms into one involving products of functions of the terms. The reverse of this is "`combine(...,trig)`", which changes an expression involving products of trigonometric functions to one involving sums.

```
> normal((tan(x)/sec(x))/(1/sec(x) + 1));
```

$$\frac{\tan(x)}{\sec(x) + 1}$$

---

(19) Remember that "`display`" is part of the "`plots`" package, so you need "`with(plots):`" if it hasn't already been entered in this session.

(20) Actually, `(cos^2)(x)` would also work, as would `(cos(x))^2`. Maple displays them all as $\cos(x)^2$.

```
> simplify(tan(x)/(sec(x)+1));
```

$$\frac{sin(x)}{\cos(x)+1}$$

```
> expand(sin(4*x) + sin(2*x));
```

$$8\sin(x)\cos(x)^3 - 2\cos(x)\sin(x)$$

`> combine(",trig);` (Release 4) or

`> combine(%, trig);` (Release 5)

$$\sin(4x) + \sin(2x)$$

## EXERCISES

1. Let $f(x) = 2x^2 - 1$ and $g(x) = 2(x-1)^2$. Find, and simplify if possible:
   **(a)** $f(3)$ **(b)** $g(t+1)$
   **(c)** $f(x)^2 - f(x^2)$ **(d)** $(f \circ g)(x+1)$
2. Let $f(x) = x^2 - x$, and plot the following on the interval $-1 \le x \le 1$:
   **(a)** $f(x)$ **(b)** $f(2x-1)$
   **(c)** $2f(x) - 1$ **(d)** $f(f(x))$
3. Consider the function
   $$f(x) = \frac{x^4 + 4x^3 + 7x^2 + 6x + 1}{x^2 + 2x + 2}$$
   **(a)** About what line $x = a$ is the graph of $f$ symmetric? First plot the graph to guess the answer. Then verify it by finding the function corresponding to the reflection of this graph across $x = a$ and simplifying.
   **(b)** Find and simplify the function corresponding to shifting the graph of $f$ to the right by 2 and down by 7.
4. **(a)** Write the function $p(x) = \sqrt{x^2+1}$ as $f \circ g \circ h$ where $f$, $g$ and $h$ are simpler functions.
   **(b)** Check your answer to (a) by defining the functions in Maple and performing the composition with "`@`".
   **(c)** What other functions do you get by composing $f$, $g$, and $h$ in different ways?
5. Consider the function $f(x) = x + 1 - \sqrt{4x+1}$.
   **(a)** What is the domain of $f$? What is the range?
   **(b)** What is the domain of $f \circ f$?
   **(c)** Plot $f \circ f$ for $x$ from $-1$ to 1. The result appears to contradict (b). Can you explain what is going on? *Hint:* what are $f(-1/2)$ and $f(f(-1/2))$, according to Maple?
   **(d)** For what values of $x$ is $f \circ f(x) = x$, according to Maple's graphs?
6. Use "`plot`" to find, accurate to three significant digits, the range of the function
   $$f(x) = \frac{x}{x^2 + 20\cos 3x + 30}$$
7. Maple has trouble plotting the function
   $$f(x) = \frac{1}{1 + 2^{1/x}}$$
   on the interval $[-1, 1]$, because it tries to take 2 to a very large power near $x = 0$.
   **(a)** Find a way to produce this graph.
   **(b)** Note the symmetry in the graph of $f$. Use "`simplify`" to show that this symmetry is correct.

**8.** **(a)** Income tax rates in Fredonia are as follows. If a person's taxable income is less than 25 000 F-marks , the tax is 22% of that income. For taxable income over 25 000 F-marks, it is 5 500 F-marks on the first 25 000 plus 33% of the part of the taxable income over 25 000. Define this as a Maple function.

**(b)** The Fredonian government will now introduce a surtax in order to generate more revenue. If the basic tax as calculated in (a) is $T$, the surtax is $.03\,T$ or $.05(T - 4000)$, whichever is greater. Define the total tax (the basic tax plus the surtax) in Maple as a function of taxable income.

**(c)** Plot the basic tax, surtax, and total tax as functions of taxable income.

**(d)** What taxable income would result in a total tax of 13 000 F-marks?

**9.** **(a)** Express $\cos 5x$ in terms of $\cos x$ and $\sin x$.

**(b)** Express $\cos^5 x$ in terms of cosines of multiples of $x$.

**(c)** Maple does not seem to have a built-in command to express trigonometric functions in terms of `tan` and `sec`. However, you can do this using "`subs`". How? Try it out on the result of expanding $\sin(3x)$.

**10.** Define a function that is $\cos(x) - 1$ when $\sin(x) > 0$ and $1 - \cos(x)$ when $\sin(x) < 0$, and plot it on the interval $[-10, 10]$, making sure that the jumps are handled correctly. Where are the jumps?

**11.** *(For those interested in technical points about Maple)* The main difference between defining functions using "`->`" and using "`unapply`" is that "`unapply`" evaluates its inputs while "`->`" does not. Can you predict the results of the following sequence of commands? Try it out. Explain what happens.

```
> a:= 1;
> f:= x -> a + x;
> g:= unapply(a + x, x);
> a:= -1;
> f(x);
> g(x);
```

CHAPTER 2

# Limits and Differentiation

## LAB 3 LIMITS

**Goals:** *To investigate the concept of a limit, with help from Maple. To use Maple in calculating some limits.*

### Investigating a Limit

There are two main, and often separate, ways to use Maple in connection with a mathematical concept: as an aid to understanding, and as a tool for calculating. This lab is a good example of the separation. Maple has a "`limit`" function that works very well (with some exceptions), but before introducing it let's begin with a look at the concept of limit. Recall the (informal) definition of limit:

> Suppose $f(x)$ is defined for all $x$ near $a$ (on both sides), but possibly not at $a$ itself. Suppose we can always ensure that $f(x)$ is as close as we want to $L$ by taking $x$ close enough to $a$ (on either side of $a$). Then $L$ is the **limit** of $f(x)$ as $x$ approaches $a$, and we write
>
> $$\lim_{x \to a} f(x) = L.$$

Let's take a (non-obvious) example:

■ **EXAMPLE 1** Find the limit, if any, of $f(x) = \sin\left(\frac{1}{x}\right) - \sin\left(\frac{1}{x} - x\right)$ as $x \to 0$.

**SOLUTION** First we define the function "`f`" in Maple.

```
> f:= x -> sin(1/x) - sin(1/x - x);
```

$$f := x \to \sin\left(\frac{1}{x}\right) - \sin\left(\frac{1}{x} - x\right)$$

Does this have a limit as $x \to 0$? Let's go through the definition clause by clause. First, is $f(x)$ defined for all $x$ near $0$? Yes: the only place a problem arises in defining $f(x)$ is excluded by the "except possibly at $a$ itself". Next: can we ensure that $f(x)$ is as close as we want to $L$? What is $L$?

Take a minute or two to think about how you could attack this without Maple. You may be able to convince yourself that the limit should be 0. But it's not really obvious, and without any other aids you might be stuck here. Let's see how Maple can help. The first rule in this type of situation is:

> **When in doubt, plot!**

```
> plot(f(x), x=-1 .. 1);
```
(Figure 2.1)

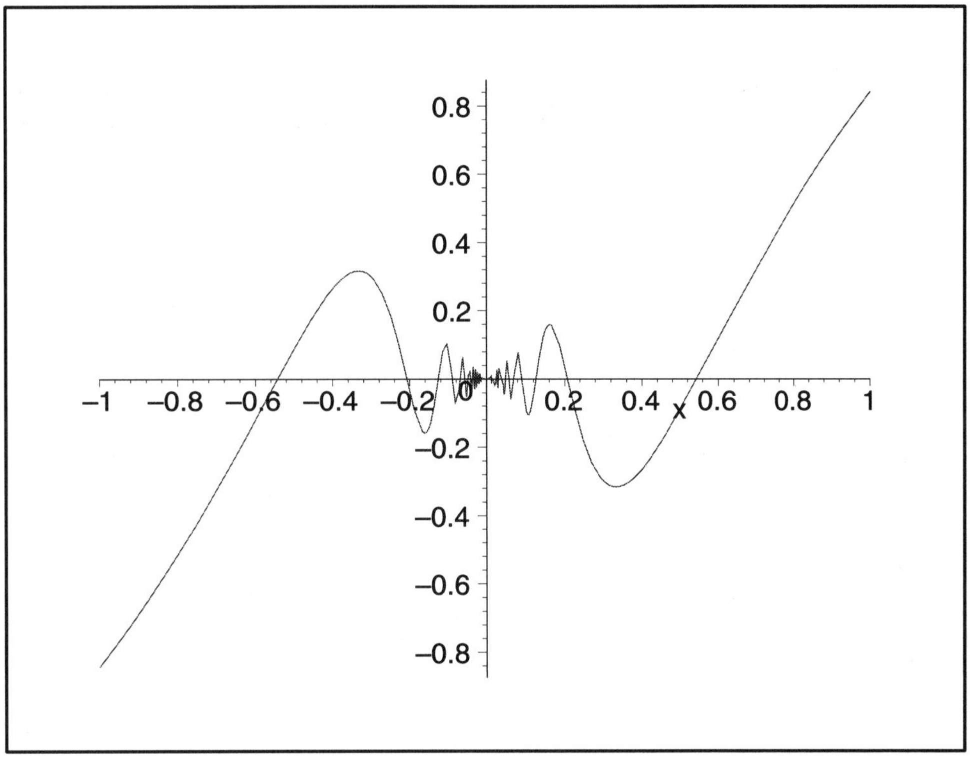


Figure 2.1. $y = \sin\left(\frac{1}{x}\right) - \sin\left(\frac{1}{x} - x\right)$

The graph does some strange "wiggling" near $x = 0$, but the important result looks fairly clear: $f(x)$ is near 0 when $x$ is near 0. If, for example, "as close as we want to 0" is anywhere between $-.2$ and $.2$, it looks like we can ensure this by taking $x$ between about $-.25$ and $.25$. That is, for every $x$ in the interval from $-.25$ to $.25$, $f(x)$ is between $-.2$ and $.2$. This is easier to see if we plot the lines $y = -.2$ and $.2$ as well as $f(x)$ on the same graph:

```
> plot({f(x), -.2, .2}, x=-1 .. 1);
```

I got the numbers $-.25$ and $.25$ by clicking the mouse cursor at the points where the curve crosses the lines closest to $x = 0$. Considering the limitations of the screen and the steadiness of my hand, the accuracy here is a bit doubtful. But there is really no need to be too precise in locating these points, as long as we err on the side of caution. So we might say, with more confidence, that we can ensure $f(x)$ is between $-.2$ and $.2$ by taking $x$ between $-.1$ and $.1$.

You might want $f(x)$ to be closer to 0, say between $-.01$ and $-.01$. Can this be ensured by taking $x$ close to 0, and if so, how close? We can't see it very well on this graph, so we might try "zooming in" toward the origin, using a smaller interval for $x$:

```
> plot({f(x), -.01, .01}, x=-.01 .. 0.01); (Figure 2.2)
```

Now the curve is doing some really serious wiggling, but it does seem to be staying between the lines $y = -.01$ and $y = .01$ for $x$ in this interval. There might be a bit of doubt near the endpoints of the interval, so to be safe we might say that we ensure $f(x)$ is between $-.01$ and $.01$ by taking $x$ between $-.005$ and $.005$.

It seems clear from the graphs that, however we define "as close as we want" to 0, we can ensure that $f(x)$ satisfies the criterion by taking $x$ sufficiently close to 0. If this is the case, we can say that 0 is the limit of $f(x)$ as $x$ approaches 0.

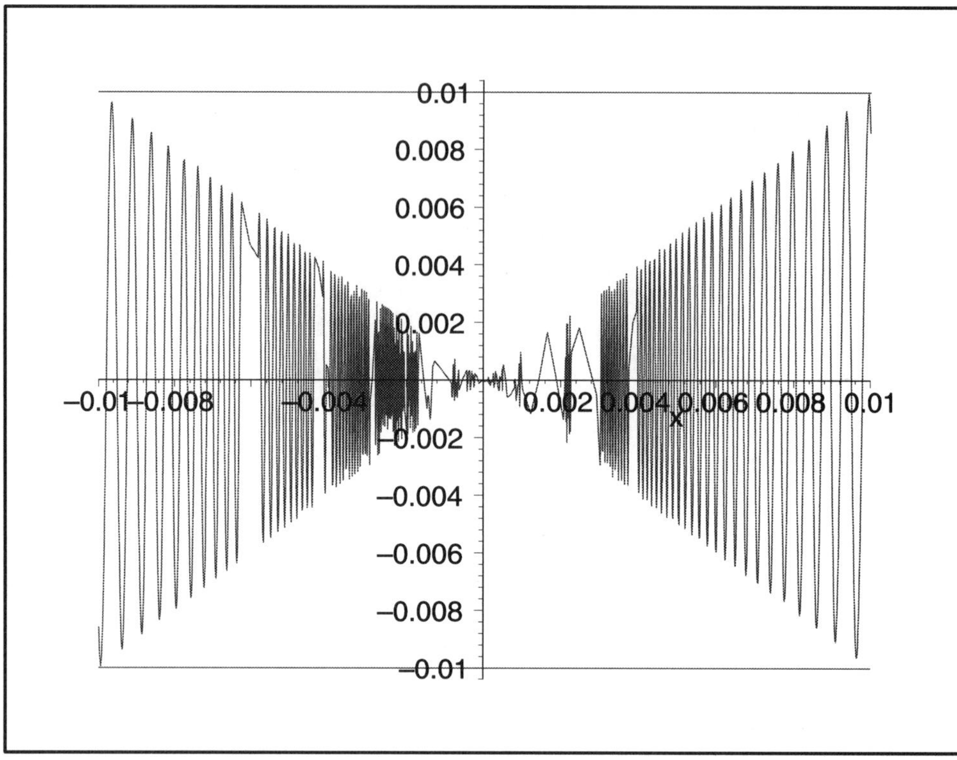


Figure 2.2. $y = \sin\left(\frac{1}{x}\right) - \sin\left(\frac{1}{x} - x\right)$

In fact, a close look at the graph seems to indicate that $f(x)$ is always closer to 0 than $x$ is, i.e., $|f(x)| < |x|$. We can test this by trying to find points where $f(x) = x$ or $f(x) = -x$:

```
> fsolve(f(x) = x);
```

$$\text{fsolve}\left(\sin\left(\frac{1}{x}\right) + \sin\left(-\frac{1}{x} + x\right) = x,\ x\right)$$

and similarly for $f(x) = -x$. The fact that Maple can't find a point where $f(x) = x$ or $f(x) = -x$ is not a proof that $|f(x)| < |x|$, but it is strong evidence for it (note that Maple doesn't have any trouble finding points with, say, $f(x) = 0.99x$). A proof that $|f(x)| < |x|$ would be the main ingredient in a proof that the limit is 0, because it would mean (in the formal definition of limit) that you could take $\delta = \epsilon$. Maple doesn't help much in constructing the proof itself, which in this case could be based on the fact that for any $t$ and $x \neq 0$, $|\sin(t) - \sin(t - x)| < |x|$. ■

## Tables of Values

An alternative to the graphical method for investigating the limit of $f(x)$ as $x \to a$ is to calculate $f(x)$ for some values of $x$ close to $a$. We should be careful to include values on both sides of $a$. If the limit is $L$, we should find that the $f(x)$ we calculate is close to $L$, and that if $x$ is very close to $a$, then $f(x)$ is very close to $L$.

To help in the presentation of these calculations, I've written a command called "`chart`". You can read its definition in to Maple from the file "`chart.def`" if that is available to you[1] or else type it in yourself: see Appendix I for the definition.

---

[1] It, as well as other files referred to in this book, is available on the Web at `http://www.geocities.com/CapeCanaveral/Campus/7705`

To illustrate the use of "chart", I'll make a table of $f(x)$ and $f(-x)$ for some values of $x$ near 0, using the $f$ from Example 1.

```
> read `chart.def`;
> chart([x, 'f(x)', 'f(-x)'], [ 0.1, 0.01, 0.001, 0.0001 ]);
```

| $x$ | $f(x)$ | $f(-x)$ |
|---|---|---|
| 0.1 | −0.0864852171 | 0.0864852171 |
| 0.01 | 0.0085977269 | −0.0085977269 |
| 0.001 | 0.0005627924 | −0.0005627924 |
| 0.0001 | −0.0000952171 | 0.0000952171 |

The two inputs to "chart" are both lists(2). The first list gives the headings $x$, $f(x)$, and $f(-x)$ for the table. The second is a list of values which will be substituted for $x$ (the first item in the headings list). For each row in the table below the headings, Maple substitutes the corresponding entry in the second list for the variable $x$ (which is the first item in the headings list).

The quotes around `'f(x)'` and `'f(-x)'` prevent Maple from evaluating these (and thus replacing them with the formulas for $f(x)$ and $f(-x)$) in the title line. Note that these are single forward quotes (apostrophes), not the back quotes that are used around file names.

From the chart, it appears that $f(x)$ is close to 0 when $x$ is close to 0. When $x$ is very close to 0, $f(x)$ is very close to 0. From this evidence we could guess that the limit is 0, although we shouldn't be too confident about this. Trying more values might increase our confidence. However, there are some pitfalls. In particular, as we will see in the next section, problems could arise if we took $x$ very much closer to 0.

## Numerical Difficulties

In some cases, numerical difficulties can arise with Maple's calculations. This is particularly common in cases where a limit involves the difference between two quantities that are very close together relative to their magnitude. When taking such a difference, there is a loss of significant digits. This loss of accuracy can be severe enough to seriously mess up our graphs and charts.

■ **EXAMPLE 2** Investigate the limit of $g(x) = (1 - \cos(x))/x^2$ as $x \to 0$.

**SOLUTION** Let's first try it with "chart". I'll put $\cos(x)$ in the table as well as $x$ and $g(x)$.

```
> g:= x -> (1-cos(x))/x^2;
> chart([x, 'g(x)', cos(x)], [seq(10^(-k), k=2 .. 7)]);
```

| $x$ | $g(x)$ | $\cos(x)$ |
|---|---|---|
| .01000000000 | .499996 | .9999500004 |
| .001000000000 | .5000 | .9999995000 |
| .0001000000000 | .50 | .9999999950 |
| .00001000000000 | 0 | 1.000000000 |
| $.1000000000\,10^{-5}$ | 0 | 1.000000000 |
| $.1000000000\,10^{-6}$ | 0 | 1.000000000 |

At first we seem to be doing quite well: $g(.01)$ is very close to .5. As we go to .001 and .0001, there are fewer digits of accuracy(3). Then at $x = .00001$, we have $\cos(x) = 1$ to the accuracy of our calculations, and so $g(x)$ is calculated as 0.

To some extent, the problem can be alleviated by taking a larger setting for "Digits"(4). Here is the same chart with 15 digits.

(2) Recall that a list in Maple is enclosed in square brackets, with items separated by commas.

(3) In our case, the limit happens to be exactly .5, but if it had been something like .54321 we would lose the last few digits.

(4) The default setting of 10 digits is really too few for most serious numerical work; a more reasonable value would be "`Digits:= 20`".

```
> Digits:= 15;
> chart([x, 'g(x)', cos(x)], [seq(10^(-k), k=2 .. 7)]);
```

| $x$ | $g(x)$ | $\cos(x)$ |
|---|---|---|
| .0100000000000000 | .49999583335 | .999950000416665 |
| .00100000000000000 | .499999958 | .999999500000042 |
| .000100000000000000 | .5000000 | .999999995000000 |
| .0000100000000000000 | .50000 | .999999999950000 |
| $.100000000000000\ 10^{-5}$ | .500 | .999999999999500 |
| $.100000000000000\ 10^{-6}$ | .5 | .999999999999995 |

This is better, but there is still a loss of digits as $x$ gets smaller, and for $x = 10^{-7}$ it looks like we would be back to computing $g(x) = 0$. No matter how many digits we use, there will always be some point at which this occurs. The definition of "limit" makes it appear that the closer $x$ is to $a$, the better $f(x)$ is as an approximation to $L$. That would be true if our calculations were perfectly accurate. In actual computations subject to roundoff error, however, it is not the case. We must avoid taking $x$ so close to $a$ that roundoff error becomes a serious problem. If we do want to take $x$ very close to $a$, we must perform the calculations with a large number of digits.

The graphical method has similar problems. Before plotting, I'll set "`Digits`" back to 10. However, the graph is actually calculated with the computer's hardware floating-point mechanism, which corresponds approximately to `Digits = 16`. Results here can vary somewhat: I made Figure 2.3 in Release 5 under Windows 95, but in Release 4 on the same machine I needed to make the interval `-10^(-8) .. 10^(-8)` to get a similar picture.

```
> Digits:= 10;
> plot(g(x), x = -10^(-7) .. 10^(-7), axes = FRAMED);
```
(Figure 2.3)

The option "`axes=FRAMED`" shows better what this does near $x = 0$, since the $y$ axis doesn't get in the way. You can also change the "`axes`" option in the plot's Axes menu, or click on the 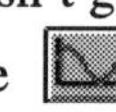 button in its context bar (click on the graph first to have these appear).

Near the edges of the graph, $g(x)$ is near .5, as it should be, but as you approach 0 the roundoff error appears in the form of wild oscillations. From about $-1.1 \times 10^{-8}$ to $1.1 \times 10^{-8}$, $\cos(x)$ is calculated as 1 so $g(x)$ is calculated as 0.

The problem is solved by taking a larger value for "`Digits`". Note that I'm specifying the $y$ axis values. Without this, Maple would choose the $y$ axis scale to fit the minimum and maximum values of $y$ it calculated (which would both be very close to 0.5), and this would magnify the effects of roundoff error so that it would still look bad.

```
> Digits:= 25:
> plot(g(x), x = -10^(-7) .. 10^(-7), 0 .. 1, axes = FRAMED);
```
(Figure 2.4)

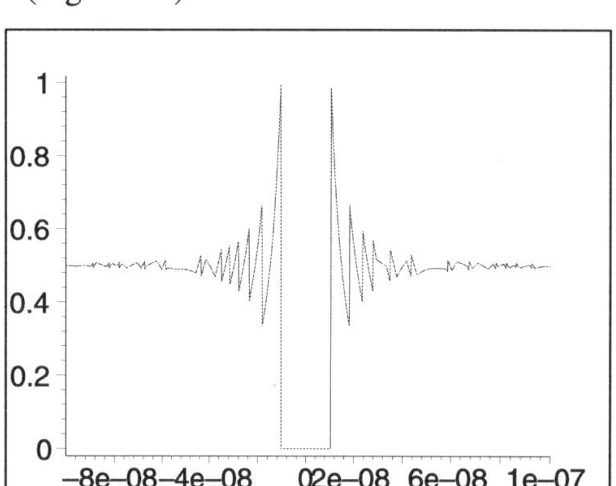


Figure 2.3. $(1 - \cos x)/x^2$ (bad)

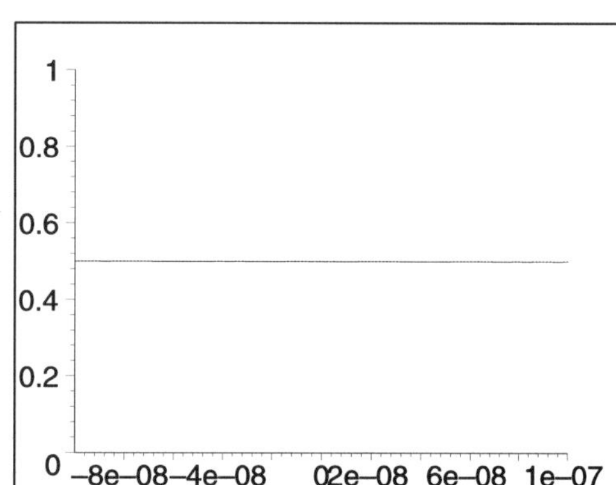


Figure 2.4. $(1 - \cos x)/x^2$ (good) ■

## Limits Involving Infinity

Our methods can be adapted for limits involving infinity.

■ **EXAMPLE 3** Investigate $\lim_{x\to\infty} \frac{x+1}{x-1}$.

**SOLUTION** The table-of-values method is quite straightforward. We try a list of large values of $x$.

```
> h:= x -> (x+1)/(x-1);
> chart([x, 'h(x)'],[10, 1000, 10^5, 10^7]);
```

| $x$ | $h(x)$ |
|---|---|
| 10. | 1.222222222 |
| 1000. | 1.002002002 |
| 100000. | 1.000020000 |
| .10000000 $10^8$ | 1.000000200 |

It seems quite clear that $h(x)$ is approaching a limit of 1.

Now let's look at it with our graphical methods. We begin with a plot on a finite interval, say $2 \le x \le 100$. This choice avoids $x = 1$, where the function "blows up".

`> plot(h(x), x=2..100);` (Figure 2.5)

We see that $h$ is decreasing on this interval. It appears that $h(x)$ is between 1 and 1.1 when $x$ is greater than 25 or so. We might actually solve $h(x) = 1.1$ to confirm this fact (the function is simple enough that we don't even need "`fsolve`"; we can even do it by hand). It is quite plausible from the graph that the limit is 1, although a skeptic might point out that we can't see what happens for values of $x$ larger than 100. For more evidence, we could try more intervals, such as $100 \le x \le 1000$ and $1000 \le x \le 10000$. But it would be better if we could somehow see the whole of $2 \le x < \infty$ all at once. We can even try an "infinity plot":

```
> plot(h(x), x=2 .. infinity);
```

This does seem to show $h(x)$ is decreasing for $2 \le x < \infty$, but doesn't provide quantitative evidence. Perhaps the best solution is to use a transformation: take $x = 1/t$.

`> plot(h(1/t), t=0..1/2);` (Figure 2.6)

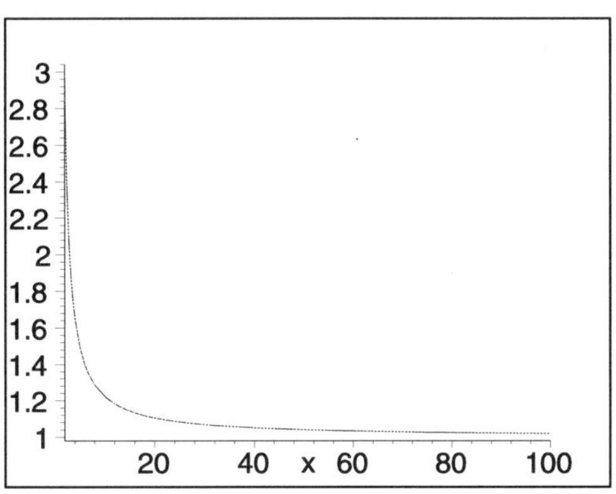


Figure 2.5. $y = \frac{x+1}{x-1}$

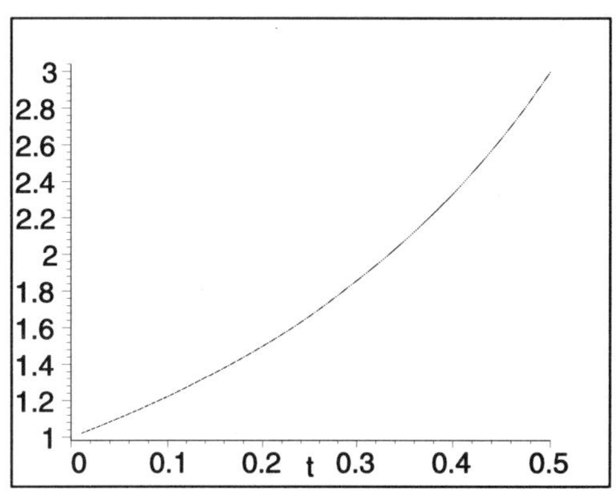


Figure 2.6. $y = h(1/t)$

Since $2 \le x < \infty$ corresponds to $0 < t \le 1/2$, this does show us the whole of $2 \le x < \infty$. It appears, for example, that $1 < h(t) < 1.1$ for $0 < t < 0.04$, which corresponds to $x > 25$. We can "zoom in" toward $t = 0$ to get more information about large values of $x$. In fact, for any function $F$, $\lim_{x\to\infty} F(x)$ and $\lim_{t\to 0+} F(1/t)$ are the same, so all our previous methods can be used. ■

In investigating a limit (say $\lim_{x\to a} F(x)$) that you suspect is infinite, it might be useful to use another transformation: $G(x) = 1/F(x)$. It is easy to see that $\lim_{x\to a} F(x) = \infty$ corresponds to $\lim_{x\to a} G(x) = 0$ with $G(x) > 0$ for all $x$ near $a$. See Exercise 6.

## Maple Calculates Limits

For actually calculating limits, Maple has the "`limit`" command: `limit(`*expr*`, x = a)` finds the limit of the expression *expr* as the variable "x" approaches "a". It uses algebraic methods rather than the numerical and graphical approach we have been trying.

```
> limit(g(x), x = 0);
```

$$\frac{1}{2}$$

Maple also knows about one-sided limits, limits at infinity, and infinite limits. For $\lim_{x\to a+} f(x)$ and $\lim_{x\to a-} f(x)$ you use "`limit(f(x), x=a, right)`" and "`limit(f(x), x=a, left)`" respectively.

```
> limit(2^(1/x), x=0);
```

*undefined*

```
> limit(2^(1/x), x=0, left);
```

$$0$$

```
> limit(2^(1/x), x=0, right);
```

$$\infty$$

```
> limit((x+1)/x, x=-infinity);
```

$$1$$

In some cases where there is no limit (or Maple can't find it) but all the values of the expression approach a certain range as $x$ approaches $a$, Maple will return that range:

```
> limit(sin(1/x), x=0);
```

$$-1..1$$

And what about the function $f(x) = \sin(1/x) - \sin(1/x - x)$ of Example 1?

```
> limit(f(x), x=0);
```

$$-2..2$$

Maple was unable to find the answer. It just made the obvious observation that each value of `sin` is between $-1$ and 1, so the difference of the two is between $-2$ and 2. This, you might say, was a hard problem, and it's not so surprising that Maple can't solve it. There are also some cases where Maple's answer is not what we might expect, due to the fact that Maple allows complex values:

```
> limit(sqrt(1-x),x=1);
```

$$0$$

We should say that the limit doesn't exist, because $\sqrt{1-x}$ is undefined when $x > 1$; only the one-sided limit as $x \to 1-$ should be 0. But Maple is perfectly happy with square roots of negative numbers, and gives the correct answer in the context of complex numbers. Maple is wrong only in the context of real numbers, which is what we are working with now.

## EXERCISES

**1.** Investigate the limit of each of the following functions $f(x)$ as $x \to 1$, using graphical methods. Does the limit exist, and if so, what is it? Compare your answer to that obtained using "`limit`". Can you ensure that $f(x)$ is within $0.1$ of some number $L$ by taking $x$ close enough to 1? What about $0.01$? What about an arbitrary $\epsilon > 0$?

**(a)** $f(x) = x^2 - 3x$

**(b)** $f(x) = \sqrt{x}$

**(c)** $f(x) = \dfrac{\sin(x-1)}{1-x^2}$

**(d)** $f(x) = \operatorname{sgn}(\sin(1/x))$

**2.** Investigate $\lim\limits_{x\to 0+} x^{\sin x}$ graphically. How small must $x$ be (with $x > 0$) to ensure that $x^{\sin x}$ is between $0.9$ and $1.1$? What about between $0.99$ and $1.01$?

**3.** Investigate the following limits graphically and using "`chart`". Determine the limit to four decimal places. Compare to Maple's answer for the limit.

**(a)** $\lim\limits_{x\to 1} \dfrac{x^2-1}{x^2+\sqrt{x}-2}$

**(b)** $\lim\limits_{x\to 0} (\cos 2x)^{1/x^2}$

**4.** A student tried to investigate $\lim\limits_{x\to 0} \sin(\pi/x)$ using "`chart`" with $x = \pm 0.1, \pm 0.01, \pm 0.001$, etc. What went wrong with this approach?

**5.** This exercise shows some of the limitations of graphical techniques for limits. Consider $\lim\limits_{x\to 0+} f(x)$ where $f(x) = \dfrac{x}{x^{\sin x}-1}$.

**(a)** Plot $f(x)$ on the intervals $0 \le x \le 0.5$, $0 \le x \le 0.05$, and $0 \le x \le 0.005$. What does the limit appear to be in each case? Compare to Maple's answer for the limit.

**(b)** Can you find an $x$ for which $|f(x)| \le 0.1$? $0.01$? *Hint:* you'll need to increase "`Digits`" to at least 50 for this one. Can you explain what is happening in (a)?

**(c)** Now consider $g(x) = |\sin(1/x)|^{1/x^2}$. Plot $g(x)$ on the same intervals used in (a). What does $\lim\limits_{x\to 0+} g(x)$ appear to be?

**(d)** What is $g(x)$ when $1/x$ is an odd multiple of $\pi/2$? What does this say about the limit in (c)?

**6.** Investigate the following limits graphically, using transformations where appropriate.

**(a)** $\lim\limits_{x\to\infty} x^{1/x}$

**(b)** $\lim\limits_{x\to 1+} \dfrac{\sqrt{x-1}}{\sqrt{x}-1}$

**(c)** $\lim\limits_{x\to -\infty} \left((x^2+1)^{3/2} + x^3\right)$

**7. (a)** Which of the following expressions is difficult to calculate accurately when $x$ is close to 0 because of roundoff error?

**(i)** $\dfrac{\sqrt{1+x^2/3}-\sqrt{1-x^2/2}}{\sin^2 x}$ **(ii)** $\dfrac{\sin x}{x+3x^2}$

**(b)** How large must "`Digits`" be in order to have Maple's value for this expression with $x = .00001$ be within $.001$ of its limit as $x \to 0$? Within $.00001$? Do you think it is ever within $10^{-20}$?

# LAB 4 SLOPES AND DERIVATIVES

**Goals:** *To use Maple as an aid to understanding the concept of derivative, by looking at the slopes of secant lines and tangent lines. To see how to calculate derivatives using Maple.*

## Secant Lines and Tangent Lines

Consider a secant line joining two points $P = (x_0, f(x_0))$ and $Q = (x_0 + h, f(x_0 + h))$ on the graph of a function $f(x)$. The slope of the secant line is

$$q = \frac{f(x_0 + h) - f(x_0)}{h}$$

(this expression is called a **Newton quotient** or **difference quotient** for $f$ at $x_0$), and so the secant line itself has equation

$$y = f(x_0) + q(x - x_0)$$

The tangent line (if it exists and is non-vertical) has slope

$$m = \lim_{h \to 0} q = \lim_{h \to 0} \frac{f(x_0 + h) - f(x_0)}{h}$$

so its equation is

$$y = f(x_0) + m(x - x_0)$$

We can do this in Maple. Note: be sure to make the following definitions before assigning values to "`f`", "`x0`" or "`h`". If any of these names has already been assigned a value in this session, you can "unassign" it:

```
> f:= 'f';
```

If there are a lot of variables to unassign and none that you want to keep, an alternative is the "`restart`" command, which will undo everything and return Maple to its initial state.

```
> q:= (f(x0 + h) - f(x0))/h;
> secline:= x -> f(x0) + q * (x - x0);
> m:= limit(q, h=0);
> tanline:= x -> f(x0) + m * (x - x0);
```

I won't bother to record Maple's response to any of these except for the "`m:=`" line, which is

$$m := \mathrm{D}(f)(x0)$$

As we will see, this is how Maple writes the derivative of the function $f$ at $x_0$. So Maple knows the definition of derivative.

We can now choose a typical function $f$ and some values for $x_0$ and $h$, and plot the function, the secant line, and the tangent line together:

```
> f:= x -> 1 + 2*x - x^3;
> x0:= 1;
> h:= 1/2;
> plot({ f(x), tanline(x), secline(x)}, x=0 .. 2); (Figure 2.7)
```

To see what happens as $h$ approaches 0, we might try $h = 1/2, 1/4, 1/8$, and $1/16$ on the same plot. Here is a bit of a trick, to get these values of $h$ as $k$ goes from 1 to 4:

```
> h:= 1/2^k;
> plot({ f(x), tanline(x),
>        seq( secline(x), k=1 .. 4) }, x = 0 .. 2);
```

We should also try negative $h$ values.

```
> h:= -1/2^k;
> plot({ f(x), tanline(x),
>          seq( secline(x), k=1 .. 4) }, x = 0 .. 2);
```

We could also try an animation:

```
> h:= 'h';
> with(plots):
> animate({ f(x), tanline(x), secline(x)},
>              x=0 .. 2, h=-1/2 .. 1/2);
```

## Tangent Lines and the Derivative

We now want to think of the slope $m$ as depending on $x_0$, so let's unassign the variable "`x0`". In the next animation, Maple shows the function $f(x)$, its derivative (written as "`D(f)(x)`"), and its tangent lines, each frame using a different $x_0$. Observe that the slope of each tangent line is the value of "`D(f)`" at the point of tangency. In particular, this is zero at the points where the tangent line is horizontal. To get both of those points in the picture, I will use the interval `-1.2 .. 1.2` this time.

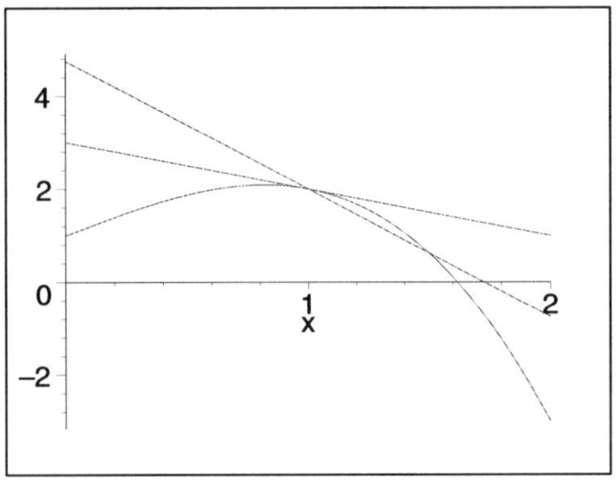


Figure 2.7. Secant and tangent lines

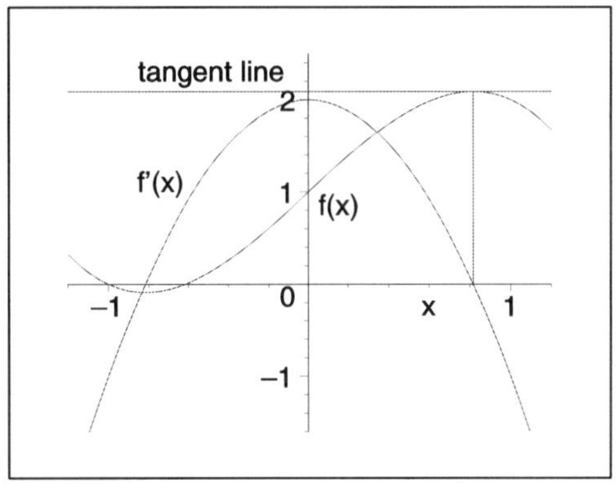


Figure 2.8. Derivative and tangent line

```
> x0:= 'x0';
> animate({ f(x), D(f)(x), tanline(x)},
>                x=-1.2 .. 1.2, x0 = -1.2 .. 1.2);
```

Now let's plot this in one of the cases where the tangent line is horizontal. I will use "`solve`" (although this example is simple enough to do by hand). More difficult examples, where "`solve`" can't find a solution, require "`fsolve`". Note that, when used on a polynomial, "`solve`" returns all of the roots. We can assign this collection of roots to a variable "`r`", and then refer to the individual roots as "`r[1]`", "`r[2]`", etc.

```
> D(f)(x);
```

$$2 - 3x^2$$

```
> r:= solve(D(f)(x) = 0, x);
```

$$r := \frac{1}{3}\sqrt{6}, -\frac{1}{3}\sqrt{6}$$

The order in which Maple prints the roots may vary from session to session, so to get the same picture as mine you should check which root you want.

```
> x0:= r[1];
```

$$x0 := \frac{1}{3}\sqrt{6}$$

```
> plot({ f(x), D(f)(x), tanline(x)}, x=-1.2 .. 1.2);
```

## An Extra Touch

There's still one improvement possible: a vertical line at $x = x_0$ to indicate the point of tangency (the tangent line is so close to the curve that it's hard to tell exactly where they touch). First I'll save the last plot by assigning it to a name. To avoid having to see the whole plot structure printed out, I'll use "`:`" instead of "`;`".

> `p1:= ":` (Release 4), or

> `p1:= %:` (Release 5)

Next I assign a plot of the vertical line to another name. The command "`plot([[a,b], [c,d]])`" draws a straight line joining points $(a, b)$ and $(c, d)$.

```
> p2:= plot([ [x0, 0], [x0, f(x0)] ]):
```

Finally, the "`display`" command[5] will put both together.

> `display({ p1, p2 });` (Figure 2.8)[6]

## Maple Does Derivatives

There are actually two differentiation operators in Maple: "`diff`", which is most useful for expressions, and "`D`", which is most useful for functions.

If "`f`" is a function of one variable, "`D(f)`" is its derivative. This is also a function of one variable. As with any function, you evaluate it at some "`p`" by putting "`p`" in parentheses after the function: "`D(f)(p)`", with two sets of parentheses.

```
> f:= x -> x^2 + a * x;
```

$$f := x \to x^2 + ax$$

```
> D(f);
```

$$x \to 2x + a$$

```
> D(f)(p);
```

$$2p + a$$

If Maple doesn't know how to differentiate a function (in particular, if that function hasn't been defined), it will leave "`D`" of that function as is.

```
> D(fnew)(p);
```

$$\mathrm{D}(fnew)(p)$$

If "`expr`" is an expression, "`diff(expr,x)`" is its derivative, where "`x`" is the independent variable. This is also an expression. The variable "`x`" must not be assigned a value when you use "`diff`", or that value would be used instead of `x` and would probably cause an error. You can use "`subs`", or assign the derivative to a name before assigning a value to "`x`".

```
> expr:= x^2 + a * x;
```

$$expr := x^2 + ax$$

```
> v:= diff(expr,x);
```

$$v := 2x + a$$

```
> subs(x=2,v);
```

$$4 + a$$

```
> x:= 2:
> diff(expr,x);
Error, wrong number (or type) of parameters in function diff
```

---

(5) In the "`plots`" package.

(6) The labels on the curves and tangent line were produced with the "`textplot`" command, which we will meet in Lab 14.

(assigning a value to "x" caused this error)

```
> v;
```

$$4 + a$$

An unevaluated "`diff(expr,x)`" is shown in Maple's output as $\frac{\partial}{\partial x}$ *expr* instead of $\frac{d}{dx}$ *expr*. You will meet "$\partial$" again in calculus of several variables; for now, you can think of it as being the same as "$d$".

Any variable, other than the independent variable, that hasn't been given a value is considered to be a constant when differentiating. An example is "a" in "`D(f)`" and "`diff(expr,x)`" above.

In "`diff(expr,x)`", Maple uses the values of any variables that occur in "`expr`". So if "a" had the value $2x$ in the example above, "`expr`" would have been taken as $3x^2$ and "`diff(expr,x)`" would have been calculated as $6x$.

On the other hand, "D" does not use the values of variables. It considers any name, other than the independent variable, occurring in the definition of "`f`" as a constant. So "`D(f)`" would be $x \to 2x + a$ regardless of the value of "a"; if "a" was $2x$ then `D(f)(x)` would be calculated as $4x$. Therefore, when using "D" you should be careful to avoid using any variables defined as expressions involving the independent variable.

"D" also does not work with functions defined in terms of other user-defined functions. Thus, in trying to avoid the problems of the last paragraph, you might be tempted to try

```
> a:= x -> 2 * x:
> f:= x -> x^2 + a(x):
> D(f);
```

$$\mathrm{D}(f)$$

Maple just refuses to differentiate this function! A way to work around this is to use "`unapply`" to define the function "`f`", so that the definition does not actually contain "a".

```
> f:= unapply(x^2 + a(x), x);
```

$$f := x \to x^2 + 2x$$

```
> D(f);
```

$$x \to 2x + 2$$

Since differentiating a formula is a purely mechanical calculation, it is not surprising that Maple can calculate the derivative of just about any function you might write down. It uses essentially the same rules that you learn in calculus (product rule, chain rule, etc.). There are just a few surprises.

```
> D(abs);
```

$$a \to \mathrm{abs}(1, a)$$

It looks a bit strange, but it has the right values: 1 if $a$ is positive, $-1$ if $a$ is negative, and undefined if $a$ is 0. In fact:

```
> D(abs)(0);
Error, (in simpl/abs) abs is not differentiable at 0
```

Similarly,

```
> D(signum);
```

$$a \to \mathrm{signum}(1, a)$$

where $\mathrm{signum}(1, a)$ is 0 if $a \neq 0$ and returns an error if $a = 0$.

For a function defined "piecewise", you can't trust "`D`" at the endpoints of the intervals. In those cases, you could use "diff" and "subs" or "eval", or resort to the definition of derivative as a limit. For example, consider the function

$$f(x) = \begin{cases} 0 & \text{for } x < 0 \\ x & \text{for } x \geq 0 \end{cases}$$

```
> f:= x -> piecewise(x < 0, 0, x);
> D(f)(0);
```

1 **(WRONG!!)**

`> eval(diff(f(x),x), x = 0);` (Release 5) or

`> eval(subs(x=0, diff(f(x),x)));` (Release 4)

*undefined* **(RIGHT)**

```
> limit((f(h) - f(0))/h, h=0);
```

*undefined*

The reason this derivative is undefined at $x = 0$ is that the limits of the Newton quotient as $h \to 0$ from the left and right are different.

```
> limit((f(h) - f(0))/h, h=0, left);
> limit((f(h) - f(0))/h, h=0, right);
```

0

1

## Exceptional Points

We must often deal with a function defined by a certain expression except for one or two points, where the original expression may be undefined but a certain value is specified, e.g.,

$$f(x) = \begin{cases} 1/x & \text{if } x \neq 0 \\ 0 & \text{if } x = 0 \end{cases}$$

The simplest way to deal with this in Maple is to define $f$ using the expression, and then define the value of $f$ at the special points:

```
> f:= x -> 1/x;
> f(0):= 0;
```

Be sure to enter the "`f:=...`" before the special values, as "`f:=`" will wipe out any special values defined previously.

## EXERCISES

1. Use Maple to find the derivatives of the following functions. Then plot the function and its derivative on the same graph, for $-2 \le x \le 2$. How can you tell from the graph which curve is the original function and which is the derivative?

   **(a)** $x^2 + x - 1$ **(b)** $x(x-1)(x+2)$

   **(c)** $\dfrac{x^2 - 1}{\cos(\pi x/2)}$ **(d)** $\sin(x) + \cos(3x)$

2. An alternative to the graphical approach in investigating slopes and derivatives is to make a table showing the slope of the secant line through $(x_0, f(x_0))$ and $(x_0 + h, f(x_0 + h))$ for various values of $h$ approaching 0. Try this, using the "`chart`" command from Lab 3, for the following functions and points. Be sure to take both positive and negative $h$ values, and watch out for roundoff error that could cause difficulty if $h$ is extremely small.

   **(a)** $\sin(x)$, $x = \pi/4$ **(b)** $1/\sqrt{x^2+1}$, $x = 1$

   **(c)** $\sqrt{1 - \sin(x)}$, $x = \pi$ **(d)** $\operatorname{sgn}(x)\sqrt{1-\sin(x)}$, $x = \pi$

3. Investigate graphically the secant lines to the graphs of the following functions through the given points. Does a tangent line exist? If so, what is it? If Maple doesn't find the slope successfully using "`D`", you should try taking the left and right limits of the Newton quotient $(f(x_0 + h) - f(x_0))/h$.

   **(a)** $(x+1)/(x^2+1)$, point $(1,1)$ **(b)** $|x^2 - 1|$, point $(1,0)$

   **(c)** $\begin{cases} -\sqrt{|x|} & \text{if } x < 0 \\ \sqrt{x} & \text{if } x \ge 0 \end{cases}$, point $(0,0)$ **(d)** $\begin{cases} x^2\sin(1/x) & \text{if } x \ne 0 \\ 0 & \text{if } x = 0 \end{cases}$, point $(0,0)$

   **(e)** $\begin{cases} x^2 & \text{if } x \le 1/2 \\ x - 1/4 & \text{if } x > 1/2 \end{cases}$, point $(1/2, 1/4)$

4. Consider the curve $y = x^2/(x-1)$. Where does it have horizontal tangents? Plot these together with the curve. Do the same for the tangents of slope $-1$. Use reasonable $x$ and $y$ intervals so that the points of tangency are visible.

5. **(a)** What are all the possible slopes of tangent lines to the curve $y = x^2/(x-1)$? *Hint:* solve the equation $y' = m$.

   **(b)** For what values of $m$ is there exactly one point on the curve where the tangent line has slope $m$?

   **(c)** What are all the possible slopes of secant lines for this curve?

6. Where is the function $f(x) = |x^3 - x| - 2|x - 1|$ non-differentiable? Where does it have a horizontal tangent? Where does it have a tangent of slope 2?

7. To see that Maple knows the differentiation rule for sums, you could enter

   ```
   > D(f+g)(x);
   ```

   where `f`, `g`, and `x` have not been given values. How does Maple express the Product Rule, Reciprocal Rule, Quotient Rule, and Chain Rule?

8. Consider the curve $y = x/(x^2 - 1)$.

   **(a)** At what other point(s) does a tangent line at $(x_1, y_1)$ intersect the curve? *Hint:* $f(x_2) = f(x_1) + f'(x_1)(x_1 - x_2)$. Discard the case $x_2 = x_1$.

   **(b)** Can the same line be tangent to this curve at two or more different points? *Hint:* this would mean $f'(x_2) = f'(x_1)$ in addition to the condition of part (a).

**9.** For **rational functions** (i.e., quotients of polynomials), you can greatly reduce difficulties with roundoff error in computing difference quotients if you use the "`normal`" command on the expression before evaluating it at a particular $h$. This essentially performs the same trick we use in calculating a derivative "from the definition", simplifying the difference quotient and canceling a factor of $h$ from the numerator and denominator. Try this for $f(x) = (x^3 - x)/(x^2 + x + 1)$ at $x_0 = 2$. Use "`chart`" to produce a table showing the difference quotient and the "`normal`" version of the difference quotient for various small values of $h$. Include values small enough that the difference quotient (in its original form) evaluates to 0.

# LAB 5 HIGHER DERIVATIVES, IMPLICIT DERIVATIVES, AND ANTIDERIVATIVES

**Goals:** *To learn to use Maple in calculating higher-order derivatives (including finding patterns for them), derivatives of implicit functions, and antiderivatives.*

## Higher-Order Derivatives

Both of Maple's differentiation operators can be used to calculate higher-order derivatives. Using "`D`", the $n$th derivative of a function "`f`" is the function "`(D@@`$n$`)(f)`". Remember to put parentheses around both the "`D@@`$n$" and the function name. If you are evaluating the function at $x$, the $x$ also goes in parentheses, so you have three sets of parentheses. Using "`diff`", the $n$th derivative of an expression "`expr`" is "`diff(expr,x$n)`", where "`x`" is the independent variable.

```
> (D@@3)(f)(x);
```

$$D^{(3)}(f)(x)$$

```
> diff(expr(x),x$3);
```

$$\frac{\partial^3}{\partial x^3} expr(x)$$

Corresponding to mathematical notation for $f^{(1)}$ and $f^{(0)}$, "`(D@@1)`" is the same as "`D`", and "`(D@@0)(f)`" is "`f`". On the other hand, while "`diff(...,x$1)`" is the same as "`diff(...,x)`", "`diff(...,x$0)`" does **not** work (it produces an error message).

■ **EXAMPLE 1** Investigate the higher-order derivatives of $f(x) = x \cos(2x)$.

**SOLUTION** We begin by defining $f$ and taking a few derivatives.

```
> f:= x -> x * cos(2*x);
> D(f)(x);
```

$$\cos(2x) - 2x \sin(2x)$$

```
> (D@@2)(f)(x);
```

$$-4 \sin(2x) - 4x \cos(2x)$$

We could proceed one derivative at a time, or save some typing by using "`seq`" to calculate many in one command. Let's go up to the sixth derivative:

```
> seq((D@@k)(f)(x), k=3..6);
```

$$-12\cos(2x) + 8x\sin(2x),$$
$$32\sin(2x) + 16x\cos(2x),$$
$$80\cos(2x) - 32x\sin(2x),$$
$$-192\sin(2x) - 64x\cos(2x)$$

Can you discern the pattern in all this? The $n$th derivative is a combination of $\cos(2x)$ and $x\sin(2x)$ (if $n$ is odd) or $x\cos(2x)$ and $\sin(2x)$ (if $n$ is even). The coefficient of $x\sin(2x)$ or $x\cos(2x)$ in $f^{(n)}(x)$ is $\pm 2^n$, $+$ if $n$ or $n+1$ is divisible by 4, $-$ otherwise. The coefficient of $\sin(2x)$ or $\cos(2x)$ is $\pm n2^{n-1}$, $+$ if $n$ or $n-1$ is divisible by 4, $-$ otherwise. Thus when $n$ is even, say $n = 2m$, we should have

$$f^{(2m)}(x) = m(-4)^m\sin(2x) + (-4)^m x\cos(2x)$$

The derivative of this should be the formula for $f^{(2m+1)}$.

```
> f2m:= x -> m * (-4)^m * sin(2*x) + (-4)^m * x * cos(2*x);
> D(f2m)(x);
```

$$2m(-4)^m\cos(2x) + (-4)^m\cos(2x) - 2(-4)^m x\sin(2x)$$

One more derivative gives us $f^{(2m+2)}$, which should be the same as the formula for $f^{(2m)}$ with $m$ replaced by $m+1$. Let's check this.

```
> a:= (D@@2)(f2m)(x);
```

$$a := -4m(-4)^m\sin(2x) - 4(-4)^m\sin(2x) - 4(-4)^m x\cos(2x)$$

```
> b:= subs(m=m+1,f2m(x));
```

$$b := (m+1)(-4)^{(m+1)}\sin(2x) + (-4)^{(m+1)}x\cos(2x)$$

```
> simplify(a-b);
```

$$0$$

After checking that the formula works for $m = 0$, we have all the ingredients for a proof of our formulas by mathematical induction. ■

## Implicit Differentiation

■ **EXAMPLE 2** Find $dy/dx$ where $xy^3 - x^3y = 1$. What is it when $x = 2$?

**SOLUTION** We want to consider the equation as implicitly defining $y$ as a function of $x$. Maple can actually solve this cubic equation, but the solution is quite complicated, so it's better to stay with the implicit definition.

```
> eq1:= x * y^3 - x^3 * y = 1;
```

It's almost always a good idea to visualize the situation first. Recall that "`implicitplot`" is in the "`plots`" package. When you only need one command from a package, it's not necessary to load the whole package using "`with`". You can refer to that command with the name of the package, followed by the name of the command in square brackets: "`plots[implicitplot]`". Or you can load just the one command from the package, with the command "`with(plots, implicitplot)`". These methods can reduce memory usage (the difference is not very much, but it might be significant on a system with very limited memory).

```
> plots[implicitplot](eq1,
>        x=-4 ..  4, y=-3 ..  3);
```

(Figure 2.9)

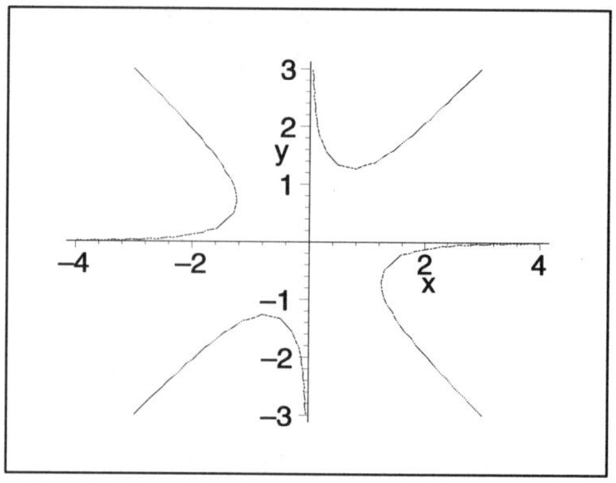

Figure 2.9. $xy^3 - x^3y = 1$

By the way, you may be wondering how I came up with the numbers `-4..4` and `-3..3`. Basically it's the result of a trial-and-error process: usually my first try would be on a fairly large scale (maybe `-10..10`), and then I would "zoom in" a bit, being careful to have any interesting features of a graph visible on the screen. I made the $y$ interval $3/4$ times the $x$ interval, because that is the height/width ratio of the printed image: with these proportions, the scale is the same in both directions.

Now we want Maple to consider $y$ as a function of $x$, so we change "`y`" to "`y(x)`" in the equation.

```
> eq2:= subs(y=y(x), eq1);
```

$$eq2 := xy(x)^3 - x^3y(x) = 1$$

When we apply "`diff`", Maple differentiates both sides of the equation.

```
> eq3:= diff(eq2, x);
```

$$eq3 := y(x)^3 + 3xy(x)^2\left(\frac{\partial}{\partial x}y(x)\right) - 3x^2y(x) - x^3\left(\frac{\partial}{\partial x}y(x)\right) = 0$$

This can now be solved for $dy/dx$:

```
> yprime:= solve(eq3, diff(y(x), x));
```

$$yprime := -\frac{y(x)^3 - 3\,x^2y(x)}{3\,xy(x)^2 - x^3}$$

As usual, our answer involves both $x$ and $y$. It is quite reasonable that to know the slope, we need to know both $x$ and $y$ coordinates of a point on the curve, because a single $x$ value can correspond to several different $y$ values: the implicit equation defines not just one but several different functions $y(x)$. Now: what happens at $x = 2$?

```
> ys:= fsolve(subs(x=2, eq1), y);
```

$$ys := -1.934297876, -.1254940943, 2.059791970$$

Since `subs(x=2,eq1)` is a polynomial, "`fsolve`" finds all its real roots. In this case, there are three of them, corresponding to the three points where the vertical line $x = 2$ intersects the curve in Figure 2.9. I put the roots in the variable "`ys`", where they can be accessed as "`ys[1]`", "`ys[2]`", and "`ys[3]`". What are the corresponding $dy/dx$ values? We want to substitute 2 for $x$ and each "`ys`" value in turn for $y(x)$ in "`yprime`". The "`subs`" command can do several substitutions at once: in the form "`subs(eqn1, eqn2, ...eqnn, expr)`" it will first substitute equation "`eqn1`" into "`expr`", then substitute "`eqn2`" into the result of this, etc.

Since "`subs`" works step-by-step, if we first substitute 2 for $x$ we must substitute the "`ys`" value for $y(2)$, not $y(x)$ (there is no more $y(x)$ when "`subs`" does the second substitution, because the $x$ has been changed to 2).

```
> seq(subs(x=2, y(2)=ys[j], yprime), j=1..3);
```

$$-1.105566345, .1902411420, .9153252040$$

■

■ **EXAMPLE 3** Where does the curve of Example 2 have horizontal and vertical tangents?

**SOLUTION** To find horizontal tangents, we want to solve simultaneously "`yprime = 0`" and our original equation. The "`solve`" command can do it:

```
> s:= solve({yprime=0,eq2},{x,y(x)});
```

$$s := \left\{x = \frac{2}{9}\text{RootOf}\left(4\_Z^8 - 27\right)^5,\ y(x) = \text{RootOf}\left(4\_Z^8 - 27\right)\right\}$$

Note the braces "`{}`" around both the set of equations to solve and the set of variables to solve for. Maple often uses "`RootOf`" to represent irrational solutions of a polynomial, with variable "`_Z`".

To get numerical values, you can use "`evalf`". This uses only one of the roots of the polynomial, which is not a big problem here (for the other real root, just change the signs of both $x$ and $y$):

```
> evalf(s);
```

$$\{x = -.7329972488,\ y(x) = -1.269588476\}$$

On the other hand, "`evalf`" will not always get you the root you want. You can use "`fsolve`" to find all real roots of the polynomial, and then substitute them for the "`RootOf`"[7].

```
> rts:= fsolve(4*_Z^8 - 27, _Z);
```

$$rts := -1.269588476,\ 1.269588476$$

```
> seq( subs(RootOf(4*_Z^8 - 27)=rts[k], s), k=1..2);
```

$$\{x = -.7329972488,\ y(x) = -1.269588476\},\ \{x = .7329972488,\ y(x) = 1.269588476\}$$

Instead of using "`solve`", we might have tried "`fsolve`" on our pair of equations.

```
> fsolve({yprime=0,eq2}, {x,y(x)});
Error, (in fsolve) should use exactly all the indeterminates
```

It looks like "`fsolve`" is confused by the "`y(x)`". It will work if we change "`y(x)`" to just "`y`".

```
> fsolve({subs(y(x)=y, yprime) = 0, eq1}, {x, y});
```

$$\{y = -1.269588476,\ x = -.7329972482\}$$

Note that "`fsolve`" returns only one real solution. However, you can find other solutions by specifying that one or more variables should be in certain intervals. In this case, we might ask for a solution with $x \geq 0$:

```
> fsolve({subs(y(x)=y,yprime)=0, eq1}, {x, y}, x = 0..infinity);
```

$$\{x = .7329972482,\ y = 1.269588476\}$$

Now that we have the horizontal tangents, what about vertical tangents? This would require that the slope has an infinite limit at some point, which can only happen if the denominator of "`yprime`" goes to 0. The "`numer`" and "`denom`" commands will extract the numerator and denominator of a fraction. So we can proceed as above, using "`denom(yprime)=0`" instead of "`yprime=0`".

```
> denom(yprime);
```

$$x\left(3y(x)^2 - x^2\right)$$

```
> solve({"=0, eq2}, {x, y(x)});
```

$$\{y(x) = \text{RootOf}(12\_Z^8 - 1),\ x = -6\,\text{RootOf}(12\_Z^8 - 1)^5\}$$ ■

The denominator of the expression for $dy/dx$ being 0 doesn't always mean a vertical tangent. Sometimes it produces a cusp. Another possibility is shown by the curve $xy^3 - x^3y = 0$ (which is really the four straight lines $x = 0$, $y = 0$, $y = x$, and $y = -x$). Note that we would get the same "`yprime`" as before. However, the point on the curve where the denominator of "`yprime`" is 0 is ($x = 0$, $y = 0$), at which the numerator is also 0. What happens in this case is that several branches of the curve intersect at the origin; each branch has its own tangent, but there is no single tangent to the curve at this point.

---

(7) Of course, if your $s$ had a different polynomial than $4\_Z^8 - 27$, you should use that polynomial in the next two commands.

**■ EXAMPLE 4** Find $d^2y/dx^2$ for the curve of Example 2.

**SOLUTION** To obtain higher derivatives of $y(x)$, we can differentiate our expression for $dy/dx$.

```
> ypp:= diff(yprime,x);
```

$$ypp := -\frac{3\,y(x)^2\,\left(\frac{\partial}{\partial x}y(x)\right) - 6\,x\,y(x) - 3\,x^2\,\left(\frac{\partial}{\partial x}y(x)\right)}{3\,x\,y(x)^2 - x^3} + \frac{\left((y(x))^3 - 3\,x^2 y(x)\right)\left(3\,(y(x))^2 + 6\,x\,y(x)\frac{\partial}{\partial x}y(x) - 3\,x^2\right)}{\left(3\,x\,(y(x))^2 - x^3\right)^2}$$

This contains $\partial y/\partial x$, for which we should substitute our value "`yprime`". The result is rather complicated, but some simplification is possible. Note that Maple sometimes shortens a complicated expression by finding pieces of the expression that occur several times and abbreviating them with "%1", "%2", etc. In this case, each "%1" stands for $y(x)^3 - 3x^2y(x)$, which occurs four times in the expression.

```
> ypp:= subs(diff(y(x),x)=yprime,ypp);
```

$$ypp := -\frac{-3\dfrac{y(x)^2\%1}{3xy(x)^2 - x^3} - 6x\,y(x) + 3\dfrac{x^2\%1}{3xy(x)^2 - x^3}}{3xy(x)^2 - x^3} + \frac{\%1\left(3y(x)^2 - 6\dfrac{xy(x)\%1}{3xy(x)^2 - x^3} - 3x^2\right)}{\left(3xy(x)^2 - x^3\right)^2}$$

$$\%1 := y(x)^3 - 3x^2y(x)$$

```
> ypp:= simplify(ypp);
```

$$ypp := \frac{12\,y(x)\left(y(x)^4 + 2\,y(x)^2x^2 + x^4\right)\left(x^2 - y(x)^2\right)}{x^2\left(x^2 - 3\,y(x)^2\right)^3}$$ ■

## Antiderivatives

Maple can find antiderivatives. To find an antiderivative of an expression *expr* involving the variable $x$, you use "`int(`*expr*`,x)`". Note that Maple does not supply the arbitrary constant "$+C$", but returns just one antiderivative (if it can find one). In some cases, Maple can't find an antiderivative, and returns only $\int expr\,dx$. This usually means that the antiderivative can't be expressed in terms of "elementary functions".

```
> int(cos(x), x);
```

$$\sin(x)$$

```
> int(sin(cos(x)), x);
```

$$\int \sin(\cos(x))\,dx$$

Sometimes Maple's answer may use some functions that you have not seen before.

```
> int(sin(x)/x, x);
```

$$\mathrm{Si}(x)$$

**■ EXAMPLE 5** Solve the initial-value problem

$$y'' = 2x^2\sin\left(x^2\right) - \cos\left(x^2\right)$$
$$y(\sqrt{\pi}) - 2\sqrt{\pi}$$
$$y'(\sqrt{\pi}) = 2$$

**SOLUTION** First, we ask Maple for an antiderivative of $y''$.

```
> v:= int(2 * x^2 * sin(x^2) - cos(x^2), x);
```

$$v := -\cos\left(x^2\right) x$$

Now we must add the appropriate constant to get $y'$ satisfying the initial condition at $x = \sqrt{\pi}$. We first subtract the value of $v$ at $\sqrt{\pi}$, and then add 2.

```
> yprime:= v - subs(x = sqrt(Pi),v) + 2;
```

$$yprime := -\cos\left(x^2\right) x + 2 + \cos(\pi)\sqrt{\pi}$$

Yes, Maple knows that $\cos(\pi) = -1$. This is one of the simplifications that Maple usually does automatically. However, as I mentioned in Lab 2, many of these simplifications are not immediately done on the output of "`subs`". The simplification will be done the next time you look at "`yprime`". Alternatives, as I mentioned, are to use "`eval`" to evaluate the output of "`subs`", or, in Release 5, to use "`eval(v, x = sqrt(Pi))`".

```
> yprime;
```

$$-\cos\left(x^2\right) x + 2 - \sqrt{\pi}$$

Another antiderivative, and adjustment of constants, gives us $y$.

```
> w:= int(yprime, x);
```

$$w := -\frac{1}{2}\sin\left(x^2\right) + 2x - \sqrt{\pi}x$$

```
> y:= w - eval(subs(x=sqrt(Pi),w)) + 2*sqrt(Pi);
```

$$y := -\frac{1}{2}\sin\left(x^2\right) + 2x - \sqrt{\pi}x + \pi$$ ■

## EXERCISES

1. Use Maple to verify that the function $f(x) = \cos(x^2)$ satisfies the equation $f'''(x) + 4x^2 f'(x) + 12x f(x) = 0$.

2. Let $f(x) = 6x^2 - x^4$. Find the largest interval on which $f'' \geq 0$. Verify (by plotting it) that $f'$ is increasing on this interval.

3. Plot the function $f(x) = (x^3 + 1)/(x^2 + x + 1)$ and its first and second derivatives on the same graph, for $-2 \leq x \leq 2$. Which curve is which?

4. **(a)** Find the first and second derivatives of $y$ with respect to $x$, where $\dfrac{x^2 + y}{1 - xy} = 2$, using implicit differentiation.

   **(b)** Find the same two derivatives by first solving for $y$ as a function of $x$. Compare the results. How can you show they are equal?

5. **(a)** Find the first to sixth derivatives of $\sqrt{x^2 - 1}$.

   **(b)** Describe the pattern. Don't worry about the values of the coefficients, for which finding a formula is not so simple, only the general form of the terms.

   **(c)** Use mathematical induction (with an assist from Maple) to prove that this pattern is valid.

6. **(a)** Find the first to sixth derivatives of $f(x) = x^2 \sin(Ax)$, where $A$ is a constant.

   **(b)** Find a formula to describe the even-numbered derivatives. *Hint:* the coefficient of $\sin(Ax)$ in $f^{(2m)}(x)$ is of the form $(a + bm + cm^2)\left(-A^2\right)^m$ for some constants $a$, $b$, and $c$.

   **(c)** Find the formula for the odd-numbered derivatives.

   **(d)** Check your answers by comparing the derivative of the answer to (c) with the answer to (b) with $m$ replaced by $m + 1$.

**7.** Consider the curve $\cos(x) - \cos(y) + xy = 1$.

**(a)** Find the slope at $(1,\ 1)$ by implicit differentiation.

**(b)** Find the two points closest to the origin at which the curve has a horizontal tangent.

**(c)** Find the two points closest to the origin at which the curve has a vertical tangent.

**(d)** Find the two points closest to the origin at which $d^2y/dx^2 = 0$.

**8. (a)** Plot the curves $y^2(a - x) = x^3$ and $(x^2 + y^2)^2 = b(2x^2 + y^2)$ for several values of the parameters $a$ and $b$, together on the same graph. Use some negative and some positive $a$'s, but only positive $b$'s (why?).

**(b)** Show that these curves always intersect at right angles.

**9. (a)** Solve the initial value problem below using Maple. Use "`simplify`" after each "`int`".

$$y'' = \frac{x(\cos(x) + 2) - 2\sin(x)}{(1 - \cos(x))^2}$$

$$y(\pi) = -\pi/2$$

$$y'(\pi) = -1/2$$

**(b)** On what interval is the solution valid?

**(c)** Plot $y$, $y'$, and $y''$ over this interval. You will need to specify an interval for the $y$ axis as well. Can you tell which curve is which on this graph?

*CHAPTER 3*

# *Applications of Differentiation*

## LAB 6 RELATED RATES AND EXTREME VALUES

**Goals:** *To use Maple as a tool in solving some fairly realistic problems involving rates of change and maximization or minimization.*

### Related Rates

The main difficulty in a related rates problem usually lies in translating the words of the problem into mathematical equations between variables. Maple often has little help to offer for this part of the solution process. Of course, once the equations are written down, Maple can be helpful in manipulating them.

Here are two sample problems. The first is a standard "calculus text" example, while the second is a bit more complicated.

**■ EXAMPLE 1** The volume of a spherical balloon is increasing at 10 $cm^3/s$. How fast is the radius increasing when the surface area is 200 $cm^2$?

**SOLUTION** We first express the volume $V$ and area $A$ in terms of the radius $r$, all of these being functions of time $t$.

```
> V := t -> 4/3 * Pi * r(t)^3;
> A := t -> 4 * Pi * r(t)^2;
```

We can find the rate of change of $r$ by differentiating the $V$ equation, using the known value of $V'$.

```
> D(V)(t);
```

$$4\pi r(t)^2 \mathrm{D}(r)(t)$$

```
> rprime:= solve(D(V)(t) = 10, D(r)(t));
```

$$rprime := \frac{5}{2}\frac{1}{\pi r(t)^2}$$

Now we need to know a value for $r(t)$. This comes from solving the $A$ equation.

```
> r1:= solve(A(t) = 200, r(t));
```

$$r1 := 5\frac{\sqrt{2}}{\sqrt{\pi}},\ -5\frac{\sqrt{2}}{\sqrt{\pi}}$$

Well, we can't blame Maple for giving us two answers: how should it know that a balloon's radius is positive? Let's take the positive solution[(1)].

---

(1) The order in which Maple returns the two solutions may vary from one session to another. You might have to use `r1[2]`.

```
> r1:= r1[1];
```

$$r1 := 5\frac{\sqrt{2}}{\sqrt{\pi}}$$

```
> subs(r(t)=r1, rprime);
```

$$\frac{1}{20}$$

The radius is increasing at $1/20$ cm/s. ■

■ **EXAMPLE 2** A small boat travels north in a straight line at a constant speed of 10 m/s. A hovering seagull drops a stone straight down from a height of 30 m as the boat passes directly beneath it. A pigeon flies in such a way that it, the stone, and the boat always form the vertices of an equilateral triangle in a north-south vertical plane. How is the pigeon moving when the stone hits the water? Neglect air resistance, and take the acceleration of gravity $g$ to be $9.8\,\mathrm{m/s^2}$.

**SOLUTION** First we might draw a picture, such as Figure 3.1, showing the relationship of the boat $B$, the stone $S$, and the pigeon $P$. Note that the pigeon could start out either in front of the boat or behind it. However, since the triangular relationship is supposed to persist until the stone hits the water, the pigeon (which is not a diving bird!) would much prefer to start out in front. For an animation of this picture, read the file `pigeon.txt` if it is available to you.

Next, we introduce coordinates so that the boat moves on the $x$ axis and the stone falls down the $y$ axis. At time $t$, measured from the release of the stone, the stone is at $(0, y_s)$, the boat is at $(x_b, 0)$, and the pigeon is at $(x_p, y_p)$. We have $x_b = 10t$ and $y_s = 30 - gt^2/2$. The pigeon, boat, and stone are equal distances from each other. We want to find $dx_p/dt$ and $dy_p/dt$ at the moment when $y_s = 0$.

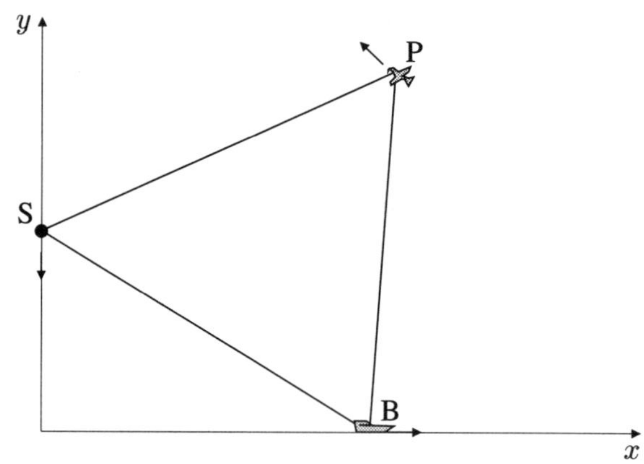

Figure 3.1. Boat, stone, and pigeon

First, I'll find $x_p$ and $y_p$ in terms of $x_b$ and $y_s$. It's usually more convenient to work with squares of distances, rather than distances themselves, so I'll let "`ps`", "`sb`", and "`bp`" be the squares of the distances from pigeon to stone, boat to stone, and boat to pigeon respectively. To find $x_p$ and $y_p$, we solve equations that say that `ps`, `sb`, and `bp` are all equal.

```
> ps:= xp^2 + (yp - ys)^2;
> bp:= (xp - xb)^2 + yp^2;
> sb:= xb^2 + ys^2;
> s:= solve({ps = bp, bp = sb}, {xp, yp});
```

$$s := \left\{xp = \frac{1}{2}\mathrm{RootOf}\left(-3ys^2 + xb^2 + \_Z^2 - 2\_Z\,xb\right),\right.$$
$$\left. yp = \frac{1}{2}\left(ys^2 + \mathrm{RootOf}\left(-3ys^2 + xb^2 + \_Z^2 - 2\_Z\,xb\right)xb - xb^2\right)/ys\right\}$$

The polynomial in the "`RootOf`" is only a quadratic. Let's actually solve it.

```
> poly:= - 3*ys^2 + xb^2 + *_Z^2 - 2*_Z*xb;
> rts:= solve(poly, _Z);
```

$$rts := xb + \sqrt{3}\,ys,\ xb - \sqrt{3}\,ys$$

Now substitute each of these roots for the "`RootOf`" in `s`.

```
> s:= seq(subs(RootOf(poly) = rts[k], s), k=1..2);
```

$$s := \left\{ xp = \frac{1}{2}xb + \frac{1}{2}\sqrt{3}\,ys,\ yp = \frac{1}{2}\frac{ys^2 + (xb + \sqrt{3}ys)xb - xb^2}{ys} \right\},$$
$$\left\{ xp = \frac{1}{2}xb - \frac{1}{2}\sqrt{3}\,ys,\ yp = \frac{1}{2}\frac{ys^2 + (xb - \sqrt{3}ys)xb - xb^2}{ys} \right\}$$

Presumably one of the two solutions corresponds to the pigeon starting out in front of the boat, and the other to the pigeon starting behind the boat and taking a dive. It's not hard to see that the solution(2) with the $+\sqrt{3}$ is the one we want (e.g., look at $x_p$ in the initial position with $x_b = 0$ and $y_s = 30$). The expression for $ys$ looks like it could use some simplification.

```
> simplify(s[1]);
```

$$\left\{ xp = \frac{1}{2}xb + \frac{1}{2}\sqrt{3}\,ys,\ yp = \frac{1}{2}ys + \frac{1}{2}xb\sqrt{3} \right\}$$

I'd like to make this into a definition of "`xp`" and "`yp`". The "`assign`" command will do it, giving "`xp`" and "`yp`" the values found in the solution.

> `assign(");` (Release 4) or

> `assign(%);` (Release 5)

This command produces no visible output, but if you examine `xp` and `yp` you'll see that they were indeed assigned those values. Now let's look at the dependence on $t$.

```
> xb:= 10 * t;
> ys:= 30 - g/2 * t^2;
```

Since "`assign`" made "`xp`" and "`yp`" into expressions in "`xb`" and "`ys`", which in turn are now defined in terms of "`t`", we should now have "`xp`" and "`yp`" as expressions in `t`.

```
> xp;
```

$$5t + \frac{1}{2}\sqrt{3}\left(30 - \frac{1}{2}gt^2\right)$$

```
> yp;
```

$$15 - \frac{1}{4}gt^2 + 5t\sqrt{3}$$

Now all we have to do is find when the stone hits the water, i.e., when $y_s = 0$.

```
> t1:= solve(ys=0, t);
```

$$t1 := 2\frac{\sqrt{15}}{\sqrt{g}},\ -2\frac{\sqrt{15}}{\sqrt{g}}$$

Again there are two solutions, and the positive one (`t1[1]` in this case) is the one we want.

```
> vx:= subs(t=t1[1], diff(xp,t));
```

$$vx := 5 - \sqrt{3}\sqrt{g}\sqrt{15}$$

```
> vy:= subs(t=t1[1], diff(yp,t));
```

$$vy := -\sqrt{15}\sqrt{g} + 5\sqrt{3}$$

We want a numerical answer. Let's substitute the numerical value 9.8 $\mathrm{m/s^2}$ for $g$. We can actually do this to both "`vx`" and "`vy`" in one command by putting them in a list.

```
> V:= subs(g=9.8, [vx, vy]);
```

$$V := [5 - 3.130495169\sqrt{3}\sqrt{15},\ -3.130495169\sqrt{15} + 5\sqrt{3}]$$

We need "`evalf`" to get rid of the square roots.

(2) This was `s[1]` in my Maple session, but you may need to use `s[2]` instead.

```
> evalf(V);
```

$$[-16.00000001, -3.464101610]$$

So here is our answer: when the stone hits the water, the pigeon is flying down and to the south, with velocity components approximately $-16$ m/s in the $x$ direction and $-3.46$ m/s in the $y$ direction. ■

## Extreme Values

One of the main differences between a "calculus text" extreme-value problem and a more realistic problem is that in the "calculus text" problem it is usually possible to locate all the critical points explicitly by solving an equation. In a more realistic problem, it is often impossible to solve this equation exactly, and we are forced to use numerical approximations ("`fsolve`" instead of "`solve`"). Moreover, it may not even be clear how many critical points there are in an interval. The latter is a question of some importance, because "`fsolve`" only produces one solution (except for a polynomial) and we have no guarantee that this solution will be the one we want. In this situation, graphing the function becomes very useful.

The procedure I recommend for finding the maximum or minimum of a function $f$ when "`solve`" won't locate all critical points is the following:

(1) Plot $f$ on the interval in question. If the interval is infinite, an "infinity plot" may be useful.
(2) Locate an interval (or perhaps several intervals) that appears to contain the absolute maximum or minimum, but no other critical points.
(3) Plot $f'$ on the interval of (2) to confirm that there are no other critical points in this interval.
(4) If the maximum or minimum is at a critical point, use "`fsolve`" to find the critical point in the interval.

■ **EXAMPLE 3** Find the absolute maximum and absolute minimum of the function $f(x) = \dfrac{|\sin(x)| + x}{1 + x^2}$.

**SOLUTION** Since we initially don't know where on the real line the maximum and minimum might be, we begin with an infinity plot.

```
> f:= x -> (abs(sin(x))+ x)/(1+x^2);
> plot(f(x),x=-infinity .. infinity); (Figure 3.2)
```

The graph shows what appears to be a single large peak for positive $x$ and a valley for negative $x$, with $f(x)$ appearing to go to 0 as $x \to \infty$ or $-\infty$. Since $|\sin(x)| \le 1$, it is easy to see that indeed $\lim_{x\to\pm\infty} f(x) = 0$ (although Maple can't do these limits directly). We might next try an interval such as `-5 .. 5` (Figure 3.3).

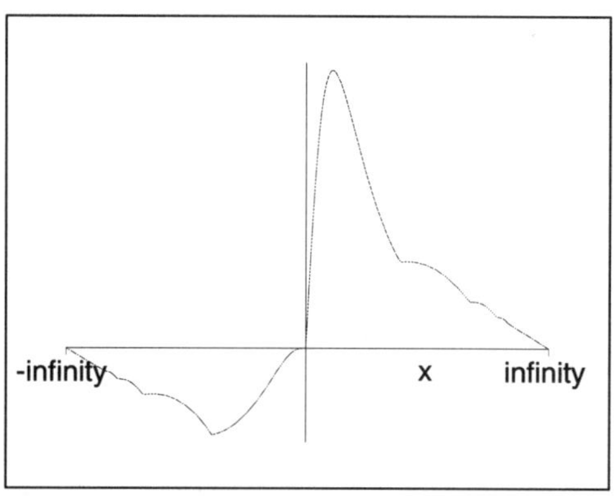


Figure 3.2. Infinity plot

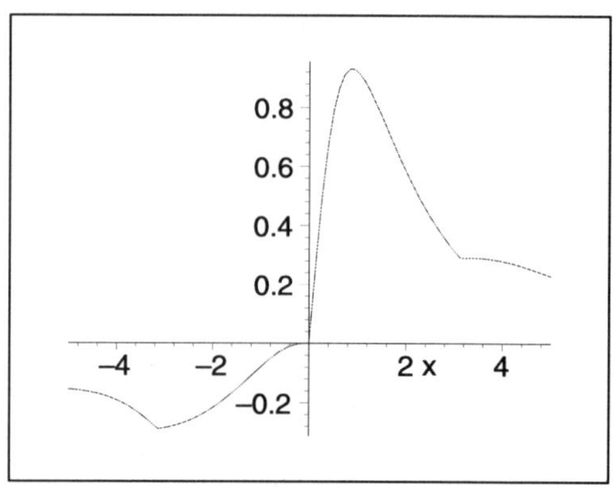


Figure 3.3. $f(x) = \dfrac{|\sin x| + x}{1 + x^2}$

In Figure 3.3 we see a peak and a valley, which apparently are the ones that showed up in the infinity plot. The maximum seems to be around $x = 0.9$, certainly between $0.8$ and $1.0$, and the minimum around $x = -3.1$, certainly between $-3.3$ and $-2.9$.

> `plot(D(f)(x), x = 0.8 .. 1.0);` (Figure 3.4)

The graph of $f'(x)$ looks like a smooth curve, evidently crossing the $x$ axis once in this interval. This confirms that there is one critical point in the interval. Since $f'$ is positive to the left of the critical point and negative to the right, the critical point is a (local) maximum.

> `xmax:= fsolve(D(f)(x)=0, x, x=0.8 .. 1.0);`

$$xmax := .8799272979$$

> `f(xmax);`

$$.9303082089$$

The maximum value, 0.9303082089, occurs at $x = .8799272979$ (to 10 digit accuracy). Next we look for the minimum.

> `plot(D(f)(x), x = -3.3 .. -2.9);` (Figure 3.5)

We see that $f'$ is negative on the left side of this interval, then near $x = -3.14$ takes a sudden jump (which Maple plots as an almost vertical line segment) to positive values. This is no critical point, it is a singular point. Indeed it is not hard to see that $x = -\pi$ is likely to be a singular point, because $\sin(-\pi) = 0$ and the absolute value function is not differentiable at 0. The main purpose of plotting Figure 3.5 is not to identify the singular point, but rather to check that there are no critical points hiding near the singular point[(3)]. Also, the facts that $f$ is continuous at the singular point with $f'$ negative on its left and positive on its right confirm that the singular point is a (local) minimum.

> `evalf(f(-Pi));`

$$-.2890254823$$

The minimum value, $-.2890254823$ (to 10 digit accuracy), occurs at the singular point $x = -\pi$. ■

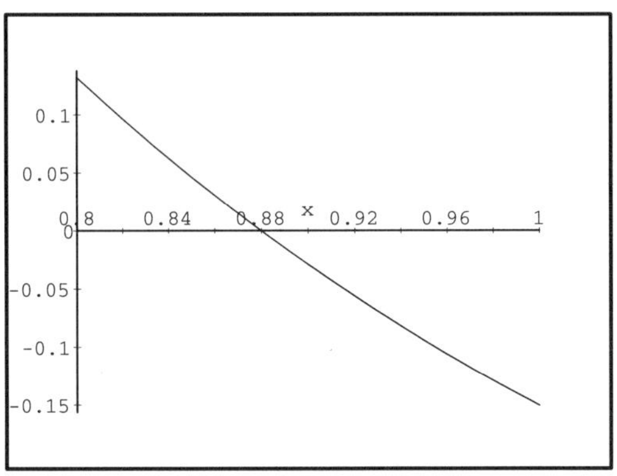

Figure 3.4. $f'$ near the maximum

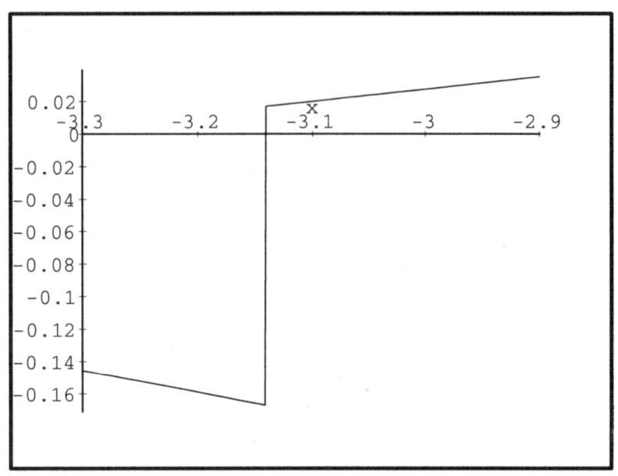

Figure 3.5. $f'$ near the minimum

The next example is a fairly routine "applied" problem. It is also a warm-up for the more difficult Example 5.

---

(3) As it turns out, $f'$ gets quite close to 0 to the right of the singular point. A very small change in the function $f$ might produce a local minimum just to the right of the singular point, which would be very hard to detect without a graph such as Figure 3.5.

**■ EXAMPLE 4** A 10 cm by 100 cm rectangular metal sheet is to be bent to form a 100 cm long trough, open at the top. The cross-section of the trough is to be a circular arc. How should this be done in order to maximize the volume of the trough?

**SOLUTION** Let $r$ be the radius of the arc and $\theta$ the angle it subtends at the centre. The circumference of the arc is the width of our rectangle, 10 cm. I think it is most natural to use $\theta$ as the variable in this problem, so I'll express $r$ in terms of $\theta$:

```
> r:= 10/theta;
```

Now the cross-sectional area of the trough is the area of sector $AOB$ ($\theta r^2/2$) minus the area of triangle $AOB$. The triangle has base $2r\sin(\theta/2)$ and height $r\cos(\theta/2)$. The volume is the length (100) times the area. The picture looks different for $\theta > \pi$, but the formula is the same.

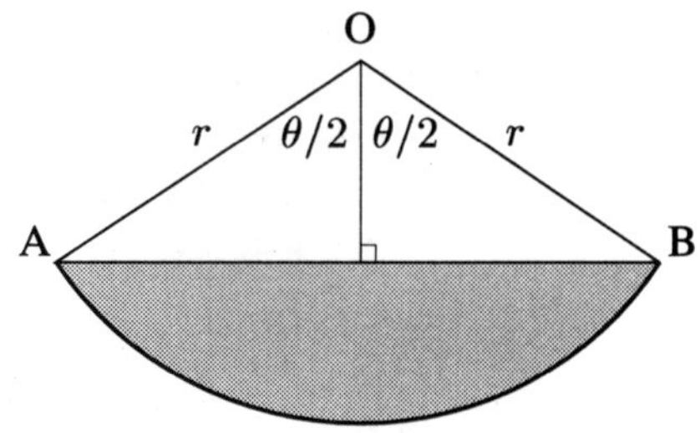


Figure 3.6. Cross-section of trough

```
> v:= 100 * (theta * r^2/2 - r^2 * sin(theta/2) * cos(theta/2));
```

$$v := 5000\frac{1}{\theta} - 10000\frac{\sin\left(\frac{1}{2}\theta\right)\cos\left(\frac{1}{2}\theta\right)}{\theta^2}$$

It may be more convenient to have a function of $\theta$, rather than just an expression.

```
> vol:= unapply(v,theta);
```

$$vol := \theta \to 5000\frac{1}{\theta} - 10000\frac{\sin\left(\frac{1}{2}\theta\right)\cos\left(\frac{1}{2}\theta\right)}{\theta^2}$$

A plot of $vol(\theta)$ for $\theta$ from 0 to $2\pi$ shows a nice smooth function (even though $vol(0)$ is undefined, the function has a limit at 0) with a single peak somewhere between $\theta = 2.9$ and 3.5. A plot of the derivative on that interval shows it crossing the $\theta$ axis once. So we should find the maximum with

```
> fsolve(D(vol)(theta)=0,theta,theta=2.9 .. 3.5);
```

$$3.141592654$$

Well, that number looks familiar! Is it $\pi$?

```
> D(vol)(Pi);
```

$$0$$

Yes, the maximum occurs at $\theta = \pi$. To find the maximum volume:

```
> vmax:= vol(Pi);
```

$$vmax := 5000\frac{1}{\pi}$$

```
> evalf(vmax);
```

$$1591.549431$$

■

**■ EXAMPLE 5** Now suppose that the ends of the trough must also be cut out of the same sheet of metal. One method would be to place the two "D" shaped pieces as shown in Figure 3.7, with their straight sides vertical and their curved sides touching each other. What is the maximum volume of the trough now?

**SOLUTION** We introduce coordinates so one corner of the sheet is at $(0, 0)$, with a long edge along the positive $x$ axis. The straight side of the lower "D" is at $x = u$, so the trough will have length $100 - u$. We need to determine $u$ as a function of $\theta$.

Let the centre $A$ of the curved side of the lower "D" be at $(x_c, y_c)$. The centre $B$ of the upper one will then be at the point $(-r\cos(\theta/2), 10 - y_c)$. Since the bottom of the lower "D" rests on the $x$ axis, we get $y_c = r\sin(\theta/2)$ if $\theta \le \pi$, but $y_c = r$ if $\theta > \pi$. We can use the "piecewise" function to represent this:

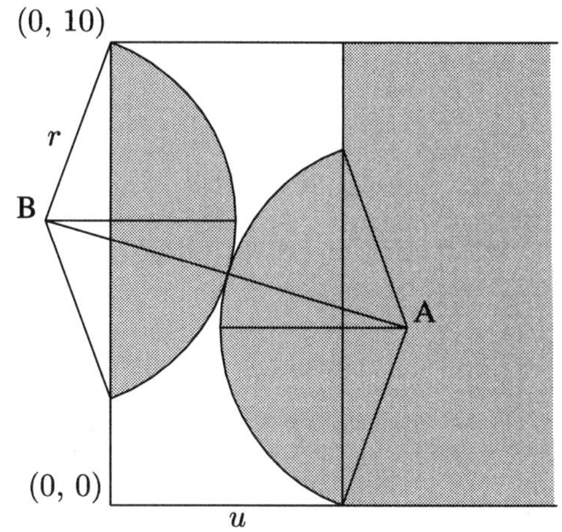


Figure 3.7. Ends of trough on sheet

```
> yc:= piecewise(theta <= Pi, r*sin(theta/2), r);
```

It is easy to see that $u = x_c - r\cos(\theta/2)$, but what is $x_c$ in terms of $\theta$? Imagine the upper "D" fixed in its place, and the lower "D" starting to the right and sliding left as far as possible. One of two things can stop its slide: it may contact the upper "D" (as in Figure 3.7) or the left edge of the sheet (the $y$ axis). We don't really know which will happen, but Maple can help us. If the lower "D" contacts the left edge of the sheet, then $x_c = r$. Let's look at the distance between the two centres $(-r\cos(\theta/2), 10 - y_c)$ and $(r, y_c)$ in that case:

```
> d:= sqrt((r + r * cos(theta/2))^2 + (2 * yc - 10)^2);
```

In order for the "D's" not to overlap, this distance must be at least $2r$. To compare $d$ to $r$, we can plot $d/r$. Since the important value is $d = 2r$, we can also plot the horizontal line $y = 2$ on the same graph.

```
> plot({d/r, 2}, theta = 0 .. 2*Pi);
```
(Figure 3.8)

It appears that $d \le 2r$ for $\theta$ from 0 to some number, call it $b$, slightly less than 4, and $d > 2r$ from $b$ to $2\pi$ (you might want to take a closer look at a range of $\theta$ near 0, where this is hard to see). We want to find the value of $b$ more exactly.

```
> b:= fsolve(d = 2 * r, theta,
>      theta = 3 ..  4);
```

$$b := 3.897922911$$

Now for $\theta > b$ we have $x_c = r$, while for $\theta < b$ it is that value that makes the distance between the two centres equal to $2r$.

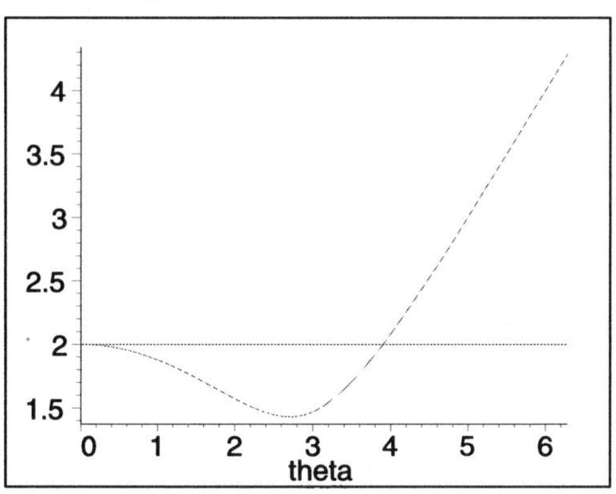


Figure 3.8. $d/r$ and 2

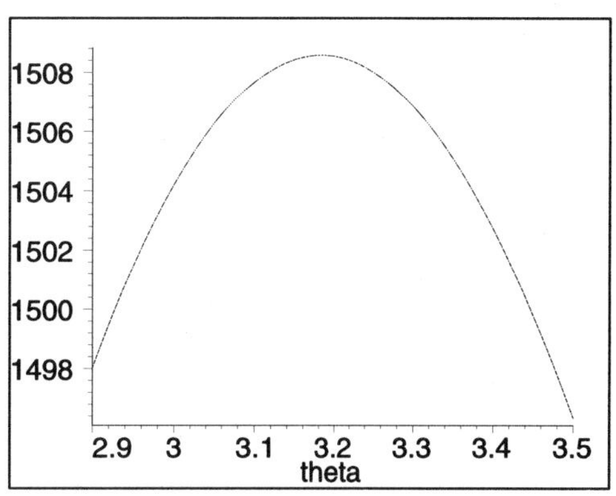


Figure 3.9. Volume near the maximum

```
> s:= solve((x + r*cos(theta/2))^2 + (2*yc - 10)^2 = (2*r)^2, x);
```

The result is a bit complicated. Since it is a quadratic equation, there are two solutions, one of the form $\ldots + \sqrt{\ldots}$ and the other $\ldots - \sqrt{\ldots}$. We want the solution with a "+" (the other one would have the lower "D" to the left of the sheet). In my Maple session this was the first solution `s[1]`, though this may vary.

```
> xc:= piecewise(theta <= b,s[1],r);
```

Now we can obtain $u$ and the volume $v$.

```
> u:= xc - r * cos(theta/2);
> v:= (100 - u) * (theta * r^2/2
>        - r^2 * sin(theta/2) * cos(theta/2));
> vol:= unapply(v, theta);
```

After all that buildup, the graph of $vol(\theta)$ is a bit of an anticlimax, looking very similar to that of Example 4. The peak may be in a slightly different place, but again it is between $\theta = 2.9$ and $\theta = 3.5$. A closer look shows that it is slightly greater than $\pi$:

```
> plot(v, theta = 2.9 .. 3.5); (Figure 3.9)
```

Unfortunately, there are problems using "`fsolve(D(vol)(theta), theta)`" here. Release 4 can't calculate "`D(vol)`" (I'm not sure why). Release 5 can, but the result is undefined at $\theta = b$ (this is correct; the function $vol$ is not differentiable there). And the presence of $undefined$ causes "`fsolve`" to fail (it seems to think that $undefined$ is another variable). Fortunately, we can use the "`assume`" command to tell Maple that $\theta$ should be in the interval $(Pi, 3.5)$. With this assumption, Maple is able to simplify $v$, choosing the branches of the "`piecewise`" expressions that correspond to $\theta$ in this interval.

```
> assume(theta > Pi, theta < 3.5);
> vol1:= unapply(v, theta);
```

$$vol1 := \theta \to \left(100 - 2\frac{-5\cos\left(\frac{1}{2}\theta\sim\right) + \sqrt{100\,\theta\sim - 25\,\theta\sim^2}}{\theta\sim} + 10\frac{\cos\left(\frac{1}{2}\theta\sim\right)}{\theta\sim}\right) \left(\frac{50}{\theta\sim} - 100\frac{\sin\left(\frac{1}{2}\theta\sim\right)\cos\left(\frac{1}{2}\theta\sim\right)}{\theta\sim^2}\right)$$

The "~" after the $\theta$ is just to remind you that an assumption has been made about this variable(4). We will see more about "`assume`" in Lab 9.

```
> tmax:= fsolve(D(vol1)(theta) = 0, theta, theta = 2.9 .. 3.5);
```

$$3.183383372$$

```
> vol(tmax);
```

$$1508.585282$$

■

## EXERCISES

1. Consider the locomotive and car of Exercise 7 from Lab 1.
   (a) What is the rate of change of the distance between them when the car is 10 metres past the crossing?
   (b) Using calculus, find the minimum distance between the car and the locomotive, and the time at which this occurs.
2. Use Maple to find the maximum and minimum of $x^3 - x^2 - x$ for $-2 \le x \le 2$.

---

(4) If you find it annoying, you can stop it from being displayed: in the Options menu, choose "Assumed Variables" and then "No Annotation".

**3.** A conical paper cup is 10 cm high and its radius at the top is 4 cm. Water is poured into the cup at the rate of 2 $\text{cm}^3/\text{s}$. What is the rate of change of the area of the (top) surface of the water after 5 seconds? Assume that the cup's axis is vertical and it is initially empty.

**4.** A triangle has one vertex at $(0, 0)$. The other two have the same $y$ coordinate and are on the ellipse $4x^2 + y^2 = 4$. What is the maximum area of such a triangle?

**5.** Some authors advocate differentiating implicitly in a related rates problem rather than solving for the quantities whose rates are to be found. I disregarded this advice in my solution of Example 2. Solve the problem by following this advice:

**(a)** Making $x_p$, $y_p$, $x_b$ and $y_s$ functions of $t$, differentiate the equations `ps=bp` and `bp=sb` and solve for the derivatives of $x_p$ and $y_p$.

**(b)** Solve `ps=bp` and `bp=sb` for the values of $x_p$ and $y_p$ at the time when the stone hits the water.

**(c)** Substitute the values from (2) into the results of (1), and simplify to find the answers.

**6.** **(a)** What is the highest point reached by the pigeon in Example 2?

**(b)** What is the fastest speed of the pigeon? *Hint:* if the $x$ and $y$ components of velocity are $v_x$ and $v_y$, the speed is $\sqrt{v_x^2 + v_y^2}$.

**7.** A balloon being inflated consists of a cylinder with hemispheres on the ends. When the balloon is 50 cm long and 10 cm in radius, its volume is increasing at a rate of 100 $\text{cm}^3/\text{s}$, and its surface area is increasing at 15 $\text{cm}^2/\text{s}$. How fast is its radius increasing at that time?

**8.** Find any local maximum or minimum of the function of Example 3 between $x = 3$ and $x = 4$.

**9.** In front of a high vertical wall is a hill, described by the equation $y = 2x - x^2$ where the $y$ axis is the wall and the $x$ axis is the ground (level except for the hill). How long is the shortest ladder that can reach over the hill from the wall to level ground?

**10.** Find the maximum and minimum of the function $\sin x + |x \sin x - \cos x|$ for $-10 \leq x \leq 10$. *Hint:* "`fsolve`" can be useful for finding singular points, as well as critical points.

**11.** Find the rectangle of largest area that can fit under the graph of $f(x) = 1 - 3x + 8x^3 - 6x^4$ for $0 \leq x \leq 1$, with one side on the $x$ axis and one on the $y$ axis. *Hint:* a rectangle with top right corner on the curve might **not** fit under it.

# LAB 7 LINEAR APPROXIMATION AND NEWTON'S METHOD

**Goals:** *To examine linear approximations to functions, and the error in such approximations. To see how Newton's method solves equations numerically, and what happens when it doesn't work.*

## Linear Approximation

The linear approximation of a function $f$ at $x = a$ is the function $L(x) = f(a) + f'(a)(x - a)$. The graph of $L$ is the straight line tangent to the graph of $f$ at $x = a$. We wish to explore how $L(x)$ is an approximation to $f(x)$ when $x$ is near $a$.

■ **EXAMPLE 1** Study the linear approximation of $f(x) = \sqrt{4 + x \sin x}$ at $x = \pi$.

**SOLUTION** We begin by defining $L$, $f$, and $a$ in Maple, and plotting $f$ and $L$ on the same graph.

```
> L:= x -> f(a) + D(f)(a) * (x - a);
> f:= x -> sqrt(4 + x * sin(x));
> a:= Pi;
> plot({ f(x), L(x) }, x = -5 .. 5);
```

The picture of the curve and its tangent line is similar to the pictures we saw in Lab 4. Notice how the graphs of $f(x)$ and $L(x)$ are very close together near $x = \pi$. Let's take a closer look.

`> plot({ f(x), L(x) }, x = 3 .. 3.2);` (Figure 3.10)

At this scale it starts to become difficult to tell the two graphs apart. If we tried, say, the interval `x = 3.14 .. 3.142`, the limited resolution of the computer screen would not be able to separate them.

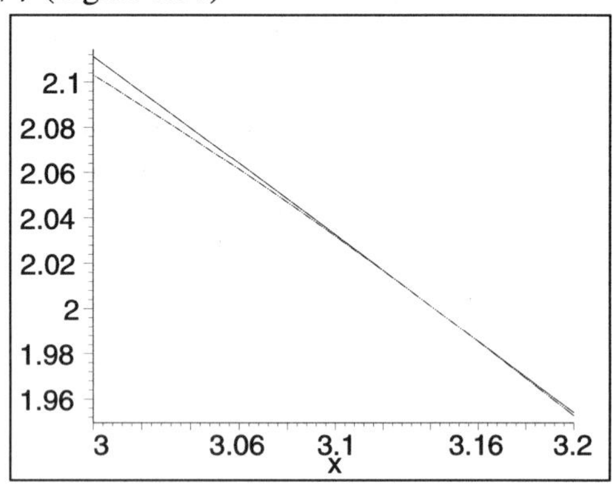


Figure 3.10. $f(x)$ and $L(x)$

The graphs of differentiable functions, you might say, are very boring when examined under a microscope: at sufficiently small scales, they look like their tangent lines. This statement must be qualified slightly in a case where $f'(a) = 0$: a close-up under Maple might not look like a straight line unless you used "constrained" scaling, which makes the scale the same in both $x$ and $y$ directions. With the default "unconstrained" scaling, the graph would be magnified so much more in $y$ than in $x$ that the deviation of $f(x)$ from its horizontal tangent line would be visible.

When the tangent line is so close to the curve, the tangent-line value $L(x)$ is a good approximation to $f(x)$. To take a numerical example, let's try $f(\pi + 0.01)$.

```
> evalf(L(Pi + 0.01));
```

1.992146018

```
> evalf(f(Pi + 0.01));
```

1.992105569

Not bad: $L(x)$ is correct to five significant digits. If you took $x$ even closer to $a$, $L(x)$ would be even more accurate. You might not find this too impressive since Maple gives us the "exact"[5] value of $f(x)$ just as easily as it does $L(x)$. The method would really prove its worth in a case where we don't know how to compute the true $f(x)$. Some examples of this are Exercises 3 and 4. ■

(5) Actually, it computes an approximation.

## Error Estimates

The error in the linear approximation for $f(x)$ is $E(x) = f(x) - L(x)$ (the true value minus the approximate value). We can estimate this error using the fact, shown in most calculus texts, that

$$E(x) = \frac{f''(X)}{2}(x-a)^2$$

where $X$ is some point between $x$ and $a$. This assumes that $f''(t)$ exists for all $t$ in an interval containing $a$ and $x$. We don't know exactly what point $X$ is. But if we know two numbers $A$ and $B$ such that $A \le f''(t) \le B$ for all $t$ in this interval, this will apply to $f''(X)$. Then we will be able to say

$$\frac{A}{2}(x-a)^2 \le E(x) \le \frac{B}{2}(x-a)^2$$

■ **EXAMPLE 2** Find error estimates for Example 1 on the interval $2.5 \le x \le 3.5$.

**SOLUTION** The function $f''(t)$ looks rather complicated:

```
> (D@@2)(f)(t);
```

$$\frac{1}{2}\frac{2\cos(t) - t\sin(t)}{\sqrt{4+t\sin(t)}} - \frac{1}{4}\frac{(\sin(t)+t\cos(t))^2}{(4+t\sin(t))^{3/2}}$$

Proceeding by hand, we might produce crude but rigorous numbers $A$ and $B$ by the following method. Step by step, we would go through the expression for $f''(t)$, getting estimates for each small part of the expression and using these to estimate larger and larger parts. For example, if $a \le x \le b$ and $c \le y \le d$, then $a + c \le x + y \le b + d$. We might start as follows:

$2.5 \le t \le 3.5$ so $-1 \le \cos(t) \le 0$ [6] and $-2 \le 2\cos(t) \le 0$
$-1 \le \sin(t) \le 1$ and $2.5 \le t \le 3.5$ so $-3.5 \le t\sin(t) \le 3.5$
Therefore $-5.5 \le 2\cos t - t\sin(t) \le 3.5$
...

This would take a lot of work. Fortunately, Maple has a command for it: "`evalr`". This takes an expression containing intervals, written in the form "`INTERVAL(a..b)`", and returns an interval. If you replaced each interval in the input with a number in that interval, the value of the expression would always be in the output interval. The "`evalr`" function must first be read in to Maple using "`readlib`".

```
> readlib(evalr):
> evalr((D@@2)(f)(INTERVAL(2.5 .. 3.5)));
```

$$\text{INTERVAL}(-2.032750990\,..\,-.1086288975)$$

Let's look at $E(x)$ and its estimates for our example $x = \pi + 0.01$.

```
> A:= -2.032750990;
> B:= -.1086288975;
> E:= x -> f(x) - L(x);
> evalf(E(Pi + 0.01));
```

$$-.000040449365$$

```
> evalf(A/2 * ( Pi + 0.01 - a)^2);
```

$$-.0001016375495$$

```
> evalf(B/2 * ( Pi + 0.01 - a)^2);
```

$$-.5431444875\,10^{-5}$$

Yes, the estimates worked: the actual error is between the two error estimates.

---

(6) Better estimates for $\cos(t)$ and $\sin(t)$ are possible, but this is to give you the idea.

Rather than looking at individual $x$ values, let's plot the error and its estimates as functions of $x$ in our interval.

```
> plotset:= { E(x), A/2 * (x-a)^2, B/2 * (x-a)^2 };
> plot(plotset, x = 2.5 .. 3.5);
```

Of the three curves in this graph, $E(x)$ is in the middle, the error estimates using $A$ and $B$ on the bottom and top respectively. Well, you shouldn't have to take my word for this. On a colour monitor, sets of curves are plotted in different colours: the first three are red, green, and yellow. The problem is that we are plotting a set of expressions. Maple can, and often does, change the order of the elements in a set. The reason I first defined "`plotset`" and then plotted it was so you could see the order that Maple is using for this set. In Maple's response to the "`plotset:=`" command, the order of the three expressions corresponds to the order of the colours, as mentioned above.

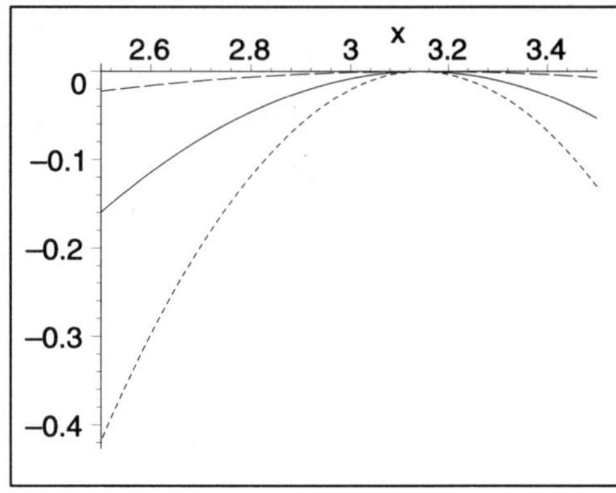


Figure 3.11. $E(x)$ and bounds

In order to control which curve is plotted in which colour, instead of a set you can plot a list of expressions:[7]

```
> plot([ E(x), A/2 * (x-a)^2, B/2 * (x-a)^2 ], x = 2.5 .. 3.5);
```

In fact, you can also specify the colours of the expressions as a list:[8]

```
> plot([ E(x), A/2 * (x-a)^2, B/2 * (x-a)^2 ],
>           x = 2.5 .. 3.5, colour = [blue, red, green]);
```

If you don't have a colour monitor, you can still distinguish the three curves: instead of "`colour`," you specify a "`linestyle`" for each. Linestyle 1 is solid, 2 is dotted, 3 is dashed, 4 is dot-dash. Again, you can use one "`plot`" command, specifying the line styles as a list "`linestyle=[1,2,3]`".

```
> plot([ E(x), A/2 * (x-a)^2, B/2 * (x-a)^2 ],
>           x = 2.5 .. 3.5, linestyle = [1,2,3]);
```
(Figure 3.11)

Another way to obtain error estimates is by finding the maximum and minimum of $f''(t)$ on the interval using the methods of Lab 6.

```
> plot((D@@2)(f)(x), x = 2.5 .. 3.5);
```

We see from the graph that $f''(x)$ is a decreasing function of $x$ on this interval. Its maximum and minimum values on the interval, therefore, are $f''(2.5)$ and $f''(3.5)$ respectively.

```
> A:= (D@@2)(f)(3.5);
```

$$A := -.9067864924$$

```
> B:= (D@@2)(f)(2.5);
```

$$B := -.6990918093$$

The graphical method produces tighter bounds. However, it is less rigorous. It requires us to believe what we see on the graph of $f''$, but graphs can sometimes be misleading. For example, we haven't ruled out the possibility that the graph of $f''$ really has a very narrow "spike" somewhere, which doesn't appear on Maple's graph because Maple didn't evaluate $f''$ at any point on the spike. ■

(7) Recall that a list is enclosed in square brackets, and the order of its items is maintained.

(8) For those who prefer U.S. spelling, "`color`" is equivalent to "`colour`".

## Newton's Method

Newton's method is one of the best methods of finding an approximate solution to an equation of the form $f(x) = 0$. It actually forms part of the programming of Maple's "`fsolve`" command. Starting from an initial "guess" $x_0$, it produces a sequence $x_1, x_2, x_3, \ldots$ that (when it works well) converges very rapidly to a solution of the equation. As we will see, however, the method does not always work well.

■ **EXAMPLE 3** Solve the equation $x = 1 + \sin(2x)$ using Newton's method.

**SOLUTION** We first put the equation in the form $f(x) = 0$. The simplest way is just to subtract one side from the other.

```
> f:= x -> 1 + sin(2*x) - x;
```

The iteration formula on which the method is based is $x_{n+1} = x_n - f(x_n)/f'(x_n)$. Let's define a function "g" so that $x_{n+1} = g(x_n)$.

```
> g:= evalf @ unapply( x - f(x)/D(f)(x), x);
```

$$g := \mathit{evalf}@\left(x \to x - \frac{1 + \sin(2x) - x}{2\cos(2x) - 1}\right)$$

I'm using "`unapply`" rather than "`->`" mainly for reasons of speed: this way, Maple does not have to differentiate the function $f$ each time it evaluates $g$. The "`evalf`" is to ensure that Maple gives you a numerical answer, not a symbolic one. In this example it wouldn't be needed if you were careful to only use $x$ values with decimal points in them, but it's better to be safe than sorry[9].

Now let's take an initial guess, and apply $g$ repeatedly to it.

```
> x[0]:= 2;
```

$$x_0 := 2$$

This is the first entry in what Maple calls a "table". If you've done some computer programming, you may be pleasantly surprised to see that you don't need to tell Maple that you're making a table, or how big it will be: the table is created automatically whenever you use "`[...]`" following a name, and will hold whatever entries you put in it. [10].

```
> x[1]:= g(x[0]); x[2]:= g(x[1]);
```

$$x_1 := 1.238585269$$

$$x_2 := 1.385414630$$

My typing fingers might get tired soon. Let's ask Maple to do several more of these in one command. The "`for`" command will repeat a statement or sequence of statements a certain number of times, with values of a variable (the loop variable) going from an initial value to a final value (increasing by 1 each time). It is used as follows:

> `for` *loop variable* `from` *initial value* `to` *final value* `do` *statements* `od;`

Don't forget the "`od`" (the reverse of "`do`") which ends the "`for`" command.

The loop variable is left with a value at the end of the loop. This could cause problems if you wanted to use the loop variable for other purposes. As a safety measure, I follow the practice of using double-letter names such as "`nn`" for loop variables.

```
> for nn from 2 to 5 do x[nn+1]:= g(x[nn]) od;
```

$$x_3 := 1.377353614$$

$$x_4 := 1.377336877$$

$$x_5 := 1.377336877$$

$$x_6 := 1.377336877$$

---

(9) In preparing this lab, I was not careful a few times, and actually had to reboot my system as a result. Without the "`evalf`", $x_{10}$ (as computed below) can be a **very** complicated expression.

(10) Maple shows "`x[0]`" in output as $x_0$.

Well, Maple seems to have settled on a value of 1.377336877. Is this the solution?

```
> f(x[6]);
```

$$0$$

This was an example where Newton's method worked very well. But with other starting points we might not be so lucky. To save the $x$ values we just calculated, I'll use a different name for the next table.

```
> w[0]:= 0:
> for nn from 0 to 49 do w[nn+1]:= g(w[nn]) od;
```

I'm leaving out the 50 lines of output, but you can examine it yourself. It shows no signs of settling down[11].

Let's look at some graphs, to see how Newton's method works. The idea is that we approximate $f(x)$ near $x = x_n$ by the tangent line at $x = x_n$, so that our next approximation $x_{n+1}$ for a root of $f$ is the point where this tangent line hits the $x$ axis. So let us plot $f(x)$ and its tangent line from $(x_n, f(x_n))$ to $(x_{n+1}, 0)$. As an added touch, we can plot the straight lines from $(x_n, 0)$ to $(x_n, f(x_n))$ and from $(x_{n+1}, 0)$ to $(x_{n+1}, f(x_{n+1}))$. I'll use separate plots for $f$ and the straight lines, and combine them with "display".

In Lab 4 we used the "plot" command to make a straight line joining two points. It can also take a list of points and join them by straight lines, one to the next. In this case, to work easily with different $x_n$, it is convenient to define a function that will take $x_n$ and draw the straight lines.

```
> p:= plot(f(x), x = 0 .. 3):
> q:= x0 -> plot([[x0,0], [x0, f(x0)], [g(x0), 0],
>                [g(x0), f(g(x0))]]);
> with(plots,display);
> display({ p, q(x[0]) }); (Figure 3.12)
```

This shows the first step in the case that worked well for us. We can also display several steps in one graph. The step from $x_1$ to $x_2$ would be practically indistinguishable from the graph of $f$ itself, but let's try the case that didn't seem to converge.

```
> p:= plot(f(x), x = -3 .. 1):
> display({ p, q(w[0]), q(w[1]), q(w[2]) }); (Figure 3.13)
```

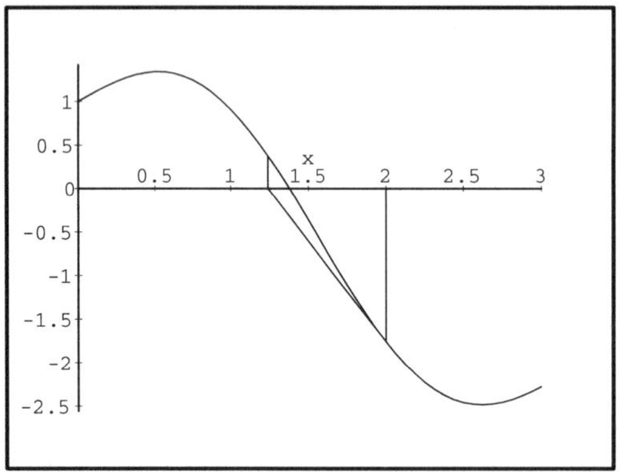


Figure 3.12. Newton's method, $x_0 = 2$

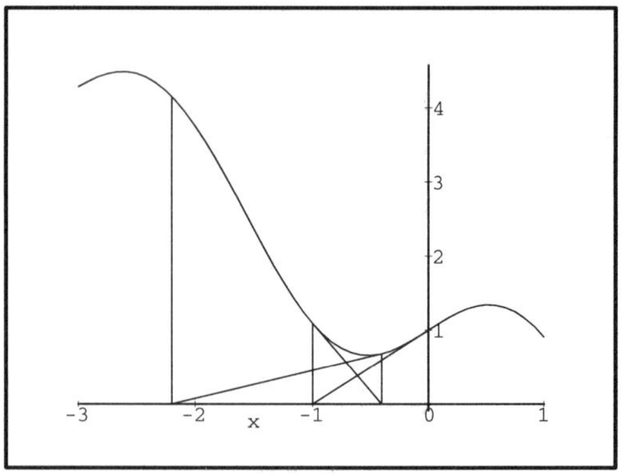


Figure 3.13. $w_0 = 0$

Why the big difference between the two cases? The tangent line at $x = x_0 (= 2)$, which of course is a very good approximation to $f(x)$ near $x = 2$, is still not too bad an approximation near the root $r = 1.377336877$. It intersects the $x$ axis at a point $x_1$ even closer to $r$, and the tangent line at $x_1$ is an even better approximation to $f(x)$ near $x = r$. On the other hand, the tangent line at

(11) Actually, in this case it does eventually settle down after about $w_{90}$.

$x = w_0 (= 0)$, although a good approximation to $f(x)$ near $x = 0$, bears no resemblance to $f(x)$ near $x = r$. Because of this there is no reason to expect $w_1$ to be close to $r$, and therefore no reason to expect Newton's method starting at $w_0$ to converge. In fact, the slope of the tangent line at $w_0$ is in the wrong direction, so that $w_1$ is even farther from $r$ than $w_0$ is. The $w_n$'s jump around, never coming near $r$.

It may be instructive to make an animation of Newton's method, with one frame per iteration. We can use the "`display`" command with option "`insequence=true`". The individual frames will themselves be "`display`" commands similar to those above. It's probably best to view this animation frame by frame.

```
> p:= plot(f(x), x=-3 .. 6):
> display([ seq(display({ p, q(w[kk])}), kk= 0 .. 10) ],
>            insequence=true);
```
■

## Error Bounds for Newton's Method

In general it is hard to determine for which starting values Newton's method will converge. Let's see if we can use error bounds for Newton's method to check that in Example 3 the method converges to the root $r$ if, say, $1 \le x_0 \le 2$. This is based on the following theorem:

> Suppose that $f$, $f'$ and $f''$ are continuous on an interval $I$ containing $x_n$, $x_{n+1}$, and a root $x = r$ of $f(x) = 0$. Suppose also that there exist constants $K$ and $L > 0$ such that for all $x$ in $I$ we have (i) $|f''(x)| \le K$, and (ii) $|f'(x)| \ge L$.
>
> Then (a) $|x_{n+1} - r| \le \dfrac{K}{2L}|x_{n+1} - x_n|^2$, and (b) $|x_{n+1} - r| \le \dfrac{K}{2L}|x_n - r|^2$.

The interval $1 \le x \le 2$ does contain the root $r = 1.377336877$. To see that $x_{n+1}$ is in the interval whenever $x_n$ is, we can plot $g(x)$ for $1 \le x \le 2$. To find $K$, we plot $|f''(x)|$ on the interval and discover that its maximum there is at $x = 1$: $K = |f''(1)| = 3.637189707$. Plotting $|f'(x)|$ on the interval, we see the minimum there is also at $x = 1$: $L = |f'(1)| = 1.832293673$. Thus $K/(2L) = .9925236770$.

Now suppose $x_0$ is in the interval. We have

$$
\begin{aligned}
|x_0 - r| &\le 2 - r = .622663123 \\
|x_1 - r| &\le \frac{K}{2L}(2-r)^2 = .3848107243 \\
|x_2 - r| &\le \frac{K}{2L}\left(\frac{K}{2L}(2-r)^2\right)^2 = \left(\frac{K}{2L}\right)^3 (2-r)^4 = .1469722049 \\
|x_3 - r| &\le \frac{K}{2L}\left(\left(\frac{K}{2L}\right)^3 (2-r)^4\right)^2 = \left(\frac{K}{2L}\right)^7 (2-r)^8 = .02143933424 \\
|x_4 - r| &\le \frac{K}{2L}\left(\left(\frac{K}{2L}\right)^7 (2-r)^8\right)^2 = \left(\frac{K}{2L}\right)^{15} (2-r)^{16} = .0004562085978
\end{aligned}
$$

and in general

$$|x_n - r| \le \left(\frac{K}{2L}\right)^{2^n - 1} (2-r)^{2^n}$$

Since $\dfrac{K}{2L}(2-r) = .6180078925 < 1$, this tends to 0 as $n \to \infty$. Thus we conclude that the sequence converges to $r$ whenever $x_0$ is in the interval. But the estimates show more than this: they show that the convergence is very fast, with the number of correct decimal places essentially doubling at each iteration.

## EXERCISES

1. Plot $x \sin x$, together with its linear approximation at $x = \pi/2$, on the interval $1 \leq x \leq 2$.
2. Use $f''$ to estimate the maximum error in using the linear approximation at $x = 1$ to approximate $f(x) = x/(1+x)$ on the interval $0.9 \leq x \leq 1.1$. Compare this to the actual maximum error.
3. A certain function $f(x)$ is known to satisfy the differential equation $f'(x) = \cos(x + f(x))$ with $f(1) = 2$.
   (a) What is the linear approximation to $f(1.2)$?
   (b) Estimate the error in this approximation. *Hint:* you can estimate $f''$ (crudely) by just using the fact that $\sin t$ and $\cos t$ are between $-1$ and 1.
4. Let $f(x)$ be the antiderivative of $\sqrt{1 + x \cos x}$ with $f(0) = 1$. Use a linear approximation to find an approximate value of $f(0.1)$. Estimate the error in this approximation.
5. Consider the function $f(x) = (3 + x)\cos(x) + x^2$.
   (a) Plot $f(x)$ and its linear approximation at $x = 0$ over the intervals $-1$ to 1 and on some smaller intervals centred at 0. Find an interval for which the two graphs appear identical on your screen.
   (b) Find error estimates for this linear approximation on the interval $-1$ to 1, using "`evalr`" to estimate $f''(x)$. Plot the actual error and the two error estimates over this interval.
   (c) Over what interval, according to the error estimates from (b), is the linear approximation accurate to within 0.001?
   (d) Repeat (b) and (c) using the graphical method to estimate $f''(x)$.
6. Consider the function
$$f(x) = \sin x + \frac{1}{1 + (1000x - 600)^{100}}$$
   (a) Use the linearization at $x = 0$ to approximate $f(0.6)$.
   (b) Show that the graphical method of estimating $f''$ on the interval $0 \leq x \leq 0.6$ produces an incorrect estimate for the error in this approximation. Explain why this happens.
7. Solve $x \cos x = \sin x + 1$ with $0 < x < 2\pi$ using Newton's method. *Hint:* draw a graph to find a good starting point.
8. Find the positive solution of $x^2 + \sin x = 8$ using Newton's method. *Hint:* use the fact that $-1 \leq \sin x \leq 1$ to guess a good starting point.
9. (a) Find an equation of the form $1 + ax + bx^3 = 0$ for which Newton's method starting at $x_0 = 0$ endlessly repeats the sequence $0, 1, 0, 1, \ldots$. *Hint:* you want $g(0) = 1$ and $g(1) = 0$. Solve these equations for $a$ and $b$.
   (b) Choose a better starting point, and solve the equation using Newton's method.
10. When Newton's method works, it usually converges very rapidly to a solution: typically the number of correct decimal places doubles at each iteration. This is related to the fact that the error estimate for $x_n$ is proportional to the square of the error in $x_{n-1}$.
   (a) Try solving the equation of Example 3 with $x_0 = 2$, using "`Digits:= 80`". How many iterations are needed until $x_{n+1} = x_n$? To how many digits is each $x_n$ correct?
   (b) On the other hand, the error estimate is no good at all if $f'(r) = 0$, and indeed the convergence may be much slower in this case. Consider Newton's method for the function $f(x) = x^3$, starting with $x_0 = 1$. How does the number of correct digits change with each iteration? How many iterations are needed to get the root to 10 decimal places? Note that you can simplify $g(x)$ and find an explicit formula for $x_n$.
   (c) For an example with even slower convergence, try
$$f(x) = \begin{cases} x^{1/x} & \text{for } x > 0 \\ 0 & \text{for } x \leq 0 \end{cases}$$

with $x_0 = 1/2$. When is $x_n < 1/10$?

**11.** The equation $x^2 = \sin x$ has two solutions, the obvious $x = 0$ and another near $x = 0.9$. Consider solving it using Newton's method.

**(a)** What can you say about $g(x)$ when $x > 0.5$? *Hint:* plot $g(x)$.

**(b)** Find an interval $I$ of length $0.3$, containing the solution $r$ near $0.9$ but not containing $0$, such that $g(x)$ is in $I$ whenever $x$ is.

**(c)** Using the error bound for $|x_{n+1} - r|$ in terms of $|x_n - r|$, how many iterations would suffice to determine $r$ to an accuracy of $10^{-100}$ starting at any point in $I$? Ignore roundoff error.

# LAB 8 ITERATION AND FIXED POINTS

**Goals:** *To examine the limiting behaviour of sequences obtained by iterating a function, including examples of attracting and repelling fixed points and cycles, and chaos. To use the derivative to decide whether a fixed point or cycle is an attractor or repeller.*

Consider a sequence $x_0, x_1, x_2, \ldots$ obtained by choosing some number $x_0$ and repeatedly calculating $x_{n+1} = f(x_n)$, where $f$ is some function. What happens as $n \to \infty$? One possibility is that $x_n$ will approach a limit $r$ which is a **fixed point** of $f$, i.e., $f(r) = r$. As we will see, there are other possibilities.

There are many interesting applications of this subject. Iteration can be used to approximate a fixed point of a function(12). On a more "applied" level, a biologist studying the population of a certain organism may model the population $x_n$ in year $n$ as some function $f$ of the population in the previous year. The result could be that the population approaches equilibrium at a fixed point of $f$. On the other hand, it might fluctuate periodically or even chaotically, which might be an even more interesting result. Investigations of this type by the mathematical biologist Robert May and others in the 1970's led to a very active area of the theory of dynamical systems, sometimes called "chaos theory"(13).

## Attractors and Repellers

■ **EXAMPLE 1** Examine the limiting behaviour of the sequences $x_{n+1} = f(x_n)$ for the function $f(x) = ax - x^2$, where $a$ is a parameter. We will take several different starting points $x_0$ and several different values of $a$. We begin with $a = 2.8$ and $x_0 = 1.5$.

(12) Newton's method is an example of this: a fixed point of $g(x) = x - f(x)/f'(x)$ is a solution of $f(x) = 0$. See Exercise 9.

(13) The connection to *Jurassic Park* is, however, a bit tenuous.

**SOLUTION**
```
> f:= x -> a * x -  x^2:
> a:= 2.8:
> x[0]:= 1.5:
> for nn from 0 to 39 do x[nn+1] := f(x[nn]) od;
```

$$x_1 := 1.95$$

$$x_2 := 1.6575$$

$$\dots$$

You may see more clearly what is going on if you plot the points $(n, x_n)$, rather than looking at a list of numbers. There's no real purpose in joining the points by lines, so I'll use the option "`style=POINT`".

`> plot([ seq( [k, x[k]], k=0..40) ], style=POINT);` (Figure 3.14)

It looks like the sequence converges to a limit of 1.8. In fact, $f(1.8) = 1.8$, so this is a fixed point.

If you try other starting points $x_0$, you will find that the sequence converges to 1.8 as long as $0 < x_0 < 2.8$. If $x_0 < 0$ or $x_0 > 2.8$, the limit is $-\infty$. There is another fixed point at $x = 0$, but the only ways to get $x_n$ to converge to 0 are $x_0 = 0$ (in which case all $x_n$ are 0) or $x_0 = 2.8$ (in which case $x_1$ and all later $x_n$ are 0). If $x_0$ is close to, but not exactly, 0, $x_n$ will be close to 0 for a while, but eventually moves away toward either 1.8 or $-\infty$. Of the two fixed points, 1.8 is said to be an **attractor**, while 0 is a **repeller**.

A fixed point $r$ of $f$ is **stable** if we can guarantee that $x_n$ always stays close to $r$ by taking the starting point $x_0$ sufficiently close to $r$. Otherwise it is **unstable**.

A stable fixed point $r$ is an **attractor** if $x_n \to r$ as $n \to \infty$ whenever $x_0$ is sufficiently close to $r$.

An unstable fixed point $r$ is a **repeller** if whenever the starting point $x_0$ is sufficiently close to, but not equal to, $r$, there will be some $x_n$ which are at least a certain distance away from $r$.

For us the main classifications of fixed points are attractors and repellers. We won't consider any examples that are neither one nor the other. The main way to tell whether a fixed point $r$ is an attractor or a repeller is by looking at $f'(r)$.

A fixed point $r$ of a differentiable function $f$ is an attractor if $|f'(r)| < 1$. It is a repeller if $|f'(r)| > 1$. If $|f'(r)| = 1$, it could be an attractor or a repeller or neither; we need more information to decide which.

This can be seen as follows: consider the slope of the secant line to the graph of $f$ through $(r, f(r)) = (r, r)$ and $(x_0, f(x_0)) = (x_0, x_1)$, where $x_0$ is close to (but not exactly) $r$. This slope is $(x_1 - r)/(x_0 - r)$. It will be arbitrarily close to $f'(r)$ if $x_0$ is close enough to $r$, since the derivative is the limit of the slopes of secant lines. In particular, if $|f'(r)| < 1$ we will also have $|(x_1 - r)/(x_0 - r)| < 1$, i.e., $|x_1 - r| < |x_0 - r|$, if $x_0$ is close enough to $r$. Thus $x_1$ will be even closer to $r$ in this case. That is in itself not enough to make the sequence converge to $r$, but we can improve the result: if $|f'(r)| < 1$, take some number $p$ with $|f'(r)| < p < 1$. If $x_0$ is close enough to $r$, we will have $|(x_1 - r)/(x_0 - r)| < p$, i.e., $|x_1 - r| < p|x_0 - r|$. Since $x_1$ will also be "close enough" to $r$ (if not exactly equal to $r$), $|x_2 - r| < p|x_1 - r| < p^2|x_0 - r|$. This will go on forever: at each iteration, the distance to $r$ decreases at least by a factor of $p$, and as $n \to \infty$ the distance goes to 0 and $x_n$ converges to $r$. Thus in this case the fixed point is an attractor. On the other hand, if $f'(r) > 1$ a similar argument shows that when $x_0$ is close enough to (but not equal to) $r$ the distance from $r$ increases at each iteration until $x_n$ is no longer "close enough" to $r$. In this case, the fixed point is a repeller.

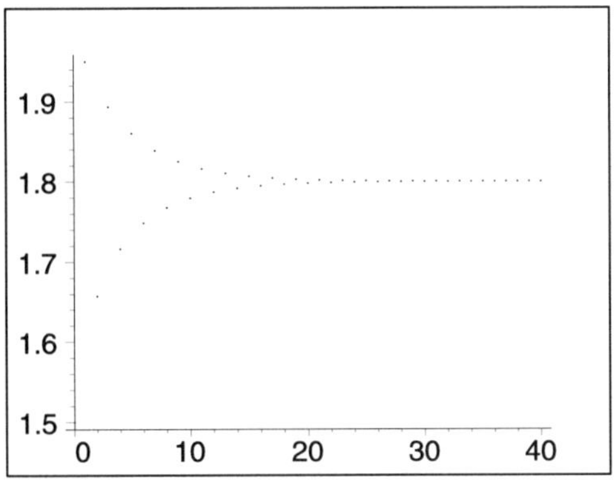


Figure 3.14. Iterations with $a = 2.8$

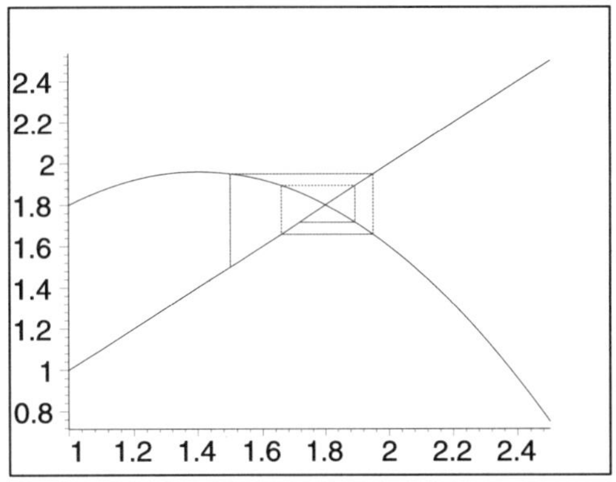


Figure 3.15. Staircase with $a = 2.8$

To help visualize what is going on here, we can have Maple draw "staircase" diagrams. We will plot the graphs of $y = f(x)$ (I'll refer to this as "the curve") and $y = x$ (I'll refer to this as "the line"). Then we will draw some line segments: starting at $(x_0, x_0)$ on the line, moving vertically to $(x_0, f(x_0)) = (x_0, x_1)$ on the curve, then horizontally to $(x_1, x_1)$ on the line, then vertically again to $(x_1, f(x_1)) = (x_1, x_2)$ on the curve, etc. For convenience, I'll define a command "`stair`" that takes a value $x$ and plots a vertical segment from $(x, x)$ to $(x, f(x))$ followed by a horizontal segment to $(f(x), f(x))$. Then we can combine the graphs of $y = f(x)$ and $y = x$ with several "stairs" to produce our diagram.

```
> stair:= x -> plot([[x, x], [x,f(x)], [f(x), f(x)]]):
> p:= plot({f(x), x}, x = 1 .. 2.5):
> with(plots):
> display({p, stair(x[0]), stair(x[1]),
>              stair(x[2]), stair(x[3]) }); (Figure 3.15)
```

The fixed point $r = 1.8$ is an attractor since $f'(r) = -.8$. Each $x_n$ is closer to $r$ than the previous one, and the "staircase" spirals inwards toward $(r, r)$. On the other hand, $f'(0) = 2.8$ so 0 is a repeller. Let's see what happens when $x_0$ is near 0.

```
> x[0]:= 0.05:
> for nn from 0 to 2 do x[nn+1] := f(x[nn]) od;
> p:= plot({f(x), x}, x = -1 .. 1):
> display({p, stair(x[0]), stair(x[1]),
>              stair(x[2]), stair(x[3]) });
```

We see a "staircase" moving away from $(0, 0)$. A few steps farther along, this would turn into a spiral and approach the attractor at $(1.8, 1.8)$. ■

## Attracting and Repelling Cycles

■ **EXAMPLE 2** Examine the limiting behaviour for the function $f(x) = ax - x^2$ with $a = 3.1$.

**SOLUTION** First we plot the points $(n, x_n)$ with $x_0 = 1.5$.

```
> a:= 3.1: x[0]:= 1.5:
> for nn from 0 to 39 do x[nn+1]:= f(x[nn]) od:
> plot([ seq( [k, x[k]], k = 0 .. 40)], style=POINT); (Figure 3.16)
```

We see a regular pattern with the even-numbered members near 1.73 and the odd-numbered ones near 2.37. What has happened to the fixed point? You can easily check that the fixed points are at 0 and 2.1, and both are repellers ($f'(2.1) = -1.1$). What happens when we start near 2.1? I'll take $x_0 = 2.05$ and draw a staircase diagram with 41 stairs, using "`seq`" to avoid having to type each "`stair`" individually.

```
> x[0]:= 2.05:
> for nn from 0 to 39 do x[nn+1] := f(x[nn]) od;
> p:= plot({f(x), x}, x = 1.5 .. 2.5):
> display({p, seq(stair(x[n]), n = 0 .. 40) }); (Figure 3.17)
```

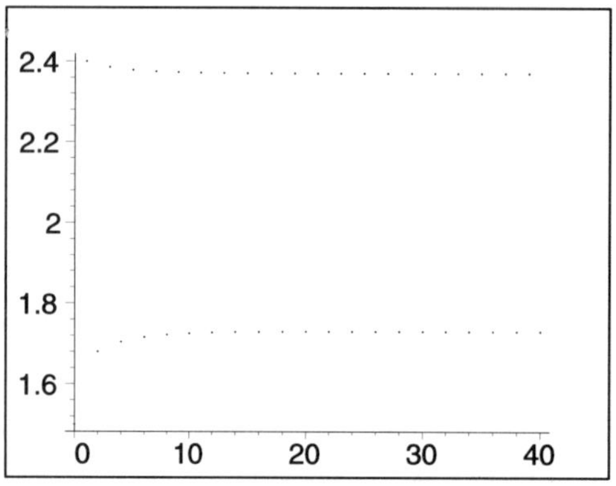


Figure 3.16. Iterations with $a = 3.1$

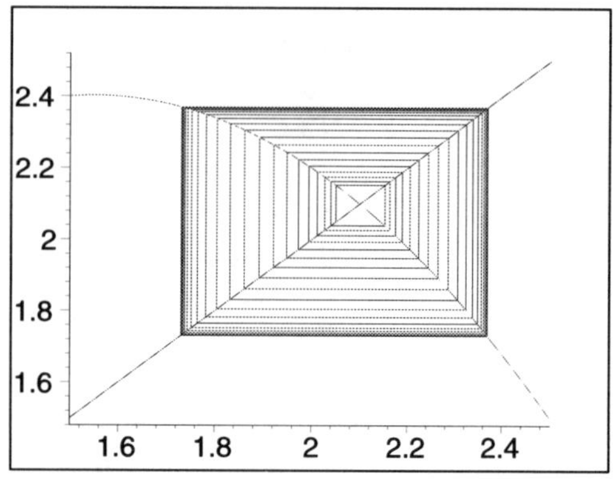


Figure 3.17. Staircase with $a = 3.1$

We see the stairs spiraling away from the fixed point, and then approaching a rectangle (if we used "constrained" scaling to make the scales on the axes equal, it would be a square). If the lower left corner of the rectangle is at $(p, p)$ on the line, the top left would be at $(p, f(p))$ on the curve, the top right at $(f(p), f(p))$ on the line, and the bottom right at $(f(p), f(f(p)))$ on the curve. But to make this a rectangle we must have $f(f(p)) = p$. We say that $p$ is a **periodic point** of **period** 2, and that $p$ and $f(p)$ form a **cycle** of period 2. We can find the periodic points using "`solve`" or "`fsolve`":

```
> s:= solve(f(f(x))=x, x);
```

$$s := 0,\ 2.100000000,\ 2.370156212,\ 1.729843788$$

Of these four solutions, the first two are fixed points, not periodic, and the last two form our cycle. You can check that $f$ takes the third solution to the fourth and the fourth to the third. Thus if you started with $x_0$ as one of these points, $x_n$ would alternate between the two.

Just as a fixed point can be an attractor or a repeller, so can a cycle. A periodic point $p$ of period 2 is a fixed point of the function $f \circ f$. If $|(f \circ f)'(p)| < 1$ it is an attractor, so if $x_0$ is close enough to $p$ the even-numbered $x_n$'s will converge to $p$ while the odd-numbered ones converge to $f(p)$. We then say the cycle is an attractor. If $|(f \circ f)'| > 1$ it is a repeller (as a fixed point of $f \circ f$), so if $x_0$ is close to $p$ the $x_n$'s will move away from $p$ and $f(p)$. We then say the cycle is a repeller. Note that by the Chain Rule, $(f \circ f)'(p) = f'(p)f'(f(p))$. In particular, it doesn't matter whether we use $p$ or $f(p)$ in deciding stability.

```
> D(f @ f)(s[3]);
```

$$.5899999997$$

This confirms that the cycle is an attractor. ■

In general, a periodic point $p$ of period $m$ is a point such that $f(\ldots f(p) \ldots) = p$ (with $m$ $f$'s, but not with any smaller positive number of $f$'s). It is a fixed point of the function (in Maple notation) `f@@m`. Together with the points `(f@@k)(p)` for $k = 1$ to $m - 1$, which are also periodic points of period $m$, it forms a cycle of period $m$. To test stability of the cycle, we look at `D(f@@m)(p)`, which is the product of the values of $f'$ at the $m$ points of the cycle. If this has absolute value less than 1 the cycle is an attractor, while if its absolute value is greater than 1 the cycle is a repeller.

■ **EXAMPLE 3** Examine the limiting behaviour for the function $f(x) = ax - x^2$ with $a = 3.8$.

**SOLUTION** Figure 3.18 shows the points $(n, x_n)$ for $n \leq 50$, with $x_0 = 1.5$. No regular pattern is apparent: $x_n$ jumps around in an unpredictable fashion. There are fixed points and periodic points at this value of $a$, but all of them are repellers. This is an example of "chaos".

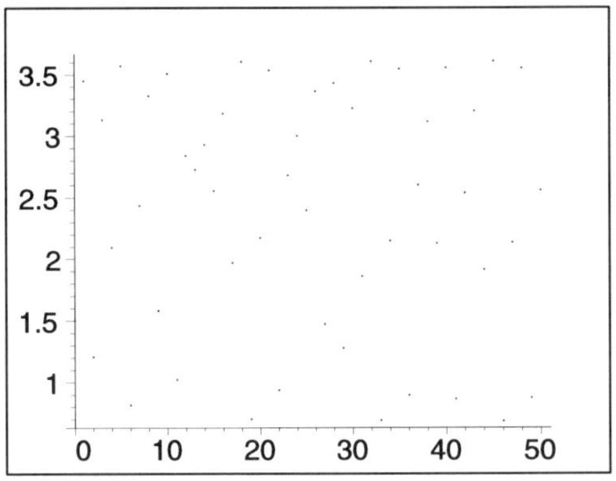


Figure 3.18. Iterations with $a = 3.8$

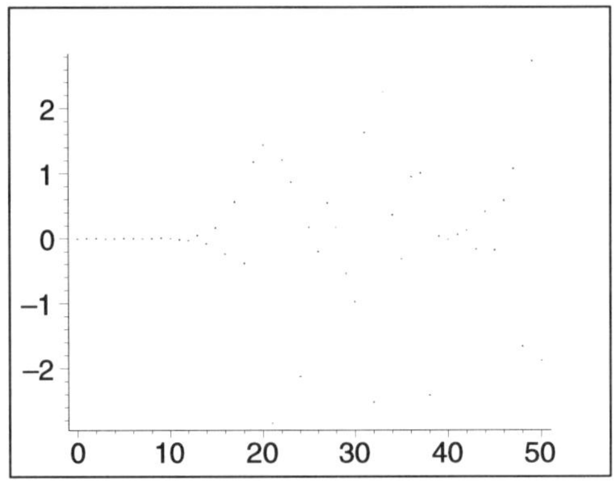


Figure 3.19. $w_n - x_n$

One of the hallmarks of a chaotic system is "sensitive dependence on initial conditions". To illustrate this, we can compare the $x_n$ calculated above with $w_n$ calculated by the same formula but with $w_0 = 1.5001$ instead of 1.5. You might think that such a slight change wouldn't make much difference, and indeed it doesn't for the first few $n$. But the difference gradually builds up, until by about $n = 20$ the two sequences are essentially unrelated. One way to see this is to plot the differences $w_n - x_n$:

```
> plot([ seq( [k, w[k] - x[k]], k = 0..50) ],
>       axes=FRAMED,  style=POINT); (Figure 3.19)
```
■

The sensitive dependence on initial conditions has practical significance for the predictability of systems in real life. In principle, if you know the function $f$ and the initial value $x_0$, this determines all the $x_n$. However, in a real system you never know $x_0$ exactly, but can only measure it to a certain degree of accuracy. In systems that exhibit sensitive dependence on initial conditions, even a very small initial inaccuracy grows over time, until eventually it completely ruins any predictions made about the system. This is believed to be the reason that accurate long-range weather forecasts are impossible.

## EXERCISES

**1.** **(a)** Find the fixed points of the function $f(x) = x(4 + x^2)/(1 + 4x^2)$, and classify them as attractors or repellers.

**(b)** What happens to the points $x_n$ with starting point $x_0 = 2$?

**2.** **(a)** Examine the limiting behaviour of the sequence $x_{n+1} = (10+x_n)/(1+10x_n^2)$ with $x_0 = 1.5$.

**(b)** Find the cycle to which this converges, and show that it is an attractor.

**3.** Consider iterating the function $f(x) = \cos(ax)$, where $a$ is a parameter.

**(a)** With $a = 1$ and any starting point, calculate and plot the first 40 points. Does the sequence appear to converge to a fixed point?

**(b)** Find the fixed point with $a = 1$ using "`fsolve`", and check whether it is an attractor or repeller.

**(c)** Repeat (a) for $a = 1.5$. Is there an attractor that is a fixed point? An attractor that is a cycle?

**(d)** Plot "staircase" diagrams for $a = 1$ and $a = 1.5$, using starting points near the fixed points.

**4.** In this exercise we try to make sense of the expression $x^{x^{x^{\cdot^{\cdot^{\cdot}}}}}$ (an infinite "tower" of $x$'s). If this is to mean anything, it must be the limit of finite "towers". Let $t_n$ be a "tower" of $n$ $x$'s. Thus we have $t_1 = x$ and $t_{n+1} = x^{t_n}$. So the question becomes: "What happens when we iterate the function $f(t) = x^t$, starting with $t_1 = x$?"

**(a)** Calculate and plot the first 40 "towers" for $x = 0.04$, for $x = 0.1$, and for $x = 1.2$. In each case, are the "towers" converging? For $x = 1.5$, you won't be able to go all the way to 40. What happens?

**(b)** Plot "staircase" diagrams for the first few iterations, with each of the $x$ values from (a). This exercise will be continued in Lab 9, as we will need exponential and logarithm functions to proceed further.

**Exercises 5 to 8 use the function $f(x) = ax - x^2$ of Examples 1 to 3.**

**5. (a)** The function $f$ has a fixed point at $a - 1$. For what range of values of $a$ is this fixed point an attractor?

**(b)** Use "`solve`" to find the cycle of period 2. For what range of values of $a$ is this valid?

**(c)** For what range of values of $a$ is the cycle of period 2 an attractor?

**6. (a)** For $a = 3.5$, calculate and plot the first 50 points, with $x_0 = 1$.

**(b)** What is the period of the attractor cycle? Verify that it is an attractor.

**(c)** Plot a "staircase" diagram with 50 stairs corresponding to the points of (a).

**7. (a)** With $a = 3.9$, find a cycle of period 3.

**(b)** Show that the cycle of (a) is a repeller.

**(c)** One way to ensure that a cycle is an attractor is to arrange for $f'$ to be 0 at one of its points. For our function, $f'(a/2) = 0$. Find an $a$ for which $f(f(f(a/2))) = a/2$, but $a/2$ is not a fixed point.

**(d)** Plot a "staircase" diagram for the $a$ of (c), with $x_0 = a/2$.

**8. (a)** Using `Digits:= 30` and $a = 3.8$, what is the least $n$ for which $|x_n - w_n| \geq 1$, where $x_0 = 1.5$ and $w_0 = 1.5 + 10^{-5}$? What about with $w_0 = 1.5 + 10^{-10}$? $1.5 + 10^{-20}$?

**(b)** With $a = 3.1$ and the same $x_0$ and $w_0$'s as in (a), find $n$ for which (to 30-digit accuracy) $x_n = w_n$. This exhibits the opposite of sensitive dependence on initial conditions: small inaccuracies eventually disappear.

**9.** Let $f$ be any smooth function and $r$ a point where $f(r) = 0$ but $f'(r) \neq 0$. Let $g(x) = x - f(x)/f'(x)$ be the function whose iteration constitutes Newton's method. Show that $r$ is an attracting fixed point of $g$.

# LAB 9 INVERSE FUNCTIONS AND TRANSCENDENTAL FUNCTIONS

**Goals:** *To use Maple in deciding whether a function is one-to-one, and in computing and graphing inverse functions. To learn how to use exponential, logarithmic, and inverse trigonometric functions in Maple while avoiding some of the pitfalls that can arise.*

## Inverse Functions

Inverse functions arise naturally when we want to solve an equation $y = f(x)$ for $x$ as a function of $y$. Usually, Maple can at least solve such an equation numerically (using "`fsolve`"). In some cases, it can do it symbolically with "`solve`".

The first question that must be asked when we want an inverse function is whether our original function $f$ is one-to-one. A one-to-one function $f$ has the property that any horizontal line meets the graph of $f$ in at most one point. In some cases, you can actually check this by solving $f(x_1) = f(x_2)$ for $x_2$. The function is one-to-one if there are no solutions with both $x_1$ and $x_2$ in the domain of $f$, except for $x_2 = x_1$. However, you should exercise caution because "`solve`" does not always report all solutions, only the ones it found. Sometimes it doesn't find any solutions.

■ **EXAMPLE 1** Decide whether the following function is one-to-one: $f(x) = \sqrt{x^2 - 3x + 2}$ with domain the interval $2 \le x < \infty$.

**SOLUTION** We try solving $f(x_1) = f(x_2)$.

```
> f:= x -> sqrt(x^2 - 3*x + 2):
> solve(f(x1) = f(x2), x2);
```

$$-x1 + 3,\ x1$$

These are all the solutions, because $f(x_1) = f(x_2)$ is equivalent to the quadratic equation (in the variable $x_2$) $x_2^2 - 3x_2 + 2 = x_1^2 - 3x_1 + 2$, and a quadratic has no more than two solutions. If $x_1 \ge 2$ then $x_2 = 3 - x_1 \le 1$ is not in the domain. Therefore $f$ is one-to-one on its domain. ■

■ **EXAMPLE 2** Decide whether the following function is one-to-one: $g(x) = x + \sin(x)$, with domain all real numbers.

**SOLUTION** We try solving $g(x_1) = g(x_2)$.

```
> g:= x -> x + sin(x);
> solve(g(x1) = g(x2), x2);
```

$$\mathrm{RootOf}(\_Z - x1 - \sin(x1) + \sin(\_Z))$$

This is no help. Maple doesn't even find the solution $x_2 = x_1$. We could use "`fsolve`" for any particular $x_1$, but this would probably just return $x_2 = x_1$ and not prove anything. We could divide the equation by $x_2 - x_1$, so that it won't be defined when $x_2 = x_1$[14]. That might prevent "`fsolve`" from returning $x_2 = x_1$, but it still only works with one particular $x_1$ at a time, while we want to know about all of them. Let's try a graph.

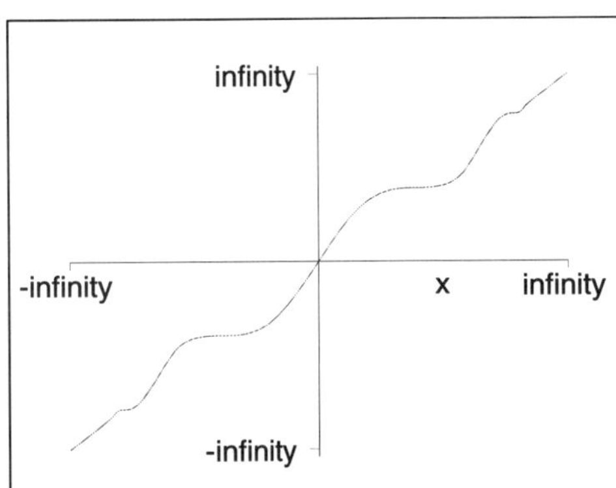


Figure 3.20. Infinity plot of $g$

[14] In Release 5, you can also use "`fsolve(g(x1)=g(x2), x2, avoid={x2=x1});`".

```
> plot(g(x),x=-infinity..infinity);
```
(Figure 3.20)

Well, this one looks promising. There are no horizontal lines that appear to cut the graph more than once. The graph does look rather flat in some places, but presumably these are not really flat intervals, and $g$ is really increasing. A closer look might be in order. We can also look at $g'$, using a result which is a consequence of the Mean Value Theorem:

> If $f'(x) \geq 0$ for all $x$ in an interval $I$, with $f'(x) = 0$ only at isolated points (if at all), then $f$ is increasing on $I$.

In this case, $g'(x) = 1 + \cos(x)$ is simple enough to study by hand. It is easy to verify, with or without Maple, that the minimum value of $g'(x)$ is 0, attained only at the isolated points $\pm\pi$, $\pm 3\pi$, .... So $g$ is increasing and therefore one-to-one. ■

Once we know our function has an inverse, the next question is how to obtain that inverse in Maple. If the function is simple enough for "`solve`" to work, that is the best way to handle it:

```
> s:= solve(x=f(y), y);
```

$$s := \frac{3}{2} + \frac{1}{2}\sqrt{1+4x^2},\ \frac{3}{2} - \frac{1}{2}\sqrt{1+4x^2}$$

Which of these do we want? The one with "+", since we're working on the interval $2 \leq y < \infty$.

```
> finv:= unapply(s[1],x);
```

$$\mathit{finv} := x- > \frac{3}{2} + \frac{1}{2}\sqrt{1+4x^2}$$

In cases such as Example 2 where "`solve`" doesn't work, we probably won't be able to represent the inverse function explicitly in terms of the standard functions of mathematics. Instead, we can use "`fsolve`":

```
> ginv:= x -> fsolve(g(y)=x, y);
```

If necessary, we can specify an interval for $y$ (the domain of the function $g$) in the "`fsolve`" to ensure that the correct value is found.

```
> finv2:= x -> fsolve(f(y)=x, y, y=2..infinity);
```

This works fine for calculating particular numerical values of the inverse function, but doesn't work with a non-numerical input:

```
> ginv(2);
```

$$1.106060158$$

```
> ginv(x);
```
Error, (in fsolve) x, is in the equation, and is not solved for

Moreover, Maple isn't able to differentiate a function defined using "`fsolve`".

```
> D(ginv);
```

$$\mathrm{D}(\mathit{ginv})$$

An alternative approach is to use "`RootOf`".

```
> ginv2:= unapply(RootOf(x=g(y), y), x);
```

$$\mathit{ginv2} := x \to \mathrm{RootOf}(x - \_Z - \sin(\_Z))$$

This can be evaluated with a symbolic input, and it can be differentiated. On the other hand, calculating numerical values is uncertain because it's less easy to specify which of several possible roots to use: in fact, sometimes "`evalf`" will choose a complex root, even though a real one is available.

```
> ginv2(x);
```

$$\mathrm{RootOf}(x - \_Z - \sin(\_Z))$$

```
> D(ginv2)(x);
```

$$-\frac{1}{-1-\cos(\text{RootOf}(x-\_Z-\sin(\_Z)))}$$

Note that this is what it's supposed to be: $1/g'\left(g^{-1}(x)\right)$.

```
> evalf(ginv2(2));
```

$$3.986898134 - 1.632344899\,I \text{ (in Release 4)}$$

For some reason (I really don't know why) Maple has chosen a complex root[(15)].

## Graphing Inverse Functions

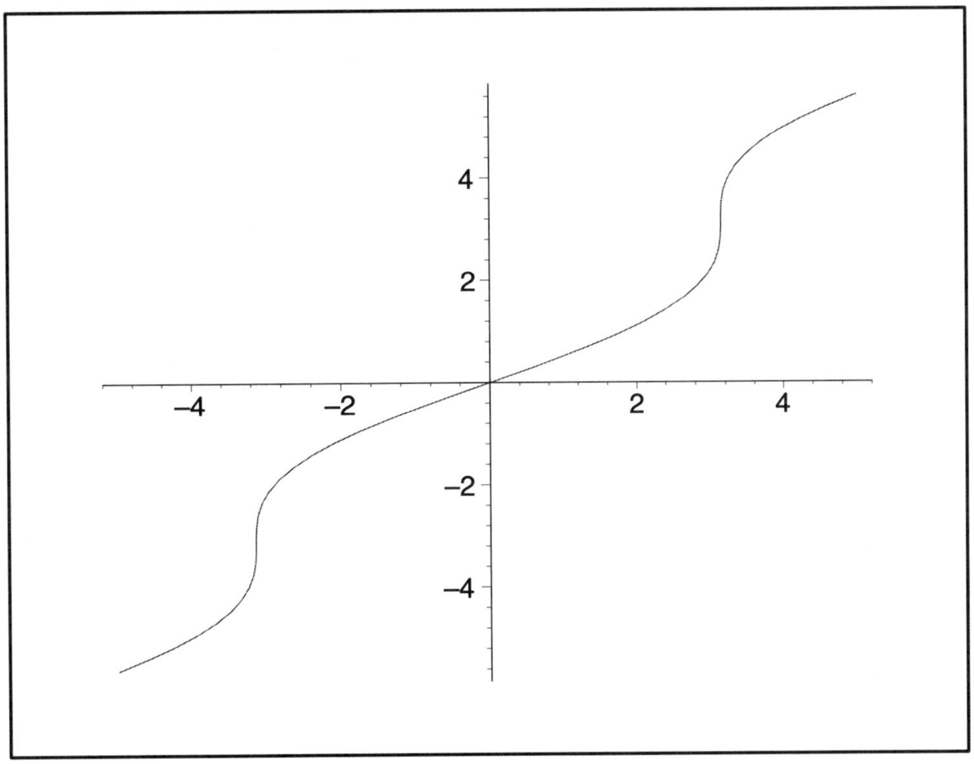


Figure 3.21. Plot of $g^{-1}$

For graphing inverse functions, there are at least four possible approaches: implicit, direct (using "`fsolve`"), parametric, and transformational.

```
> plots[implicitplot](x = g(y), x=-5..5, y=ginv(-5)..ginv(5));
```

This can take some time, and often produces rather poor-quality results. You can improve the quality if you use the `grid` option to increase the number of points evaluated, but the time needed to produce the plot increases with the number of points.

```
> plot('ginv(x)',x=-5..5);
```

Note that we use the "`fsolve`" version, avoiding the possibility of complex values. We put the `ginv(x)` in quotes to prevent Maple from trying to evaluate `ginv` with the symbolic input "x", which as we saw above would produce an error. The results from this method are usually good, but it could be very slow: each point calculated requires an "`fsolve`", and each of those can take a noticeable amount of time.

(15) Recall that $I$ is $\sqrt{-1}$, which is not a real number. Release 5 produces a real root in this example.

```
> plot([ g(y), y, y=ginv(-5)..ginv(5)]);
```

This is an example of a "parametric" plot. Note the square brackets. Instead of choosing values of $x$ and plotting the corresponding points $(x, g^{-1}(x))$ (as we would ordinarily plot a function, and as the second method does) we choose the values of $y$ and plot $(g(y), y)$. Since $x = g(y)$ means $y = g^{-1}(x)$, this does produce the desired results.

```
> plottools[transform]((x,y) -> [y,x])
>     (plot(g(y), y = ginv(-5) .. ginv(5))); (Figure 3.21)
```

This is the transformational approach, where we take an ordinary plot of $g(y)$ as a function of $y$ and interchange the $x$ and $y$ axes, using "`transform`" from the "`plottools`" package. Both the parametric and transformational approaches are usually quite fast and produce good-quality results.

## Exponential and Logarithmic Functions

Maple uses "`exp`" for the exponential function. In output $\exp(t)$ is written as $\mathbf{e}^t$. However, this is merely cosmetic: if you use "`e`" in input Maple will just treat it as a variable. To refer to the constant $e$ you must use "`exp(1)`". Of course, if you wish you can make a definition "`e:= exp(1);`". However, it's still best to use "`exp(t)`" rather than "`e^t`".

Maple uses "`ln`" for the natural logarithm (you can also use "`log`" as a synonym). The logarithm of $x$ to an arbitrary base $a$ is written as "`log[a](x)`". Since Maple uses the complex numbers rather than just the real numbers, it will happily take the "`ln`" of a negative number and return a complex result. On the other hand, `ln(0)` will produce an error message.

The laws of exponents, such as $(a^x)^y = a^{xy}$, are valid for positive $a$, but not always for negative $a$ (e.g., $((-1)^2)^{1/2} \neq (-1)^1$). At one time Maple was too eager to apply these laws even when $a$ was not positive. Thus the following "simplification" occurred in Release 2:

```
> sqrt(a^2);
```

$$a$$

If $a$ can be negative, this is wrong. In later releases this "bug" (i.e., programming error) was fixed, and Maple won't make such simplifications except in cases where they are known to be safe.

```
> S:= sqrt(a^2);
```

$$S := \sqrt{a^2}$$

```
> simplify(S);
```

$$\text{csgn}(a)\, a$$

It won't say $|a|$, because that would not be true for complex numbers. The "`csgn`" function is a version of the signum function, $-1$ for $a < 0$, $+1$ for $a > 0$ and 0 for $a = 0$.

## Assumptions About Variables

This is fine, you might say, but what about when I do want $a$ to be positive? How can I tell Maple that the laws of exponents are OK again? The answer is the "`assume`" command.

```
> assume(a > 0);
> sqrt(a^2);
```

$$a \sim$$

The "$\sim$" after a name is to remind you that some assumption has been made about this variable, You don't need to use it yourself when referring to the variable. There are a number of other assumptions that can be made about a variable. For example,

```
> assume(b, real);
> assume(c, RealRange(2,4));
```

(this assumes that $2 \leq c \leq 4$)

```
> assume(d, RealRange(Open(2),Open(4)));
```

(this assumes that $2 < d < 4$)

You can find out what assumptions have been made about a variable with the "about" command:

```
> about(b);
        Originally b, renamed b~:
  is assumed to be: real
```

The "assume" command is not cumulative: if you first assume $a > 0$ and then later $a < 1$, Maple forgets the first assumption. To add new assumptions while keeping the old ones, you can use "additionally".

```
> additionally(a < 1);
> about(a);
        Originally a, renamed a~:
  is assumed to be: RealRange(Open(0),Open(1))
```

To remove all assumptions about a variable, "unassign" the variable:

```
> a:= 'a';
```

## Inverse Trigonometric Functions

The inverse trigonometric functions are all standard Maple functions: "arcsin", "arccos", "arctan", "arcsec", "arccsc", and "arccot". You can't use the "$\sin^{-1}$" notation.

```
> arccsc(2/sqrt(3));
```

$$\frac{1}{3}\pi$$

There is also a two-variable version of "arctan": `arctan(y,x)` is a number $\theta$ with $-\pi < \theta \leq \pi$ such that the point $(x, y)$ lies on the ray from the origin at counterclockwise angle $\theta$ from the positive $x$ axis. Note that the $y$ comes before the $x$. This form is especially useful in calculating the angle for a point when you don't know which quadrant the point is in.

```
> arctan(1,0);
```

$$\frac{1}{2}\pi$$

Simplification of expressions involving inverse trigonometric functions is complicated by the fact that these are only inverses of the restrictions of the trigonometric functions to particular intervals. Thus although $\sin(\arcsin(x)) = x$ is always valid,(16) $\arcsin(\sin(x)) = x$ is not. Maple will not make the simplification unless $x$ is known to be in the correct interval.

```
> sin(arcsin(x));
```

$$x$$

```
> simplify(arcsin(sin(x)));
```

$$\arcsin(\sin(x))$$

```
> assume(x>=-Pi/2, x <= Pi/2); simplify(arcsin(sin(x)));
```

$$x\sim$$

**■ EXAMPLE 3** Study the direct and retrograde motions of the planets.

The ancients were puzzled by the apparently erratic motions of the planets against the background of the stars. Sometimes a planet moves from west to east ("direct" motion), and sometimes from east to west ("retrograde" motion). The ancients lacked the theories of Copernicus, Kepler, and Newton, but Maple would have come in handy too! We are particularly interested in the times at which a planet switches between direct and retrograde motion.

---

(16) You might object that $x$ should be restricted to the interval $[-1, 1]$ here, as $\arcsin(x)$ is undefined elsewhere. The answer is that outside this interval $\arcsin(x)$ is defined in terms of complex numbers.

**SOLUTION** We can approximate the Earth and planets as moving in the same plane, counterclockwise in circles centred on the sun. The orbits are really ellipses and not in the same plane, but this is quite a good approximation except for Mercury and Pluto. We will measure length in "astronomical units", where 1 a.u. is the radius of the Earth's orbit, and time in years. We can take the Sun as the origin of an $xy$ coordinate system, and suppose that at time $t = 0$ the Earth is on the positive $x$ axis. Thus the Earth's coordinates at time $t$ are as follows:

```
> xe:= t -> cos(2*Pi*t);
> ye:= t -> sin(2*Pi*t);
```

Next, consider a planet whose orbit has radius $r_p$ and period $t_p$. We suppose that at $t = 0$ it is also on the positive $x$ axis. An astronomer would call this configuration "opposition" for one of the outer planets (i.e., the planet is opposite the sun) or "inferior conjunction" for one of the inner planets. Here are the planet's coordinates:

```
> xp:= t -> rp * cos(2*Pi*t/tp);
> yp:= t -> rp * sin(2*Pi*t/tp);
```

Now for the position where we (on Earth) see the planet against the background of stars: this is the angle, measured counterclockwise, from the positive $x$ axis to the point $(x = x_p(t) - x_e(t),\ y = y_p(t) - y_e(t))$. Since at different times this can be in any quadrant, it is easiest to use the two-variable form of "`arctan`".

```
> theta:= unapply(arctan(yp(t)-ye(t), xp(t)-xe(t)),t);
```

$$\theta := t -> \arctan\left(rp \sin\left(2\frac{\pi t}{tp}\right) - \sin(2\pi t),\ rp \cos\left(2\frac{\pi t}{tp}\right) - \cos(2\pi t)\right)$$

"Direct" motion corresponds to $\theta(t)$ increasing and "retrograde" motion to $\theta(t)$ decreasing. We expect that the times when the planet appears to change direction will be when $\theta'(t) = 0$. Our first thought is simply to solve this:

```
> t0:= solve(D(theta)(t)=0,t);
```

Maple's "solution" is not very helpful (expressed using "`RootOf`" a complicated expression). That is not too surprising when you look at what a complicated expression "`D(theta)(t)`" turns out to be (it's so complicated I won't even reproduce it here). Now for any particular planet, we could plug in numerical values for $r_p$ and $t_p$ and use "`fsolve`", but I'm hunting for a general formula. Let's try to simplify the expression.

```
> v:= simplify(D(theta)(t));
```

$$v := 2\pi\left(-rp\cos\left(2\frac{\pi t}{tp}\right)\cos(2\pi t) - \cos(2\pi t)\,tp\,rp\cos\left(2\frac{\pi t}{tp}\right) + rp^2 - rp\sin\left(2\frac{\pi t}{tp}\right)\sin(2\pi t)\,tp - rp\sin\left(2\frac{\pi t}{tp}\right)\sin(2\pi t) + tp\right)\Big/\left(tp\left(-2\,rp\cos\left(2\frac{\pi t}{tp}\right)\cos(2\pi t) + rp^2 - 2\,rp\sin\left(2\frac{\pi t}{tp}\right)\sin(2\pi t) + 1\right)\right)$$

Well, maybe some more simplification is possible: there are trigonometric identities that express a product such as $\sin A\ \sin B$ in terms of trigonometric functions of $A+B$ and $A-B$. The "`combine`" command with the "`trig`" option does this for us (while "`expand`" goes the other way, expressing functions of $A + B$ and $A - B$ as products of sines and cosines of $A$ and $B$).

```
> v:= combine(v,trig);
```

$$v := \left(-2\pi\,rp\cos\left(2\frac{\pi t(tp-1)}{tp}\right) - 2\pi\,tp\,rp\cos\left(2\frac{\pi t(tp-1)}{tp}\right) + 2\pi rp^2 + 2\pi tp\right)\Big/\left(-2\,tp\,rp\cos\left(2\frac{\pi t(tp-1)}{tp}\right) + rp^2\,tp + tp\right)$$

Note that this depends on $t$ only through $\cos(2\pi t(tp-1)/tp)$. We should therefore be able to solve this version of $v = 0$ for $t$.

```
> t0:= solve(v=0,t);
```

$$\frac{1}{2}\frac{\arccos\left(\frac{tp+rp^2}{rp(1+tp)}\right)tp}{\pi(tp-1)}$$

This is not the only solution, of course. It is easy to see that if $t_0$ is a solution, so are $-t_0$ and $t_0 + \dfrac{1}{1/t_p - 1}$.

Let's try the planet Jupiter.

```
> tp:= 11.86;
> rp:= 5.203;
> evalf(t0);
```

$$.1650861959$$

```
> plot(v,t=-1..1);
```
(Figure 3.22)

The graph shows that $v < 0$ over the interval $-t_0 < v < t_0$, i.e., Jupiter's retrograde motion lasts for $t_0$ years (approximately 60.3 days) before and the same time after opposition. ■

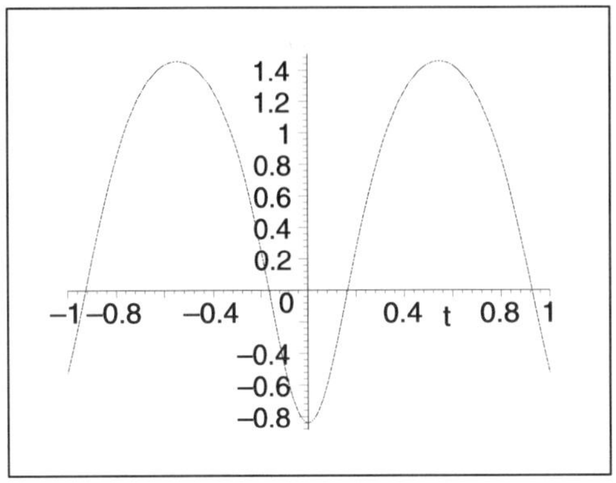


Figure 3.22. $d\theta/dt$ for Jupiter

## EXERCISES

1. Which of the following identities are correct? Verify by plotting.
   - **(a)** $\sin(\arcsin(x)) = x$
   - **(b)** $\arcsin(\sin(x)) = x$
   - **(c)** $\sin(\arcsin(\sin(x))) = \sin(x)$
   - **(d)** $\arcsin(\sin(\arcsin(x))) = \arcsin(x)$
2. Let $f(t) = t \ln t$, and let $g$ be its inverse function.
   - **(a)** Find a formula and a numerical value for $g'(3)$.
   - **(b)** Plot $g(x)$ for $0 < x \le 5$, using the various approaches discussed in this lab. *Hint:* $f(0)$ is undefined, so you may want to replace 0 by some small positive number.
3. **(a)** Find the largest interval containing 0 on which $f(x) = 3x - x^3$ is one-to-one.
   - **(b)** Maple can solve cubic equations. Use "`solve`" to find the inverse of $f$ on this interval. *Hint:* to see which of the three solutions is the one you want, evaluate them at 0.
   - **(c)** Roundoff error produces a very small imaginary component when evaluating $f^{-1}$ using the solution from (b). This must be removed in order to plot the function: use "`Re(finv(...))`" instead of "`finv(...)`". Plot $f^{-1}(f(x))$ for $-2 \le x \le 2$. Explain the result.
4. **(a)** Let $f(x) = x^2 + \sin(x)$ with domain $0 \le x < \infty$. Show that $f$ is one-to-one on this domain.
   - **(b)** Plot $f(x)$ and its inverse on the interval $0 \le x \le 5$, using "constrained" scaling. Observe that the graphs are each other's reflection across $x = y$.
5. This is the continuation of Exercise 4 of Lab 8.
   - **(a)** Solve the equation $x^t = t$ for $x$ as a function of $t$. What is the limit of $x$ as $t \to \infty$? What is the range of this function for $t > 0$?
   - **(b)** For what values of $x$ does the function $f(t) = x^t$ have a fixed point? How many fixed points does it have? *Hint:* look at the answer to (a).
   - **(c)** Which of these fixed points are attractors? For which $x$ is there a fixed point that is an attractor? *Hint:* find $f'(t)$, and use `subs` with the result of (a) to make this a function of the fixed point $t$ alone. Start by assuming $t > 0$.
   - **(d)** For $x = 0.04$, find a cycle of period 2 and show it is an attractor.

**6.** When a power and an exponential compete, the eventual winner is the exponential. When a power and a logarithm compete, the eventual winner is the power. Of course, it might take a long time to win...

**(a)** Find $x > 2$ such that $x^{100} < 1.01^x$.

**(b)** Find $x > 3$ such that $x^{0.01} > \ln(x)$.

**7.** In a baseball game, the batter hits a home run. Under the influence of gravity and air resistance, the ball's flight is described by the equations

$$x = \frac{1 - e^{-rt}}{r} v_0 \cos\alpha, \qquad y = 3 + \frac{1 - e^{-rt}}{r}\left(v_0 \sin\alpha + \frac{g}{r}\right) - \frac{gt}{r}$$

where $t$ is time in seconds from the time the ball is hit, $x$ and $y$ the horizontal and vertical coordinates in feet[(17)] with home plate at $(0, 0)$, $r$ the coefficient of air resistance, which we take to be 0.25/s, $g$ the acceleration of gravity (32 ft/s$^2$), $v_0$ the initial speed, and $\alpha$ the initial angle of the ball.

**(a)** The ball hits the ground at $t = 5$ seconds, 360 feet from home. What were $v_0$ and $\alpha$?

**(b)** The ball passes directly over the shortstop, who is standing 120 feet from home. His eyes are 5.5 feet off the ground. What is the maximum rate of change of the angle of elevation of the ball as seen by the shortstop? What is the angle of elevation when this maximum occurs?

**8.** What are the maximum and minimum of $d\theta/dt$ for Jupiter in Example 3? When do they occur?

**9. (a)** Find the largest interval containing 0 on which the function $g(x) = xe^x$ is one-to-one.

**(b)** Maple has a name for the inverse of $g(x)$ on the interval of (a). What is it? *Hint:* use "`solve`".

**(c)** Maple won't simplify $g^{-1}(g(x))$ to $x$ (with $g^{-1}$ replaced by the answer to (b)). Under what circumstances would this simplification be valid?

**(d)** The inverse of $g$ can be used to find the fixed points of $f(t) = x^t$ (see Exercise 5). How? Which fixed point does it find, when there are more than one?

**10.** Find the time intervals for retrograde motion of the following planets:

| Planet | $r_p$ | $t_p$ |
|---|---|---|
| Venus | 0.723 | 0.6152 |
| Mars | 1.524 | 1.881 |
| Saturn | 9.522 | 29.46 |

---

(17) Baseball is not a metric game. You can, if you wish, convert the measurements to metric, one foot = 0.3048 m.

# LAB 10 APPLICATIONS OF THE MEAN VALUE THEOREM

**Goals:** *To use Maple in studying the features of graphs of functions, and in proving inequalities by examining derivatives of functions.*

## Sketching Graphs

As we have seen, Maple's "`plot`" command will produce graphs of any functions we might give it. Sketching graphs by hand, it would seem, is as obsolete as slide rules and log tables. Why, then, do we need this section? The idea is to learn how to discover and recognize the main features of graphs. These features include the following:

1. symmetry
2. intercepts
3. critical points
4. singular points
5. inflection points
6. asymptotes
7. intervals of increase and decrease
8. intervals of concavity up and down

Often, we will want to discover these features before plotting, so that we can choose appropriate $x$ and $y$ intervals that will show the features.

■ **EXAMPLE 1** Plot the graph of $f(x) = \dfrac{x^4 + 3x^2 - 4}{x^3 + 2x^2 - 4}$, noting all features.

**SOLUTION** There is no obvious symmetry (the numerator is an even function, but the denominator is neither odd nor even). You don't need Maple to find the $y$ intercept: $(0, 1)$. For the $x$ intercepts we can use "`solve`".

```
> f:= x -> (x^4 + 3*x^2 - 4)/(x^3 + 2*x^2 - 4);
> solve(f(x)=0, x);
```

$$1, -1, 2I, -2I$$

Since $f$ is a rational function, "`solve`" has found all the roots, including the complex ones, $2I$ and $-2I$, which we disregard. The $x$ intercepts are $(1, 0)$ and $(-1, 0)$.

Next we find the critical points.

```
> S:= solve(D(f)(x)=0, x);
```

$$S := 0, \mathrm{RootOf}\left(\_Z^5 + 4\,\_Z^4 - 16\,\_Z^2 - 3\,\_Z^3 - 8 + 12\,\_Z\right)$$

One critical point is $x = 0$, and any others are in the "`RootOf`". We can use "`fsolve`" to find the numerical values. I'll use a trick to avoid typing in the polynomial in the "`RootOf`": change the "`RootOf`" to "`fsolve`" using "`subs`". I'll use "`eval`" to evaluate the result, which causes Maple to perform the "`fsolve`".

```
> cp:= eval(subs(RootOf=fsolve,S[2]));
```

$$cp := 1.718148949$$

Let's use the Second Derivative Test to classify these critical points.

```
> (D@@2)(f)(cp);
```

$$1.947983120$$

```
> (D@@2)(f)(0);
```

$$\frac{-1}{2}$$

This shows that $x = 0$ is a local maximum and $x = 1.718148949$ is a local minimum.

There are no singular points, but there may be vertical asymptotes, which would occur at roots of the denominator. The denominator being a cubic, "`solve`" finds its roots exactly, but they are quite complicated-looking (and I won't bother including them here). I'll use "`fsolve`" instead to find the numerical values.

```
> va:= fsolve(denom(f(x)) = 0, x);
```

$$va := 1.130395435$$

Thus there is one vertical asymptote, at $x = 1.130395435$. What does $f$ do as $x$ approaches this value? Note that we can't get the answer by asking for left and right limits at $x = 1.130395435$, because this is only an approximation. We would need the exact value (which we could have found using "`solve`"). Instead, let's check on the sign of the numerator at the (approximate) root, and of the denominator on each side of it.

```
> subs(x= va, numer(f(x)));
```

$$1.466138613$$

```
> subs(x=1, denom(f(x)));
```

$$-1$$

```
> subs(x=2, denom(f(x)));
```

$$12$$

Since $va$ is the only real root of the denominator, the fact that the denominator is negative at $x = 1$ means it is negative as $x$ approaches the vertical asymptote from the left; since the numerator is positive, we must have $f(x) \to -\infty$. Similarly, since the denominator is positive at $x = 2$, $f(x) \to +\infty$ as $x$ approaches the vertical asymptote from the right.

The two critical points 0 and $cp$ and the vertical asymptote $va$ divide the real line into four intervals: $-\infty \le x \le 0$, $0 \le x < va$, $va < x \le cp$, and $cp \le x < \infty$. In each of these, $f$ should be either increasing or decreasing. Since $x = 0$ is a local maximum, $f$ must be increasing on the first interval and decreasing on the second. Since $cp$ is a local minimum, $f$ must be decreasing on the third interval and increasing on the fourth.

Now to find inflection points:

```
> S:=solve((D@@2)(f)(x) = 0, x);
```

$$S := \mathrm{RootOf}\left(16 - 48\_Z + 20\_Z^3 + 120\_Z^2 - 48\_Z^4 + 12\_Z^5 + 7\_Z^6\right)$$

```
> infl:= eval(subs(RootOf=fsolve,S));
```

$$infl := -3.466559909, -1.589279030$$

These two candidates for inflection points, together with the vertical asymptote, divide the real line into four intervals. We check the sign of $f''$ on each interval.

```
> (D@@2)(f)(-4), (D@@2)(f)(-2), (D@@2)(f)(0), (D@@2)(f)(2);
```

$$\frac{-103}{486}, \frac{17}{2}, \frac{-1}{2}, \frac{13}{18}$$

The second derivative does change sign at the zeros of $f''$, so they are inflection points. The function $f$ is concave down on the intervals $-\infty < x \le -3.466559909$ and $-1.589279030 \le x < 1.130395435$, and concave up on $-3.466559909 \le x \le -1.589279030$ and $1.130395435 < x < \infty$.

Finally, what about horizontal or oblique asymptotes? Since this is a rational function and the degree of the numerator is one more than that of the denominator, we should have a two-sided oblique asymptote $y = ax + b$, which is the quotient when the numerator is divided by the denominator (as polynomials). Polynomial division can be done by the "quo" function: "`quo(p,q,x)`" produces the quotient[18] when "p" is divided by "q" (as polynomials in the variable "x").

```
> quo(numer(f(x)), denom(f(x)), x);
```

$$x - 2$$

Thus $y = x - 2$ is an oblique asymptote for this graph. Another method of finding oblique asymptotes, which is not restricted to rational functions, is to use limits. The graph of $f(x)$ will have an oblique asymptote $y = ax + b$ as $x \to \pm\infty$ if $\lim_{x\to\pm\infty} \frac{f(x)}{x} = a$ and $\lim_{x\to\pm\infty} (f(x) - ax) = b$.

```
> limit(f(x)/x, x=+infinity);
```

$$1$$

```
> limit(f(x) - x, x = +infinity);
```

$$-2$$

```
> limit(f(x)/x, x=-infinity);
```

$$1$$

```
> limit(f(x) - x, x = -infinity);
```

$$-2$$

When plotting the graph of $f$, we will want all the intercepts, critical and inflection points, and the vertical asymptote to be included, so we need the $x$ interval to include at least $-3.466559909 \le x \le 1.718148949$, with some room to spare on both sides. We might try $-5 \le x \le 3$. Since there is a vertical asymptote, we must choose the $y$ interval as well. We want to include at least the inflection points and critical points. The values of $f$ at the inflection points are $f(-3.466559909) = -8.160502595$ and $f(-1.589279030) = -3.360945208$. As for the critical points, we already know $f(0) = 1$, and $f(1.718148949) = 1.945299758$. So $-9 \le y \le 3$ might seem reasonable. Actually, I'd prefer to show a bit more of the approach to the vertical asymptote from the right, so I'll take $-9 \le y \le 5$. To avoid having an almost-vertical segment crossing the vertical asymptote (from the last point calculated on the left to the first one on the right), I'll use the "`discont=true`" option[19]. I'll also show the asymptotes as dotted lines. Finally, I'd like to place diamonds at the critical points and circles at the inflection points. This can be done with "`plot`" using a list of points, `style=POINT`, and `symbol=CIRCLE` or `DIAMOND`.

---

(18) You can obtain the remainder as well, if you include as a fourth input a name `'r'` in single quotes. The remainder will be assigned to this name.

(19) In Release 4, a bug in Maple prevents "`discont=true`" from working here. For a temporary patch, enter the following commands first:

```
> readlib(`plot/discontplot`):
> `plot/discontplot`:= subs(infinity=anything,
>      eval(`plot/discontplot`));
```

The bug has been corrected in Release 5.

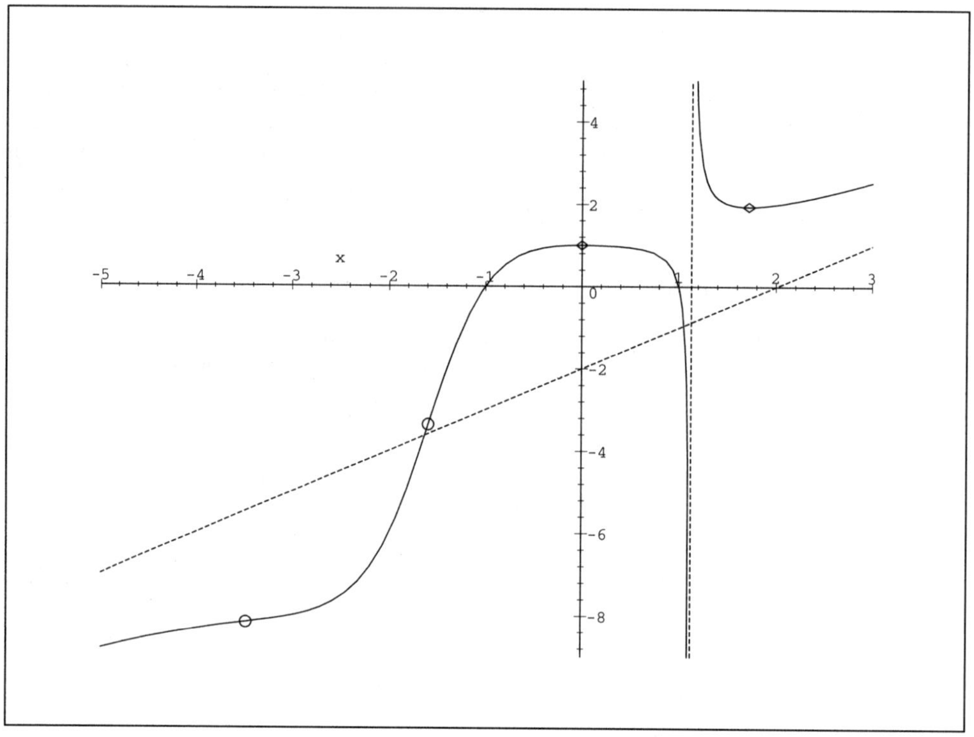


Figure 3.23. $y = \dfrac{x^4 + 3x^2 - 4}{x^3 + 2x^2 - 4}$

```
> plots[display]({plot(f(x), x=-5 .. 3, -9 .. 5, discont=true),
>           plot(x-2, x=-5 .. 3, linestyle=3),
>           plot([[va,-9], [va,5]], linestyle=3),
>           plot([[0,f(0)], [cp, f(cp)]],
>                style=POINT, symbol=DIAMOND),
>           plot([[infl[1], f(infl[1])], [infl[2], f(infl[2])]],
>                style=POINT, symbol=CIRCLE) }); (Figure 3.23)
```

■

■ **EXAMPLE 2** Identify some of the features of the graph of $f(x) = \dfrac{x^2}{1+|x|} + \ln|\cos x|$.

**SOLUTION** The symmetry is fairly easy to see: $f$ is an even function, so the graph is symmetric about the $y$ axis.

Complications begin with the $x$ intercepts.

```
> f:= x -> x^2/(1 + abs(x)) + ln(abs(cos(x)));
> solve(f(x)=0, x);
```

There is no result. Since $f$ is not a rational function, we don't expect "solve" to produce all solutions. We might have hoped for the solution $x = 0$.[20] "fsolve" produces one solution. Which one it finds may vary.

```
> fsolve(f(x) = 0, x);
```

$$0$$

Are there other solutions? How many? An "infinity plot" is often helpful in these circumstances. Because of the symmetry, we only need $x \geq 0$.

[20] Release 3 finds this solution. Not every change is an improvement.

```
> plot(f(x), x=0..infinity);
```
(Figure 3.24)

This graph can't really be taken at face value. It seems to coincide with the $x$ axis for a while: we shouldn't expect that to be exactly true, unless there is a very strange identity that we don't know about! Then there is a series of sharp, downward-pointing "spikes", of which only the first two appear to extend below the $x$ axis. Plotting on the interval $0 \le x \le 10$ shows the third spike also extending just below the axis, and there may be a hint that the graph goes slightly above the $x$ axis before the first spike. The first three spikes appear to come at about $x = 1.6, 4.7$ and $7.9$.

Enlightenment comes when we realize what is causing the spikes. The first term $x^2/(1+|x|)$ should be smooth except perhaps at $x = 0$ (and even there it turns out to be twice differentiable), so it must be the second term. The problem here is evidently that as $\cos x \to 0$, $\ln|\cos x| \to -\infty$. This will happen as $x \to \pm\pi/2, \pm 3\pi/2, \ldots$. These are our "spikes", which are actually vertical asymptotes. In particular, despite appearances, they all extend below the $x$ axis, and $f(x) = 0$ will have infinitely many solutions.

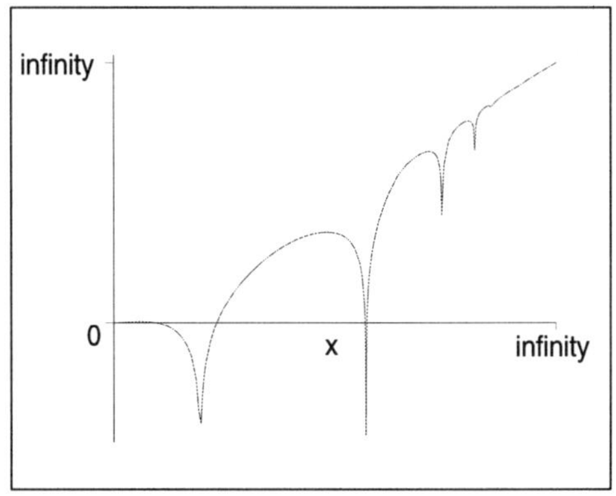

Figure 3.24. $\dfrac{x^2}{1+|x|} + \ln|\cos x|$

Now that we know where to look for these solutions, "fsolve" may be able to find them. For example, to find the solution just to the left of the vertical asymptote at $5\pi/2$:

```
> fsolve(f(x)=0, x, x=5*Pi/2 - 1 .. 5*Pi/2);
```

$$7.853038209$$

On the other hand, "fsolve" seems unable to find the solution just to the left of $23\pi/2$. Let's see: if $x \approx 23\pi/2$, we need $\ln|\cos x| \approx -35.15524914$, or $|\cos x| \approx 5.4 \times 10^{-16}$, and so $|x - 23\pi/2| \approx 5.4 \times 10^{-16}$ as well (according to the linear approximation). But with only 10 digits we can't tell the difference between $23\pi/2$ and a number this close to it, so it's not surprising that "fsolve" failed. We can find the solution if we use a higher setting of "Digits", and use the "fulldigits" option to make "fsolve" use all these digits internally.

```
> Digits:= 20:
> fsolve(f(x)=0, x, x=23*Pi/2 - 1 .. 23*Pi/2, fulldigits);
```

$$36.128315516282621702$$

The moral of this story:

| |
|---|
| Maple does not replace insight. |

More of the features of this example are explored in Exercise 3. ■

## Proving Inequalities

One of the most useful techniques for proving inequalities between functions is based on the criteria for functions to be nondecreasing or nonincreasing:

> Suppose $f$ and $g$ are continuous on the interval $a \le x \le b$ and differentiable for $a < x < b$, with $f'(x) \le g'(x)$ for $a < x < b$.
>
> (a) If $f(a) \le g(a)$, then $f(x) \le g(x)$ for $a \le x \le b$.
>
> (b) If $f(b) \ge g(b)$, then $f(x) \ge g(x)$ for $a \le x \le b$.

This is because $f(x) - g(x)$ is a nonincreasing function on the interval $a \le x \le b$.

**■ EXAMPLE 3** Show that $3\sin x + \exp(x - \pi) \geq 0$ for $0 \leq x \leq 4$.

**SOLUTION** We start with a look at the graph.

```
> f:= x -> 3 * sin(x) + exp(x - Pi);
> plot(f(x),x=0..4);
```
(Figure 3.25)

The plot certainly looks convincing. But we don't want to trust it, because the graphical method can be fooled (see Exercise 6 of Lab 7). Let's differentiate:

```
> D(f)(x);
```

$$3\cos(x) + e^{x-\pi}$$

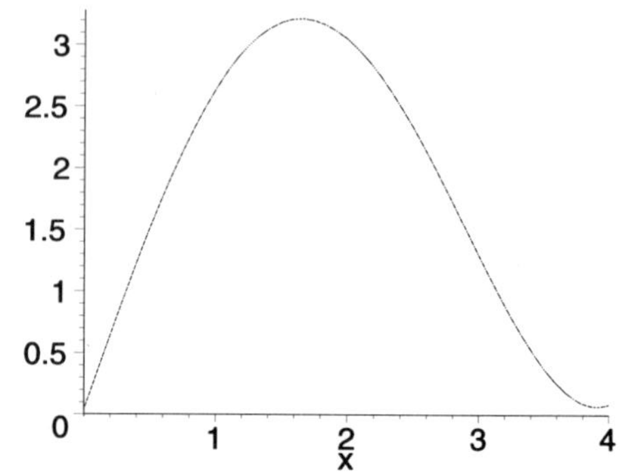


Figure 3.25. $3\sin x + \exp(x - \pi)$

Now this is certainly not always positive. But how negative can it be? Since $\exp(t) \geq 0$ for all $t$ and $\cos x \geq -1$, we certainly have $f'(x) \geq -3$. Now take $g(x) = f(0) - 3x = \exp(-\pi) - 3x$. Since $g(0) = f(0)$ and $f'(x) \geq -3 = g'(x)$ for $0 \leq x \leq 4$, we have $f(x) \geq \exp(-\pi) - 3x$ on this interval. In particular, $f(x) \geq 0$ when $0 \leq x \leq \exp(-\pi)/3 = 0.01440463942$ (approximately).

We have proven the inequality on only a small part of the interval we wanted. However, the process can be repeated. The next step would say that (with $x_1 = \exp(-\pi)/3$) $f(x) \geq f(x_1) - 3(x - x_1)$, so $f(x) \geq 0$ for $x_1 \leq x \leq x_1 + f(x_1)/3 = .04342241538$. We can take the latter as $x_2$ and continue. The computer is quite good at this sort of thing. I'll use a "`for`" loop, with 20 repetitions to begin with. In the style of Lab 7, I could use a table for $x_1, x_2, \ldots$, but since I'm not so interested in the actual numbers $x_n$ (just that they eventually reach 4 or more), I'll just use a single variable, `p`. Don't forget the "`evalf`", which ensures that we get a decimal number rather than some complicated expression involving $\pi$, and the "`od`" which terminates the loop. I've set "`Digits`" back to 10 from the 20 used in Example 2.

```
> Digits:= 10; p:= 0;
> for nn from 1 to 20 do p:= evalf(p + f(p)/3) od;
```

$$p := .01440463942$$

$$p := .04342241538$$

$$p := .1018750889$$

$$\ldots$$

$$p := 3.940064323$$

$$p := 3.964487922$$

We're almost at the target of 4. Hopefully just a couple more repetitions will get us there. Just to be safe, we can try 20 more.

```
> for nn from 1 to 20 do p:= evalf(p + f(p)/3) od;
```

$$p := 3.990397636$$

$$p := 4.018857326$$

$$\ldots$$

$$p := .5350538989\ 10^9$$

$$p := \text{Float}(9764639952, 232370944)$$

```
Error, (in evalf/sin/general) argument too large
```

Well, we did get past 4, and after a while into numbers so large that Maple can't handle them[21]. In

---

(21) "Float(9764639952, 232370944)" means $9764639952 \times 10^{232370944}$. That's big!

fact, the inequality is true for $0 \le x < \infty$. It's easy to prove that, but for this lab we want to stick with finite intervals. ■

**■ EXAMPLE 4** Show that $\sin^2(x) \le \sin(x^2)$ for $0 \le x \le 1.36$.

**SOLUTION** Let's take $f(x) = \sin^2(x) - \sin(x^2)$, so we want to show $f(x) \le 0$. Again we begin by plotting a graph: even though we won't trust it, it will show us what's going on.

```
> f:= x -> sin(x)^2 - sin(x^2);
> plot(f(x), x = 0 .. 1.4);
```

(Figure 3.26)

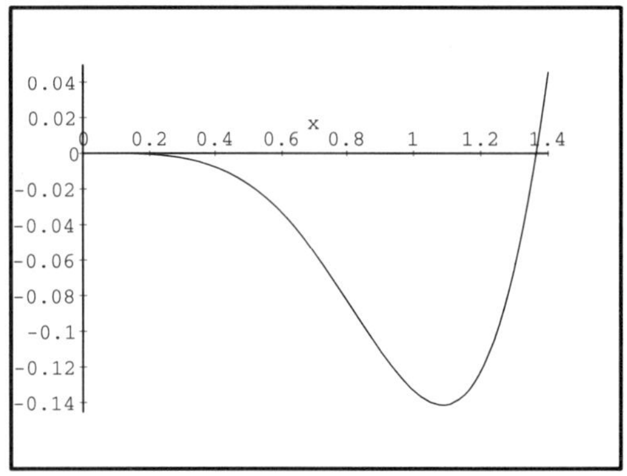


Figure 3.26. $\sin^2(x) - \sin(x^2)$

It does look like $f(x) \le 0$, at least for $0.2 < x < 1.36$. For $0 \le x \le 0.2$, $f(x)$ is almost indistinguishable from 0 on the graph. Let's differentiate:

```
> D(f)(x);
```

$$2\,\sin(x)\,\cos(x) - 2\,\cos(x^2)\,x$$

Since $f(0) = 0$, if we knew that $f'(x) \le 0$ on some interval $0 \le x \le b$, our criterion (b) would tell us that $f(x) \le 0$ on that interval. Now, as a matter of fact, it is not too hard to see that $f'(x) \le 0$ at least for $0 \le x \le 1$: on that interval, $0 \le \sin x \le x$ and (since $0 \le x^2 \le x$ and cos is decreasing) $0 < \cos(x) \le \cos(x^2)$. So this would take care of the interval $0 \le x \le 1$ (we'd need a different argument to take care of the interval $1 \le x \le 1.36$). However, let's suppose we don't notice this argument. What can we do? We can't proceed exactly as in Example 3 because $f(0) = 0$. Let's try higher derivatives.

```
> (D@@2)(f)(x);
```

$$2\,\cos(x)^2 - 2\,\sin(x)^2 + 4\,\sin(x^2)\,x^2 - 2\cos(x^2)$$

```
> (D@@@3)(f)(x);
```

$$-8\,\sin(x)\,\cos(x) + 8\,\cos(x^2)\,x^3 + 12\,\sin(x^2)\,x$$

```
> (D@@4)(f)(x);
```

$$-8\,\cos(x)^2 + 8\,\sin(x)^2 - 16\,\sin(x^2)\,x^4 + 48\,\cos(x^2)\,x^2 + 12\,\sin(x^2)$$

This last one has a negative value at $x = 0$, while all the previous ones are 0 there. So we can at least expect that $f^{(iv)}(x) \le 0$ in some interval near $x = 0$. How big an interval? We need to look at one more derivative to see this.

```
> (D@@5)(f)(x);
```

$$32\,\sin(x)\cos(x) - 32\,\cos(x^2)\,x^5 - 160\,\sin(x^2)\,x^3 + 120\,\cos(x^2)\,x$$

For $0 \le x \le 1$, the most this can be is 136: note that $32\sin(x)\cos(x) = 16\sin(2x) \le 16$, the second and third terms are negative, and the last is at most 120. Thus for $0 \le x \le 1$, we have

$$\begin{aligned}
f^{(v)}(x) &\le 136 \\
f^{(iv)}(x) &\le -8 + 136\,x \\
f'''(x) &\le -8\,x + 68\,x^2 \\
f''(x) &\le -4\,x^2 + 68/3\,x^3 \\
f'(x) &\le -4/3\,x^3 + 17/3\,x^4 \\
f(x) &\le -1/3\,x^4 + 17/15\,x^5
\end{aligned}$$

In each of these equations, we are applying our criterion. Thus to get the inequality for $f^{(iv)}(x)$, we say the following: take $g(x) = -8 + 136x$, which is the antiderivative of 136 with $g(0) = -8$. Since $f^{(v)}(x) \le g'(x)$ for $0 \le x \le 1$ and $f^{(iv)}(0) = g(0) = -8$, we conclude that $f^{(iv)}(x) \le g(x)$ for $0 \le x \le 1$.

Now our last inequality implies that $f(x) \le 0$ for $0 \le x \le 5/17$. Let's see: can we use this method again? We must check that $f^{(n)}(5/17) \le 0$ for $n = 0, 1, 2, 3$ and 4:

```
> seq(evalf((D@@k)(f)(5/17)), k = 0 .. 4);
```

$$-.00235794129, \ -.0311419488, \ -.298783847, \ -1.711862837, \ -1.492222993$$

Now for $5/17 \le x \le 1$, we have

$$f^{(iv)}(x) \le f^{(iv)}(5/17) + 136\,(x - 5/17)$$

$$\dots$$

$$f(x) \le f^{(iv)}(5/17)/24\,(x - 5/17)^4 + 17/15\,(x - 5/17)^5$$

so that $f(x) \le 0$ for $5/17 \le x \le 5/17 - (15/17) f^{(iv)}(5/17)/24$.

```
> p:= evalf(5/17 - (15/17) * (D@@4)(f)(5/17)/24);
```

$$p := .3489787866$$

```
> seq(evalf((D@@k)(f)(p)), k = 0 .. 4);
```

$$-.0045635430, \ -.0501337466, \ -.393692767, \ -1.724379031, \ 1.102183015$$

Since $f^{(iv)}(p) > 0$, we can't repeat this step again. However, $f(p) < 0$, so we can continue from this point by the method of Example 3. We have $2 \sin x \cos x = \sin(2x) \le 1$ and $2\cos(x^2)x \ge -2\cos(1.36^2)(1.36) > -.75$, so $f'(x) \le 1.75$ for $0 \le x \le 1.36$. Thus $f(x) \le f(p) + 1.75(x - p) \le 0$ if $p \le x \le p - f(p)/1.75$.

```
> for nn from 1 to 20 do p:= evalf(p - f(p)/1.75) od;
```

$$p := .3515865255$$
$$p := .3542697383$$
$$\dots$$
$$p := .4165859337$$
$$p := .4216944530$$

Progress is rather slow, because $f(p)$ is still close to 0, but we do seem to be speeding up (from a change of about .0027 to .0051). Let's continue with another 50 iterations.

```
> for nn from 21 to 70 do p:= evalf(p - f(p)/1.75) od;
```

$$p := .4270420075$$
$$p := .4326479756$$
$$\dots$$
$$p := 1.364412961$$
$$p := 1.364412963$$

Since $p$ has finally passed 1.36, we now have $f(x) \le 0$ for $0 \le x \le 1.36$, as required. ■

## EXERCISES

**1.** Plot graphs of the following functions, showing all intercepts, critical and inflection points, and asymptotes.

**(a)** $x^3 - 3x^2 + x$ **(b)** $\arctan(x^3)$

**(c)** $\ln(1 + x^2)$ **(d)** $\dfrac{x^2 + 1}{x + 1}$

**(e)** $x^{\ln x}$ **(f)** $\dfrac{1 + x^3}{1 + x + x^6}$

**2.** Plot a graph of the function $f(x) = \dfrac{x^3 + 6\sqrt{1 + x^2}}{2x - \sqrt{1 + x^2}}$, showing all intercepts, critical and inflection points, and asymptotes. Note: Due to a bug, Release 4 can't solve $f(x) = 0$. You can use "`fsolve`" (but convince yourself that you have all solutions). In other cases, "`solve`" produces results of the form "`RootOf(p(_Z), a)`", which means the root of the polynomial "`p(_Z)`" near `a`.

**3. (a)** For the function of Example 2, "`D(f)(0)`" produces an error. What should the answer be? Explain.

**(b)** Find and classify the critical points of this function for $-\pi/2 < x < \pi/2$.

**(c)** Find the inflection points for $-\pi/2 < x < \pi/2$.

**4. (a)** The graph of the function $f(x) = 5\sin x + \sqrt{3}\cos x - 16\sin x\,\cos^2 x$ is symmetric about a line $x = a$. Find $a$. *Hint:* using the graph of the function, identify points that should correspond to each other under the symmetry. Then $a$ should be about halfway between these. Use "`fsolve`" on $f(2a - u) = f(u)$ (with a particular value for $u$), specifying an interval in which you know $a$ lies. A good value for $u$ would be one with $f'(u) \neq 0$.

**(b)** This graph is also symmetric about a point $(b, 0)$. Find $b$ using a similar method.

**(c)** The exact $a$ and $b$ are rational multiples of $\pi$. Find them.

**(d)** Verify the symmetries using "`expand`" and "`simplify`".

**5. (a)** Plot a graph of the function $f(x) = (x^3 - x^2 - x + 1)^{1/3}$, showing all intercepts, critical, singular and inflection points, and asymptotes. *Hint:* Maple returns a complex number for $t^{1/3}$ when $t < 0$, which is not what we want. The real cube root of $t$ is "`surd(t, 3)`".

**(b)** On what intervals is $f(x)$ increasing? decreasing? concave up? concave down?

**6.** Establish the following inequalities:

**(a)** $\ln(1 - x) + x/(1 - x) \geq 0$ for $-\infty < x < 1$

**(b)** $\sin(\tan(x)) \geq x$ for $0 \leq x \leq 0.9999$ (it doesn't quite make it all the way to 1).

**7. (a)** Suppose $f(a) \geq 0$ and $f'(x) \geq -A$ for $b \leq x \leq c$, where $b \leq a \leq c$ and $A \geq 0$. On what interval (depending on $f(a)$ and $f'(a)$) can we be sure that $f(x) \geq 0$?

**(b)** Use this to prove the inequality of Example 3.

**(c)** Use it to prove the inequality of Example 4 for $5/17 \leq x \leq 1.36$.

**8.** Show that the inequality of Example 3 is true for all $x \geq 4$.

## CHAPTER 4

# Integration

## LAB 11 SUMS, AREAS, AND INTEGRALS

**Goals:** *To use Maple as an aid to understanding the concept of the definite integral as a limit of Riemann sums.*

### Sums

Maple performs sums using the "sum" command. If *expr* is an expression involving the variable $i$, then `sum(`*expr*`, i=m..n)` is $\sum_{i=m}^{n} expr$. You should be careful that no value is assigned to the index variable $i$. If in doubt, you can "unassign" it.

```
> i:= 'i';
> sum(sin(i), i=2 .. 4);
```

$$\sin(2) + \sin(3) + \sin(4)$$

If Maple can evaluate the sum in closed form, it will do so. Otherwise it just returns the sum, using sigma notation.

```
> sum(i^2, i=1..n);
```

$$\frac{1}{3}(n+1)^3 - \frac{1}{2}(n+1)^2 + \frac{1}{6}n + \frac{1}{6}$$

```
> sum(1/sin(k), k=1 .. m);
```

$$\sum_{k=1}^{m} \frac{1}{\sin(k)}$$

The "`Sum`" command is an "inert" form of "`sum`". You can use it to produce a sum in sigma notation without evaluating it.

```
> Sum(i^2, i=1..n);
```

$$\sum_{i=1}^{n} i^2$$

The "`student`" package, which you should load now, provides the ability to manipulate a "`Sum`" in various ways. For example, "`combine`" takes two sums with the same index variable and the same upper and lower limits and writes them as one sum. It also brings a coefficient inside the sum. This is reversed by "`expand`".

```
> with(student):
> S := combine( a * Sum(i^2, i=1..n) + Sum(i^3, i=1..n) );
```

$$S := \sum_{i=1}^{n} \left(a i^2 + i^3\right)$$

```
> S := expand(S);
```

$$S := a\left(\sum_{i=1}^{n} i^2\right) + \left(\sum_{i=1}^{n} i^3\right)$$

The "`changevar`" command performs a change of the index variable. The new variable should not already be present in the expression. We'll take the last expression and change $i$ into $j+1$, making $j$ the new index variable.

```
> changevar(i=j+1,S,j);
```

$$a\left(\sum_{j=0}^{n-1}(j+1)^2\right) + \left(\sum_{j=0}^{n-1}(j+1)^3\right)$$

When you do want to evaluate a "`Sum`", you can use the "`value`" command.

```
> value(S);
```

$$a\left(\frac{1}{3}n^3 + \frac{1}{2}n^2 + \frac{1}{6}n\right) + \frac{1}{4}n^4 + \frac{1}{2}n^3 + \frac{1}{4}n^2$$

## Areas Under Curves

We wish to study the area of the region $R$ under the graph of $y = f(x)$ between $x = a$ and $x = b$, where $a < b$ and $f(x) \geq 0$ for $a \leq x \leq b$. Various approximations are possible. Suppose we divide the interval $a \leq x \leq b$ into $n$ equal subintervals. Above each subinterval we place a rectangle to approximate the area of the part of $R$ over this subinterval. The "`student`" package[1] has commands to do this in three different ways. You can use "`leftsum`", which makes the height of the rectangle the value of $f$ at the left end of the subinterval, "`rightsum`", which uses the right end of the subinterval, or "`middlesum`", which uses the midpoint of the subinterval.

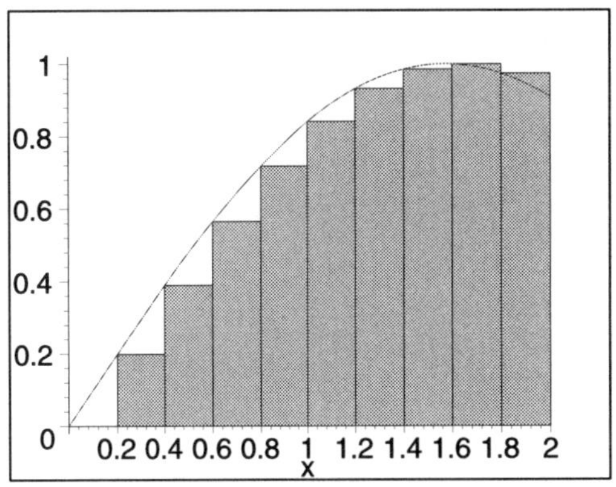


Figure 4.1. Left sum for $\sin(x)$

```
> leftsum(f(x), x=a..b, n);
```

$$\frac{(b-a)\left(\sum_{i=0}^{n-1} f\left(a + \frac{i(b-a)}{n}\right)\right)}{n}$$

```
> rightsum(f(x), x=a..b, n);
```

$$\frac{(b-a)\left(\sum_{i=1}^{n} f\left(a + \frac{i(b-a)}{n}\right)\right)}{n}$$

```
> middlesum(f(x), x=a..b, n);
```

$$\frac{(b-a)\left(\sum_{i=0}^{n-1} f\left(a + \frac{\left(i+\frac{1}{2}\right)(b-a)}{n}\right)\right)}{n}$$

---

(1) We loaded this in the last section.

■ **EXAMPLE 1** Study the area under $y = \sin x$, above the $x$ axis and between $x = 0$ and $x = 2$, using left, right, and middle sums.

**SOLUTION** First we produce pictures of the region and the approximating rectangles, using the commands "`leftbox`", "`middlebox`", and "`rightbox`", all from the "`student`" package.

```
> leftbox(sin(x), x=0 .. 2, 10);
```
(Figure 4.1)

You can specify a "`colour`", which will be used for the curve, and a "`shading`", which specifies the colour for the rectangles. Other options (as for "`plot`") can also be used.

```
> rightbox(sin(x), x=0 .. 2, 10, colour=blue, shading=pink);
> middlebox(sin(x), x=0 .. 2, 10, colour=red);
```

Now, what are those sums?

```
> L:= leftsum(sin(x), x=0 .. 2, 10);
```

$$L := \frac{1}{5}\left(\sum_{i=0}^{9} \sin\left(\frac{1}{5}i\right)\right)$$

The functions "`leftsum`", "`rightsum`", and "`middlesum`" are defined using "`Sum`" rather than "`sum`". We can use "`value`" to actually perform the sum, and "`evalf`" (either on the original sum or on the result of "`value`") to get a numerical result.

```
> value(L);
```

$$\frac{1}{5}\sin\left(\frac{1}{5}\right) + \frac{1}{5}\sin\left(\frac{2}{5}\right) + \frac{1}{5}\sin\left(\frac{3}{5}\right) + \frac{1}{5}\sin\left(\frac{4}{5}\right) + \frac{1}{5}\sin(1) + \frac{1}{5}\sin\left(\frac{6}{5}\right) + \frac{1}{5}\sin\left(\frac{7}{5}\right) + \frac{1}{5}\sin\left(\frac{8}{5}\right) + \frac{1}{5}\sin\left(\frac{9}{5}\right)$$

```
> evalf(L);
```

1.320493454

```
> R:=value(rightsum(sin(x), x=0 .. 2, 10));
```

$$R := \frac{1}{5}\sin\left(\frac{1}{5}\right) + \frac{1}{5}\sin\left(\frac{2}{5}\right) + \frac{1}{5}\sin\left(\frac{3}{5}\right) + \frac{1}{5}\sin\left(\frac{4}{5}\right) + \frac{1}{5}\sin(1) + \frac{1}{5}\sin\left(\frac{6}{5}\right) + \frac{1}{5}\sin\left(\frac{7}{5}\right) + \frac{1}{5}\sin\left(\frac{8}{5}\right) + \frac{1}{5}\sin\left(\frac{9}{5}\right) + \frac{1}{5}\sin(2)$$

```
> evalf(R);
```

1.502352939

```
> evalf(middlesum(sin(x), x=0 .. 2, 10));
```

1.418509838

The left, middle, and right sums here were not particularly close together, which is not too unexpected since $n = 10$ is rather small. They would be somewhat closer if we took a larger $n$.

```
> evalf(leftsum(sin(x), x=0 .. 2, 100));
```

1.407006657

```
> evalf(rightsum(sin(x), x=0 .. 2, 100));
```

1.425192606

```
> evalf(middlesum(sin(x), x=0 .. 2, 100));
```

1.416170439

Well, they agree to two significant digits. In fact, for any $n$ the difference between our "`rightsum`" and "`leftsum`" is $2\sin(2)/n$ (the $i = n$ term minus the $i = 0$ term). As $n \to \infty$, the difference between the two goes to 0, and both (in fact all three) sums should approach a common limit, which would be the area we want to find. Maple can actually perform these sums symbolically, and then find the limit.

```
> L:= value(leftsum(sin(x), x=0 .. 2, n));
```

$$L := 2\left(-\sin(1)\cos(1) + \frac{\sin\left(\frac{1}{n}\right)\cos\left(\frac{1}{n}\right)\cos(1)^2}{\cos\left(\frac{1}{n}\right)^2 - 1} - \frac{\sin\left(\frac{1}{n}\right)\cos\left(\frac{1}{n}\right)}{\cos\left(\frac{1}{n}\right)^2 - 1}\right)/n$$

```
> L:= limit(L, n=infinity);
```

$$L := 2 - 2\cos(1)^2$$

```
> evalf(L);
```

$$1.416146836$$

```
> R:= value(rightsum(sin(x), x=0 .. 2, n));
```

$$R := 2\left(-\sin\left(\frac{n+1}{n}\right)\cos\left(\frac{n+1}{n}\right) + \frac{\sin\left(\frac{1}{n}\right)\cos\left(\frac{1}{n}\right)\cos\left(\frac{n+1}{n}\right)^2}{\cos\left(\frac{1}{n}\right)^2 - 1} + \sin\left(\frac{1}{n}\right)\cos\left(\frac{1}{n}\right) - \frac{\sin\left(\frac{1}{n}\right)\cos\left(\frac{1}{n}\right)^3}{\cos\left(\frac{1}{n}\right)^2 - 1}\right)/n$$

```
> limit(R, n=infinity);
```

$$2 - 2\cos(1)^2$$

```
> M:= value(middlesum(sin(x), x=0 .. 2, n));
```

$$M := 2\left(\frac{\sin\left(\frac{1}{n}\right)\cos(1)^2}{\cos\left(\frac{1}{n}\right)^2 - 1} - \frac{\sin\left(\frac{1}{n}\right)}{\cos\left(\frac{1}{n}\right)^2 - 1}\right)/n$$

```
> limit(M, n=infinity);
```

$$2 - 2\cos(1)^2$$

■

## Integrals

The limits as $n \to \infty$ of all three sums coincided, as they should. This limit should be the definite integral of $\sin(x)$ on the interval $[0, 2]$. Maple can calculate the integral directly. The answer is the same (using the formula $\cos(2\theta) = 2\cos^2\theta - 1$).

```
> int(sin(x), x=0..2);
```

$$-\cos(2) + 1$$

Maple obtained this answer using an antiderivative of $\sin(x)$ and the Fundamental Theorem of Calculus. The antiderivative is also found using the "`int`" function, but without specifying an interval:

```
> J:= int(sin(x), x);
```

$$J := -\cos(x)$$

```
> subs(x=2, J) - subs(x=0, J);
```

$$-\cos(2) + \cos(0)$$

In cases where Maple can't find an antiderivative, or evaluate a definite integral by other methods, it will return the integral or antiderivative unevaluated.

```
> J:= int(sin(sin(x)), x=0..2);
```

$$\int_0^2 \sin(\sin(x))\, dx$$

It's not really surprising that Maple can't find an antiderivative here: as far as I know, this antiderivative can't be expressed in terms of the functions Maple has available. Nevertheless, Maple can still come up with a numerical answer (in Lab 13 we will discuss how this could be done):

```
> evalf(J);
```

$$1.247056058$$

## Upper and Lower Sums

The definite integral is defined using lower and upper Riemann sums $L(f, P)$ and $U(f, P)$ corresponding to a partition $P = \{x_0, x_1, \ldots, x_n\}$ with $a = x_0 < x_1 < \ldots < x_n = b$. These sums are defined as $L(f, P) = \sum_{i=1}^{n} f(l_i)(x_i - x_{i-1})$ and $U(f, P) = \sum_{i=1}^{n} f(u_i)(x_i - x_{i-1})$, where $f(l_i)$ and $f(u_i)$ are the minimum and maximum values, respectively, of $f$ on the interval $[x_{i-1}, x_i]$.

Maple doesn't have built-in functions analogous to "`leftsum`", "`leftbox`", etc. for these, so I've written some. The definitions are contained in the file "`riemann.def`", which may be available to you. If so, you can read it in now. If not, you can type it in as we go along.

```
> read `riemann.def`;
```

I'll go through the definitions step-by-step, as there are some useful points to be learned from them. If you have used the file "`riemann.def`", you don't need to enter any of the Maple commands from here to Example 2, but you should still read this section to learn what these commands do.

For convenience, I will make the function $f$, the endpoints $a$ and $b$, and the number of subintervals $n$ into variables rather than inputs to the functions. We will also use a list of $x$ values that may be local maxima or minima, and thus may be used in the upper or lower sum: I'll call it "`cp`". This will not be calculated automatically: you must define it before calculating or plotting the upper and lower sums. As usual, the minimum or maximum of $f$ on a subinterval may occur at a critical point, a singular point, or an endpoint of the subinterval. My commands will automatically look at the endpoints, but the critical and singular points should be placed in "`cp`".

The interval $[a, b]$ will be partitioned into $n$ equal subintervals. We let $h$ be the length of each.

```
> h := ('b'-'a')/'n';
```

The quotes are there so that the variable names "`a`", "`b`", and "`n`", rather than their current values, are used in the definition of "`h`". This way, if you later change the value of one of those variables, "`h`" also changes automatically.

Next we define the points $x_i$ dividing the subintervals. It will be convenient to use a function rather than a table. Since the letter "`x`" will be used elsewhere, I'll use "`X`" for the name of this function. Thus $x_i$ will be obtained as "`X(i)`".

```
> X := i -> a + i * h;
```

We'll want to know when one of the $x$ values in "`cp`" is in the $i$'th subinterval. For this purpose, it's useful to have a "`between`" function, so that "`between(x,L,R)`" is "`true`" if the number "`x`" is between "`L`" and "`R`", and "`false`" if it isn't. We need to use "`evalf`" in this definition so that it will work with expressions defined using constants such as "`Pi`" as well as with explicit numbers.

```
> between := (x,L,R) -> evalf(x - L) > 0 and evalf(R - x) > 0;
```

Now the following bit of Maple magic finds the maximum value of $f$ on the $i$'th subinterval.

```
> upper:= i -> max(f(X(i-1)), f(X(i)),
>     op(map(f,select(between,cp,X(i-1),X(i)))));
```

It's not important for you to understand the details of this construction, but here's what goes on in it: the "`select`" function extracts the members of the list `cp` that are in the subinterval, i.e., are between $x_{i-1}$ and $x_i$. The "`map`" converts this list into a list of $f$ values at these points. The "`op`" removes the "`[`" and "`]`" from the beginning and end of this list[2]. Finally, the "`max`" selects the largest of $f(x_{i-1})$, $f(x_i)$, and the $f$ values coming from `cp`, and this should be the maximum value $f(u_i)$. A similar construction, using "`min`" instead of "`max`", will produce $f(l_i)$.

```
> lower:= i -> min(f(X(i-1)), f(X(i)),
>    op(map(f,select(between,cp,X(i-1),X(i)))));
```

Now we can construct the upper and lower sums. I'll write a single function "`rsum`" so that "`rsum(upper)`" will produce the upper sum while "`rsum(lower)`" will produce the lower sum. The quotes in `'g(i)'` are needed to prevent "premature evaluation" since "`upper`" and "`lower`" require a numeric input rather than a name "`i`" (see "Premature evaluation" in Appendix II).

```
> rsum:= g -> h * sum('g(i)', i=1..n);
```

Next for something analogous to "`leftbox`", "`rightbox`", and "`middlebox`". The "`rect`" function just makes a list of four points, the corners of a rectangle bounded by the $x$ axis, two $x$ values, and a $y$ value.

```
> rect:= (x1, x2, y) -> [[x1,0], [x1,y], [x2,y], [x2,0]];
```

The "`rbox`" function plots a sequence of these rectangles, one for each subinterval, getting the $y$ values from a function that is its input. Thus "`rbox(upper)`" will plot the rectangles for an upper sum while "`rbox(lower)`" does it for a lower sum. You may, if you wish, put in other plot options such as a colour.

```
> rbox:= proc(g) local ii;
>    plot({ f(x),
>          [ seq(op(rect(X(i-1),X(i),g(i))),i=1..n)]},
>          x = a..b, args[2..nargs]);
>    end;
```

If you don't already have the file "`riemann.def`", you should save these definitions for future use. They can then be recalled using "`read`".

```
> save h, X, between, upper, lower, rsum, rect, rbox, `riemann.def`;
```

**■ EXAMPLE 2** Calculate the upper and lower Riemann sums for $f(x) = \sqrt{|x^3 - x^2 + 1|}$ on the interval $[-1, 2]$ for a partition into 10 equal subintervals. Plot the curve $y = f(x)$ with the rectangles corresponding to these sums on one graph.

**SOLUTION** We begin by defining "`a`", "`b`", "`n`", and the function "`f`".

```
> f:= x -> sqrt(abs(x^3 - x^2 + 1));
> a:= -1; b:= 2; n:= 10;
```

Next we need the candidates for local maxima and minima. These are the critical points and singular points of $f$. A singular point will occur when $x^3 - x^2 + 1 = 0$.

```
> s:= solve(x^3 - x^2 + 1 = 0, x);
```

$$s := -\frac{1}{6}\%2 - \frac{2}{3}\%1 + \frac{1}{3}, \frac{1}{12}\%2 + \frac{1}{3}\%1 + \frac{1}{3} + \frac{1}{2}I\sqrt{3}\left(-\frac{1}{6}\%2 + \frac{2}{3}\%1\right),$$
$$\frac{1}{12}\%2 + \frac{1}{3}\%1 + \frac{1}{3} - \frac{1}{2}I\sqrt{3}\left(-\frac{1}{6}\%2 + \frac{2}{3}\%1\right)$$
$$\%1 := \frac{1}{(100 + 12\sqrt{69})^{1/3}}$$
$$\%2 := (100 + 12\sqrt{69})^{1/3}$$

(2) More technically, it converts the list to a sequence.

The first of these looks real, the other two complex. We can confirm this with "evalf". Only the real root should be included in "cp". Complex roots would result in an error, because "between" wouldn't work.

```
> evalf([s]);
```

$$[-.7548776667, .8774388331 - .7448617670I, .8774388331 + .7448617670I]$$

Now for the critical points:

```
> c:= solve(D(f)(x)=0, x);
```

$$c := 0, \frac{2}{3}$$

Putting together these critical points with the singular point, we have "cp".

```
> cp:= [ s[1], c ];
```

We now have all the necessary ingredients. Here are the upper and lower Riemann sums.

```
> U:= rsum(upper);
```

$$U := \frac{3}{5}+\frac{3}{250}\sqrt{485}+\frac{3}{1000}\sqrt{9890}+\frac{54}{125}\sqrt{5}+\frac{3}{40}\sqrt{14}+\frac{3}{1000}\sqrt{11210}+\frac{3}{250}\sqrt{1115}+\frac{3}{1000}\sqrt{30230}$$

```
> evalf(U);
```

$$3.649158244$$

```
> evalf(rsum(lower));
```

$$2.759619596$$

To plot the upper and lower rectangles together, we can use "display" from the "plots" package.

```
> plots[display]({rbox(upper),
>     rbox(lower)}, axes=FRAMED);
```

(Figure 4.2)

Note that in the interval on the left the lower rectangle has height 0 since $f$ is 0 at the singular point. The upper and lower sums are rather far apart, as you might expect, since $f(u_i)$ and $f(l_i)$ are quite far apart for the first two subintervals and the last two or three. However, for a large $n$ they would be closer together. ■

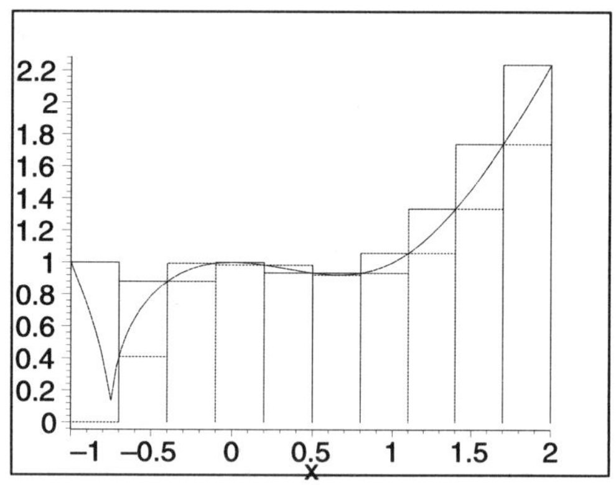


Figure 4.2. Upper and lower boxes

## EXERCISES

**1.** Evaluate the following using "sum".

**(a)** $1^2 + 2^2 + 3^2 + \ldots + 10^2$ **(b)** $\displaystyle\sum_{i=1}^{5} \frac{2i+1}{2i-1}$

**(c)** $a_5b_1 + a_4b_2 + a_3b_3 + a_2b_4 + a_1b_5$, where $a_j = 2j + 1$ and $b_j = 2j - 1$

**2.** In using "changevar" on $s = \displaystyle\sum_{i=a}^{b} f(i)$, there must be a one-to-one correspondence between the values of the original index variable (integers from $a$ to $b$) and integer values of the new index variable in some interval. Which of the following are valid uses of "changevar"?

**(a)** `changevar(i=j+1,s,j);`

**(b)** `changevar(i=j^2,s,j);`

**(c)** `changevar(j=b-i,s,j);`

**3.** Maple can evaluate lots of sums by itself, but here's one it can't.

**(a)** Use Maple to show that $\sec((n+1)a)\sec(na) = \dfrac{\tan((n+1)a) - \tan(na)}{\sin a}$.

**(b)** What does this imply about $\sum_{j=1}^{n} \sec((j+1)a)\sec(ja)$? Can Maple evaluate this sum?

**(c)** Verify the sum for $n = 5$ using Maple.

**4.** Calculate the left, right, and middle sums for $f(x) = x^4$ on the interval $[a, b]$. Find the limits as $n \to \infty$, and compare to $\int_a^b f(x)\,dx$.

**5.** Consider the function $f(x) = x^3 - 2x^2 - 3x + 10$ on the interval $[-2, 3]$.

**(a)** Calculate numerically the left, middle, right, lower, and upper sums for $n = 10$ and 100.

**(b)** Find the limit of the left sum as $n \to \infty$. Compare to the area as found using "`int`".

**(c)** Plot the graph of $f$ and the rectangles for its left and right sums for $n = 20$ (in one graph).

**(d)** The same as (c), but using the upper and lower sums.

**6.** Consider the function $f(x) = \sqrt{1 - x^2}$ on the interval $[-1, 1]$.

**(a)** Express the upper and lower sums $U(f, P)$ and $L(f, P)$, where $P$ is the partition of $[-1, 1]$ into $2n$ equal subintervals, in terms of left and right sums for $n$ equal subintervals of $[0, 1]$. *Hint:* use symmetry.

**(b)** Find some $n$ for which the difference between $U(f, P)$ and $L(f, P)$ is at most 0.01.

**(c)** Compare $U(f, P)$ and $L(f, P)$ for this $n$ with the area under the curve $y = f(x)$ for $-1 \le x \le 1$. *Hint:* use geometry to find this area. What is the curve $y = f(x)$?

**7.** Consider the function $f(x) = x + \cos x$ on the interval $[-1, 2]$.

**(a)** Of the left, right, middle, lower, and upper sums for any given $n$, which are the same?

**(b)** What is the least $n$ for which the difference between the left sum and the right sum is less than .01? *Hint:* you don't need to know the values of the sums.

**(c)** For $n = 100$, how close are the left, middle and right sums to the definite integral? What can you conclude about the practical usefulness of these sums as approximations to the area?

**8.** So far we have only used partitions into equal subintervals. However, using the "`sum`" function we can handle any partition. In this exercise we will let Maple do a rather complicated example: the area under $y = x^k$ for $a \le x \le b$, where $0 < a < b$ and $k \ne -1$.

**(a)** Find the upper sum for the partition $x_i = at^i$, $0 \le i \le n$, where $t = (b/a)^{1/n}$. *Hint:* we want $n$ to be a variable, so use a function "`xi(i)`" rather than a table "`x[i]`".

**(b)** Take the limit as $n \to \infty$, and compare to the definite integral.

**(c)** Do the same for the lower sum.

**(d)** Now do it for $k = -1$.

**9.** Let $L_n$, $R_n$ and $M_n$ be the left, right, and middle sums, respectively, for $f(x) = x^2$ on $[a, b]$ with a partition into $n$ equal intervals. Find

$$\lim_{n\to\infty} n\left(L_n - \int_a^b f(x)\,dx\right)$$

and the limits with $L_n$ replaced by $R_n$ and by $M_n$. Try it with some other $f$ for which Maple can find formulas for $L_n$, $R_n$, and $M_n$. Does this help to explain why $M_n$ is a better approximation to the definite integral?

**10.** It is easy to show that if $|f(s) - f(t)| < \epsilon$ whenever $a \le s \le t \le b$ with $t - s < h$, then $U(f, P) - L(f, P) < \epsilon(b - a)$, where $P$ is the partition of $[a, b]$ into $n$ equal subintervals of length $h = (b - a)/n$.

**(a)** Find an $n$ for which this condition is true, where $f(x) = \cos(x)$, $a = -1$, $b = 1$, and $\epsilon = 0.01$. *Hint:* use the Mean Value Theorem.

**(b)** Compute the upper and lower sums with this $n$. How close are they to the definite integral?

# LAB 12 INTEGRATION TECHNIQUES

**Goals:** *To learn some of the strategies of integration, with Maple doing the algebraic manipulations*

Maple is very good at integration. If there is an antiderivative that can be expressed in terms of the ordinary functions of calculus, Maple is likely to find it. Even when an antiderivative can't be found, Maple can evaluate definite integrals numerically to any desired degree of accuracy (we will study techniques of numerical integration in Lab 13). But this lab is not aimed at using Maple to do the integration automatically, but rather at letting you direct the integration, using the techniques you learn in your Calculus course, with Maple acting as your assistant and carrying out the detailed manipulations.

As we have seen, the "`int`" command is what is normally used in Maple to find an antiderivative or definite integral:

```
> int(x*sin(x), x);
```

$$\sin(x) - x\cos(x)$$

```
> int(x*sin(x), x=Pi.. 2*Pi);
```

$$-3\pi$$

If you don't want Maple to try to do the integration itself (as we don't in this lab), you use "`Int`" instead.

```
> Int(x*sin(x), x);
```

$$\int x\sin(x)\,dx$$

```
> Int(x*sin(x), x=Pi.. 2*Pi);
```

$$\int_{\pi}^{2\pi} x\sin(x)\,dx$$

Just like "`Sum`", which we saw in Lab 11, "`Int`" is an "inert" function, producing an expression that can be manipulated in various ways. Most of the commands for doing this manipulation are in the "`student`" package, which you should load before doing anything else in this lab.

```
> with(student):
```

For example, "`expand`" will write the integral of a sum of terms times constant coefficients as a sum of integrals times coefficients, while "`combine`" will reverse this.

```
> J:= expand(Int(x + 2*cos(x), x));
```

$$J := \int x\,dx + 2\int \cos(x)\,dx$$

```
> combine(J);
```

$$\int x + 2\cos(x)\,dx$$

Note that this is slightly different notation from standard mathematics, which would insert parentheses: $\int (x + 2\cos x)\,dx$.

The basic framework for this lab is as follows: start with an assigned integral expression, write it using "`Int`", and manipulate it in various ways. When each integrand in our expression is one of the standard integrals that every calculus student should have memorized, or a constant multiple of one, we will allow "`value`" to perform the integration. Then we will perform back-substitution and simplification as necessary, to produce the final answer.

## Substitution

Our first technique is substitution, which can be done using the "changevar" command. Suppose you have an expression $expr = \int f(x)\,dx$. To change the variable of integration from $x$ to $u$, where these are related by an equation $g(x) = h(u)$, you use

```
> changevar(g(x)=h(u), expr, u);
```

This includes both "direct" substitution $g(x) = u$ and "inverse" substitution $x = h(u)$. The third input is the name of the new variable, which should not already be present in the expression. The integral may be either indefinite or definite: in the latter case, the limits of integration are transformed as required.

■ **EXAMPLE 1** Evaluate $\int \frac{x\,dx}{x^2+1}$.

**SOLUTION** This is an easy one for a "direct" substitution: $x^2 + 1 = t$.

```
> e1:= Int(x/(x^2+1), x);
> g:= changevar(x^2+1=t, e1, t);
```

$$\int \frac{1}{2}\frac{1}{t}\,dt$$

This is now a constant multiple of an "elementary integral", so we let ourselves use "value".

```
> v:= value(g);
```

$$v := \frac{1}{2}\ln(t)$$

Note that Maple uses $\ln(t)$ rather than $\ln(|t|)$. It won't cause a problem here because $t = x^2 + 1 > 0$. We'll discuss this more in Example 4. Now we substitute back to get this in terms of $x$. We can use "subs" here.

```
> subs(t=x^2+1, v);
```

$$\frac{1}{2}\ln\left(x^2+1\right)$$ ■

■ **EXAMPLE 2** Evaluate $\int_0^1 \left(4 - x^2\right)^{-3/2}\,dt$.

**SOLUTION** Since it involves $\sqrt{4 - x^2}$, this one is a candidate for the "inverse sine substitution" $x = 2\sin t$.

```
> e2:= Int((4-x^2)^(-3/2), x=0..1);
> g:= changevar(x = 2 * sin(t), e2, t);
```

$$g := \int_0^{\frac{1}{6}\pi} 2\frac{\cos(t)}{\left(4 - 4\sin(t)^2\right)^{3/2}}\,dt$$

This should be possible to simplify: $4 - 4\sin^2 t = 4\cos^2 t$.

```
> g:= simplify(g);
```

$$g := \frac{1}{4}\int_0^{\frac{1}{6}\pi} \frac{\operatorname{csgn}(\cos(t))}{\cos(t)^2}\,dt$$

"csgn" is a version of the "sgn" function: Maple is aware that $\sqrt{\cos^2 t}$ could be either $\cos t$ or $-\cos t$, depending on what interval $t$ is in. However, in this case, the naive approach would have worked, since $t$ is in an interval where cos is positive[3]. So we are justified in getting rid of the "csgn". But how? Of course we could just retype the expression, but that might be impractical if we had a very complicated expression. One way is with a substitution:

---

(3) Unfortunately, when simplifying a definite integral Maple does not use the fact that the variable of integration should be between the endpoints.

```
> subs(csgn(cos(t))=1,g);
```

$$\frac{1}{4}\int_0^{\frac{1}{6}\pi}\frac{1}{\cos(t)^2}\,dt$$

Another method is to use the "`symbolic`" option in the "`simplify`" command. This would use $\sqrt{x^2} = x$ without any "`csgn`".

With the "`symbolic`" option, an expression of the form $\sqrt{f^2}$ may sometimes simplify as $f$ and sometimes as $-f$. Which one it will be is unpredictable. You should only use "`symbolic`" when you either don't care about signs, or know what the sign should be and are prepared to change it if necessary (as is the case here).

```
> simplify(g,symbolic);
```

$$\frac{1}{4}\int_0^{\frac{1}{6}\pi}\frac{1}{\cos(t)^2}\,dt$$

Still another method is to use "`assume`" to tell Maple that the variable is in the interval between the endpoints. With this information, Maple simplifies the expression correctly.

```
> assume(t >= 0, t <= Pi/6);
> g:= simplify(g);
```

$$g := \frac{1}{4}\int_0^{\frac{1}{6}\pi}\frac{1}{\cos(t\sim)^2}\,dt\sim$$

One of the "elementary integrals" is $\displaystyle\int \sec^2(ax)\,dx$. Recognizing that $1/\cos x = \sec x$, we can now use "`value`":

```
> value(g);
```

$$\frac{1}{12}\sqrt{3}$$ ■

## Integration by Parts

For integration by parts, the "`student`" package has the "`intparts`" function. Suppose "`expr`" is an expression containing an "`Int`" which can be written as $\int U\,dV$. Then to produce $UV - \int V\,dU$, you use

> intparts(expr,$U$);

Of course, Maple must be able to find an antiderivative of $dV$ (and you should be sure that you can as well!). This works for both definite and indefinite integrals.

```
>  g:= Int(x^2*exp(x), x=2..3);
```

$$g := \int_2^3 x^2\,e^x\,dx$$

```
> g:= intparts(g, x^2);
```

$$g := 9\,e^3 - 4\,e^2 - \int_2^3 2\,x\,e^x\,dx$$

```
> g:= intparts(g, x);
```

$$g := 3e^3 + \int_2^3 2\,e^x\,dx$$

```
> value(g);
```

$$5e^2 - 2e^2$$

Maple is often unable to do integrals involving a function to a power, where the exponent is a parameter rather than an explicit number. These may be candidates for reduction formulas, typically obtained by integration by parts.

**EXAMPLE 3** Find a reduction formula for $\int \sin^n(x)\, e^x\, dx$, and use it in the case $n = 5$.

**SOLUTION** Since this depends on the parameter $n$, it will be convenient to write it as a function of $n$.

```
> e3:= n -> Int(sin(x)^n*exp(x), x);
```

$$e3 := n \to \int \sin(x)^n e^n\, dx$$

We know how to integrate $e^x$, so we'll integrate by parts with $U = \sin^n(x)$.

```
> g:= intparts(e3(n), sin(x)^n);
```

$$g := \sin(x)^n\, e^x - \int \frac{\sin(x)^n\, n\, \cos(x)\, e^x}{\sin(x)}\, dx$$

Now we integrate by parts a second time.

```
> g:= intparts(g, sin(x)^(n-1)*cos(x));
```

$$g := \sin(x)^n\, e^x - \sin(x)^{(n-1)}\, \cos(x)\, e^x\, n + \int \left( \frac{\sin(x)^{(n-1)}\, (n-1)\, \cos(x)^2}{\sin(x)} - \sin(x)^{(n-1)}\, \sin(x) \right) e^x\, n\, dx$$

To put the integrals on the right into a similar form to $e3(n)$, we want to express the $\cos^2 x$ in terms of $\sin x$.

```
> g:= subs(cos(x)^2 = 1-sin(x)^2, g);
```

$$g := \sin(x)^n\, e^x - \sin(x)^{(n-1)}\, \cos(x)\, e^x\, n + \int \left( \frac{\sin(x)^{(n-1)}\, (n-1)\, (1-\sin(x)^2)}{\sin(x)} - \sin(x)^{(n-1)}\, \sin(x) \right) e^x\, n\, dx$$

We want to simplify this, but it's a bit tricky (I must confess that it took me several tries before I got this one right). If we just used "`simplify`" now, the $1 - \sin^2 x$ would become $\cos^2 x$ again. We need to use "`expand`" first, to separate the terms with 1 and the terms with $\sin^2 x$.

```
> g:= expand(g);
```

$$g := \sin(x)^n\, e^x - \frac{\sin(x)^n\, n\, \cos(x)\, e^x}{\sin(x)} + n^2 \int \frac{e^x\, \sin(x)^n}{\sin(x)^2}\, dx - n^2 \int \sin(x)^n\, e^x\, dx - n \int \frac{e^x\, \sin(x)^n}{\sin(x)^2}\, dx$$

Now we can try "`simplify`". But we only want to combine the powers of $\sin x$, not change $\sin^2 x$ to $1 - \cos^2 x$. A second input for "`simplify`" tells it to apply only simplifications involving powers.

```
> g:= simplify(g, power);
```

$$g := \sin(x)^n\, e^x - \sin(x)^{(n-1)}\, \cos(x)\, e^x\, n + n^2 \int e^x\, \sin(x)^{(n-2)}\, dx - n^2 \int \sin(x)^n\, e^x\, dx - n \int e^x\, \sin(x)^{(n-2)}\, dx$$

We recognize $e3(n)$ and $e3(n-2)$ here. Let's write them in terms of a new function $s(n)$, which is to give the solution to the problem.

```
> g:= subs(e3(n)=s(n), e3(n-2)=s(n-2), g);
```

$$g := \sin(x)^n\, e^x - \sin(x)^{(n-1)}\, \cos(x)\, e^x\, n + n^2\, s(n-2) - n^2\, s(n) - n\, s(n-2)$$

This is supposed to be $s(n)$, but we see "$s(n)$" on the right side of the equation too. No problem — we just solve for $s(n)$.

```
> g:= solve(s(n)=g, s(n));
```

$$g := \frac{\sin(x)^n\, e^x - \sin(x)^{(n-1)}\cos(x)\, e^x\, n + n^2\, s(n-2) - ns(n-2)}{1+n^2}$$

This is the right-hand side of a reduction formula for $s(n)$. It might be better to collect the two terms involving $s(n-2)$ into one. Otherwise, when calculating $s(n)$ with the reduction formula Maple would try to evaluate $s(n-2)$ twice, then $s(n-4)$ four times (once for each $s(n-2)$), .... That would cause the calculations to slow down tremendously for large $n$.

```
> g:= collect(g, s(n-2));
```

$$g := \frac{(n^2-n)\; s(n-2)}{1+n^2} + \frac{\sin(x)^n\, e^x - \sin(x)^{(n-1)}\cos(x)\, e^x\, n}{1+n^2}$$

```
> s:= unapply(g, n);
```

$$s := n \to \frac{(n^2-n)\; s(n-2)}{1+n^2} + \frac{\sin(x)^n\, e^x - \sin(x)^{(n-1)}\cos(x)\, e^x\, n}{1+n^2}$$

What?? Any of you who have any experience of computer programming should object loudly to this. We have just set up an infinite loop: to calculate $s(n)$ this function needs $s(n-2)$. But for that it needs $s(n-4)$, then $s(n-6)$, ... without any end in sight. And indeed, if at this point you were to ask Maple for any $s(n)$ the result would be

```
Error, (in s) too many levels of recursion
```

because all these calculations would quickly exhaust all available memory. We must tell Maple to stop at $n = 0$ or $n = 1$, and supply values for $s(0)$ and $s(1)$.[4] But we can do that in a very simple way: simply assign values to $s(0)$ and $s(1)$.

```
> s(0):= value(e3(0));
```

$$s(0) := e^x$$

```
> s(1):= value(e3(1));
```

$$s(1) := -\frac{1}{2} e^x \cos(x) + \frac{1}{2}\sin(x)\, e^x$$

Technically, what we are doing is putting values in the "remember table" for the function "s". Whenever Maple evaluates a function, it first checks for values in the remember table before using the definition of the function. So now if we ask for $s(n)$ where $n$ is a positive integer, Maple will use the reduction formula until it comes to $s(0)$ or $s(1)$, and then use the remembered value for that. Let's try it for $n = 5$.

```
> s(5);
```

$$-\frac{3}{13} e^x \cos(x) + \frac{3}{13}\sin(x)\, e^x + \frac{1}{13}\sin(x)^3\, e^x - \frac{3}{13}\sin(x)^2\cos(x)\, e^x + \frac{1}{26}\sin(x)^5\, e^x - \frac{5}{26}\sin(x)^4\cos(x)\, e^x$$

Of course, if we ask for $s(n)$ where $n$ is negative or not an integer, we will still get "`too many levels of recursion`". ∎

## Integrating Rational Functions

Our third main integration technique is the partial fraction decomposition, which is used to integrate rational functions.

---

[4] Since the reduction formula expresses $s(n)$ in terms of $s(n-2)$, we need both of these in order to deal with all positive integers $n$.

**EXAMPLE 4** Evaluate $\displaystyle\int \frac{x^7 + 2x^6 - 21x^2 - 233x - 1809}{x^6 - 2x^4 + 48x^3 + x^2 - 48x + 576}\, dx.$

**SOLUTION** We begin by examining the integrand.

```
> R:= (x^7 + 2*x^6 - 21*x^2 - 233*x - 1809)/
>     (x^6 - 2*x^4 + 48*x^3 + x^2 - 48*x + 576);
```

The "`convert`" command with option "`parfrac`" converts a rational function to partial fractions when the factors of the denominator have rational coefficients. The third input of "`convert`" tells it the name of the variable being used.

```
> F:= convert(R, parfrac, x);
```

$$F := x + 2 - 3\frac{1}{(x+3)^2} + \frac{1}{x+3} + \frac{-41 + x}{x^2 - 3x + 8} + \frac{2x - 1}{(x^2 - 3x + 8)^2}$$

Our numerator had a higher degree than the denominator, so there is a polynomial part $(x + 2)$ in the decomposition. The usual first step of long division is done automatically by "`convert`". Evidently, Maple did factor the denominator into linear and quadratic factors, and both were repeated.

Now the first four terms of the decomposition are simple enough to integrate, but the next two will require some work. We'd like to break `F` up into pieces. This can be done using the "`op`" command. A typical Maple expression is composed of a certain number of "operands", each of which might be a name, a number, or another expression made up of operands. The "`op`" command extracts one of the operands making up the expression. In this case, `F` is a sum of six terms, each of which is an operand of `F`, and these can be obtained by `op(1,F)`, `op(2,F)`, `...op(6,F)`. There is also an `op(0,F)`, which is "+" (i.e., this is what says that `F` is a sum). So let's first take the sum of the first four operands of `F`, and integrate it.

```
> g1:= op(1,F) + op(2,F)+ op(3,F) + op(4,F);
```

$$g1 := x + 2 - 3\frac{1}{(x+3)^2} + \frac{1}{x+3}$$

```
> I1:= int(g1, x);
```

$$I1 := \frac{1}{2}x^2 + 2x + 3\frac{1}{x+3} + \ln(x+3)$$

There's one thing to watch out for here: Maple wrote the antiderivative of $1/(x + 3)$ as $\ln(x + 3)$ rather than $\ln|x + 3|$. This is another example of the fact that Maple works with complex numbers rather than just real numbers. In the context of complex numbers, Maple is correct, but in calculus this is not the answer we want. We can change it with "`subs`":

```
> I1:= subs(ln(x+3)=ln(abs(x+3)),I1);
```

$$I1 := \frac{1}{2}x^2 + 2x + 3\frac{1}{x+3} + \ln(|x+3|)$$

Now on the last two operands, we want to use "`Int`".

```
> I2:= Int(op(5,F)+op(6,F), x);
```

$$I2 := \int \frac{-41 + x}{x^2 - 3x + 8} + \frac{2x - 1}{(x^2 - 3x + 8)^2}\, dx$$

The next step is "completing the square". The "`completesquare`" function will do this: `completesquare(expr,v)` will complete the square in any quadratic in the variable "`v`" that occurs in "`expr`".

```
> I2:= completesquare(I2,x);
```

$$I2 := \int \frac{-41 + x}{\left(x - \frac{3}{2}\right)^2 + \frac{23}{4}} + \frac{2x - 1}{\left(\left(x - \frac{3}{2}\right)^2 + \frac{23}{4}\right)^2}\, dx$$

Completing the square shows us which change of variables to use. The same one will work for both terms.

```
> I2:= changevar(x-3/2=t, I2, t);
```

$$I2 := \int \frac{-\frac{79}{2}+t}{t^2+\frac{23}{4}} + \frac{2+2t}{\left(t^2+\frac{23}{4}\right)^2}\,dt$$

Now let's expand this, so we can deal with the terms separately.

```
> I2:= expand(I2);
```

$$I2 := -\frac{79}{2}\int \frac{1}{t^2+\frac{23}{4}}\,dt + \int \frac{t}{t^2+\frac{23}{4}}\,dt + 2\int \frac{1}{\left(t^2+\frac{23}{4}\right)^2}\,dt + 2\int \frac{t}{\left(t^2+\frac{23}{4}\right)^2}\,dt$$

The first term is a multiple of an elementary integral.

```
> I2a:= value(op(1,I2));
```

$$I2a := -\frac{79}{23}\sqrt{23}\arctan\left(\frac{2}{23}t\sqrt{23}\right)$$

On the second and fourth terms of `I2` we can use the substitution $t^2+23/4=s$, which makes them into multiples of elementary integrals.

```
> I2b:= changevar(t^2+23/4=s, op(2,I2)+op(4,I2), s);
```

$$I2b := \int \frac{1}{2}\frac{1}{s}\,ds + 2\int \frac{1}{2}\frac{1}{s^2}\,ds$$

```
> I2b:= value(I2b);
```

$$I2b := \frac{1}{2}\ln(s) - \frac{1}{s}$$

Again we have $\ln(s)$ rather than $\ln(|s|)$, but this time it won't cause a problem because $s=t^2+23/4$ is always positive.

```
> I2b:= subs(s=t^2+23/4,I2b);
```

$$I2b := \frac{1}{2}\ln\left(t^2+\frac{23}{4}\right) - \frac{1}{t^2+\frac{23}{4}}$$

The third term of `I2` is a candidate for the inverse tangent substitution $t=\sqrt{23/4}\tan s$.

```
> I2c:= changevar(t=sqrt(23/4)*tan(s), op(3,I2), s);
```

$$I2c := 2\int 8\frac{\sqrt{23}}{529+529\tan(s)^2}\,ds$$

```
> I2c:= simplify(I2c);
```

$$I2c := \frac{16}{529}\sqrt{23}\int \cos(s)^2\,ds$$

This is an integral you've probably seen before. If not, you could use "`combine(",trig)`" to write $\cos^2 s$ as $(\cos(2s)+1)/2$ before using "`value`" on it.

```
> I2c:= value(I2c);
```

$$I2c := \frac{16}{529}\sqrt{23}\left(\frac{1}{2}\cos(s)\,\sin(s)+\frac{1}{2}s\right)$$

Now we substitute back to write this in terms of $t$. I'll also use "`eval`" to cause the automatic simplifications that are not ordinarily done after "`subs`".

```
> I2c:= eval(subs(s=arctan(t/sqrt(23/4)), I2c));
```

$$I2c := \frac{16}{529}\sqrt{23}\left(23\frac{t\sqrt{23}}{529+92t^2}+\frac{1}{2}\arctan\left(\frac{2}{23}t\sqrt{23}\right)\right)$$

In `I2a`, `I2b`, and `I2c` we must still change variables from $t$ back to $x$ before we can put everything together. While we're at it, let's expand `I2c` so that the two "`arctan`" terms can go together.

```
> e4:= I1 + subs(t=x-3/2, I2a + I2b + expand(I2c));
```

$$e4 := \frac{1}{2}x^2 + 2x + 3\frac{1}{x+3} + \ln(|x+3|) - \frac{1809}{529}\sqrt{23}\arctan\left(\frac{2}{23}\left(x-\frac{3}{2}\right)\sqrt{23}\right) + \frac{1}{2}\ln\left(\left(x-\frac{3}{2}\right)^2+\frac{23}{4}\right) - \frac{1}{\left(x-\frac{3}{2}\right)^2+\frac{23}{4}} + 16\frac{x-\frac{3}{2}}{92\left(x-\frac{3}{2}\right)^2+529}$$

Some simplification is still possible in the 6th term, and in the 7th and 8th (which can be simplified together). Here is our final answer:

```
> e4:= sum(op(i,e4),i=1..5) + simplify(op(6,e4))
>      + simplify(op(7,e4)+op(8,e4));
```

$$e4 := \frac{1}{2}x^2 + 2x + 3\frac{1}{x+3} + \ln(|x+3|) - \frac{1809}{529}\sqrt{23}\arctan\left(\frac{2}{23}\left(x-\frac{3}{2}\right)\sqrt{23}\right) + \frac{1}{2}\ln\left(x^2-3x+8\right) + \frac{1}{23}\frac{-29+4x}{x^2-3x+8}$$

■

**■ EXAMPLE 5** Evaluate $\displaystyle\int \frac{x^2+1}{x^4-x^2+1}$.

**SOLUTION** Again we try converting the integrand to partial fractions.

```
> R:= (x^2 + 1)/(x^4 - x^2 + 1);
> convert(R, parfrac, x);
```

$$\frac{x^2+1}{x^4-x^2+1}$$

The problem here is that there are no linear or quadratic factors with rational-number coefficients. The factors exist, but their coefficients are not all rational numbers. To see what kind of coefficients they might have, we can use "`solve`" to find the roots of the denominator.

```
> solve(denom(R)=0,x);
```

$$\frac{1}{2}\sqrt{2-2I\sqrt{3}},\ -\frac{1}{2}\sqrt{2-2I\sqrt{3}},\ \frac{1}{2}\sqrt{2+2I\sqrt{3}},\ -\frac{1}{2}\sqrt{2+2I\sqrt{3}}$$

The roots are all complex, so the factors we are looking for will be quadratic. But for our purpose, the important thing is that they involve $\sqrt{3}$. This means that we should look for factors whose coefficients can involve $\sqrt{3}$ as well as rational numbers. We can do this by including $\sqrt{3}$ as an additional input to "`convert`":(5)

```
> R:= convert(R, parfrac, x, sqrt(3));
```

$$R := \frac{1}{2}\frac{1}{x^2-\sqrt{3}x+1} + \frac{1}{2}\frac{1}{x^2+\sqrt{3}x+1}$$

The rest of the integration is routine: we use "`completesquare`", then change variables in each of the terms. The final answer is

$$\arctan(2x-\sqrt{3}) + \arctan(2x+\sqrt{3})$$

■

(5) You can also make the fourth input to "`convert`" a set or list of square (or higher) roots, which will allow factors involving those roots.

## EXERCISES

In exercises 1 to 7, use "`int`" or "`value`" only for "elementary" integrals that you have memorized.

**1.** Find $\displaystyle\int \frac{x^3\,dx}{x^4+1}$. *Hint:* use a substitution.

**2.** Find $\displaystyle\int \frac{\sin(3x)\,dx}{1-\cos x}$. *Hint:* expand $\sin(3x)$ and then use a substitution.

**3.** Find $\displaystyle\int \frac{dx}{x+2\sqrt{x}-3}$.

**4.** Find $\displaystyle\int \arctan(\sqrt{x})\,dx$. *Hint:* integrate by parts.

**5.** Find $\displaystyle\int \sqrt{\sec^2 x + a^2}\,dx$. This is quite a difficult one: in fact, "`int`" won't do it. *Hint:* start with the substitution $\tan x = t$. Be careful about signs when simplifying.

**6.** Find $\displaystyle\int_1^3 \frac{\sqrt{x-1}}{\sqrt{x+1}}dx$. *Hint:* start with a substitution that gets rid of one of the square roots.

**7.** Find $\displaystyle\int_0^{\pi/4} \sqrt{\tan x}\,dx$. *Hint:* a substitution will make this into the integral of a rational function.

**8.** Find a reduction formula for $a_m = \displaystyle\int \frac{x^m\,dx}{\sqrt{x^2+1}}$. *Hint:* integrate by parts $a_m + a_{m+2} = \displaystyle\int x^m\sqrt{x^2+1}\,dx$.

**9.** Find a reduction formula for $a_n = \displaystyle\int x^n \arctan x\,dx$. *Hint:* examine $a_n$ for $n = 1$ to 6 using "`int`". Find $c_n$ (which doesn't depend on $x$) so all but three terms cancel in $a_n + c_n a_{n-2}$. Integrate $a_n + c_n a_{n-2}$ by parts.

# LAB 13 NUMERICAL INTEGRATION

**Goals:** *To study some methods of numerically approximating integrals, and how to estimate and control the errors in these approximations.*

In addition to finding exact values for antiderivatives and definite integrals, Maple contains very sophisticated methods for finding numerical approximations to definite integrals. To find a numerical value for $\displaystyle\int_a^b f(x)\,dx$, you can use

```
> evalf(int(f(x), x=a..b));
```

This will only work if the answer should be a pure number: $f(x)$, $a$, and $b$ should not contain any variables (other than "$x$" in $f(x)$) that don't have numerical values.

Note that this first applies "`int`" (which will try to find an exact value), then applies "`evalf`" to the result. If you want to be sure that Maple is actually doing numerical integration rather than finding a numerical value for an exact expression for the answer, you can use

```
> evalf(Int(f(x), x=a..b));
```

One reason you might want to do this is that it can save time: "`int`" sometimes spends a lot of time trying to find the exact value before giving up. If you already know (or strongly suspect) that "`int`" won't be able to find the exact value, there's not much point in letting it try.

In this lab we will be looking at some other methods of numerical integration: the Trapezoid Rule, Midpoint Rule, Simpson's Rule, and the Romberg method. These are less sophisticated than Maple's method (which mainly uses "Curtis-Clenshaw Quadrature"), but easier to understand, and the goal in this lab is to understand how the methods work. For practical calculations, "`evalf(Int(...))`" would almost always be much better.

Most of the commands we use in this lab are in the "`student`" package, which you should load before doing any work in this lab.

```
> with(student):
```

## Midpoint and Trapezoid Rules

Let the definite integral we want to approximate be $A = \int_a^b f(x)\,dx$. The Midpoint Rule approximation based on $n$ intervals is

$$M_n = h\sum_{j=1}^{n} f(m_j)$$

where $h = (b-a)/n$ and $m_j = a + (j - 1/2)h$ for $1 \le j \le n$. I'm starting with this one because we've seen it before: it's nothing other than the "middle sum" of Lab 11.

```
> middlesum(f(x), x=a..b, n);
```

$$\frac{(b-a)\left(\sum_{i=0}^{n-1} f\left(a + \frac{\left(i+\frac{1}{2}\right)(b-a)}{n}\right)\right)}{n}$$

We also had a "`leftsum`" and a "`rightsum`", but we saw (Exercises 7 and 9 of Lab 11) that these could be much poorer approximations to $A$ than "`middlesum`". Let us examine this in more detail.

**EXAMPLE 1** Study the limiting forms of the error in the left, middle, and right sums for a "typical"(6) smooth function $f(x) = x^5 - 5x^2$.

**SOLUTION** We define $A$ as the true definite integral, and $L(n)$, $M(n)$, and $R(n)$ the left, middle, and right sums as functions of $n$, the number of subintervals.

```
> f:= x -> x^5 - 5*x^2;
> A:= int(f(x), x=a..b);
```

$$A := \frac{1}{6}b^6 - \frac{5}{3}b^3 - \frac{1}{6}a^6 + \frac{5}{3}a^3$$

```
> L:= unapply(value(leftsum(  f(x), x=a..b, n)), n);
> M:= unapply(value(middlesum(f(x), x=a..b, n)), n);
> R:= unapply(value(rightsum( f(x), x=a..b, n)), n);
```

Maple produces some rather complicated formulas for $L$, $M$, and $R$, which I won't bother to reproduce.

The error in the approximation is the true value $A$ minus the approximate value. I'll call these errors $LE(n)$, $ME(n)$, and $RE(n)$ for left, middle, and right sums respectively.

(6) Actually not that typical, in that Maple can find explicit formulas for the sums we calculate, but the behaviour we will find is typical.

```
> LE:= unapply(A - L(n), n);
> ME:= unapply(A - M(n), n);
> RE:= unapply(A - R(n), n);
```

Now we could look at these functions at particular values of $a$, $b$, and $n$, but it's more interesting to see how they behave for very large $n$. The simplest question is, what is the limit as $n \to \infty$? But the answer is simple, too: all three have limit 0. That just says that as $n \to \infty$, the (left, middle, or right) sum approaches the definite integral. We know that is true for **all** Riemann sums. A more subtle question is, what is the limit of $n\,LE(n)$ (or $n\,RE(n)$ or $n\,ME(n)$) as $n \to \infty$?

```
> limit(n*LE(n), n=infinity);
```

$$-\frac{5}{2}b^3 - \frac{5}{2}a^3 + \frac{1}{2}b^6 + \frac{1}{2}a^6 - \frac{1}{2}ab^5 + \frac{5}{2}ab^2 - \frac{1}{2}ba^5 + \frac{5}{2}ba^2$$

```
> limit(n*RE(n), n=infinity);
```

$$\frac{5}{2}b^3 + \frac{5}{2}a^3 - \frac{1}{2}b^6 - \frac{1}{2}a^6 + \frac{1}{2}ab^5 - \frac{5}{2}ab^2 + \frac{1}{2}ba^5 - \frac{5}{2}ba^2$$

Both limits exist, as rather complicated functions of $a$ and $b$. When the limit is some nonzero number $C$ it means that $LE(n)$ or $RE(n)$ is approximately $C/n$ for large $n$ (and this is a closer "approximately" than just saying $LE(n)$ or $RE(n)$ is near 0). What about $ME(n)$?

```
> limit(n*ME(n), n=infinity);
```

$$0$$

Well, this is significant: it means that $ME(n)$ goes to 0 faster than any $C/n$. Could it be $C/n^2$?

```
> C:= limit(n^2*ME(n), n=infinity);
```

$$C := \frac{5}{12}a^3 - \frac{5}{12}b^3 - \frac{5}{12}ab^5 - \frac{5}{24}a^4b^2 + \frac{5}{24}a^2b^4 + \frac{5}{4}ab^2 + \frac{5}{12}ba^5 - \frac{5}{4}ba^2 - \frac{5}{24}a^6 + \frac{5}{24}b^6$$

This confirms that when $n$ is large, $ME(n)$ is approximately $C/n^2$ for some constant $C$ depending on $a$ and $b$. $C$ may be 0 for some $a$ and $b$ values, but that would be unusual — a "lucky coincidence" in which positive errors on one part of the interval $[a, b]$ almost exactly cancel negative errors on another part of the interval.

Although $C$ looks complicated, it turns out to be $(f'(a) - f'(b))(b - a)^2/24$. This is true for any smooth function. We won't prove that here, but we can verify it for our $f$.

```
> simplify(C - (D(f)(b) - D(f)(a))*(b-a)^2/24);
```

$$0$$

If you look closely at the limits of $nLE(n)$ and $nRE(n)$, you will notice that one is $-1$ times the other. This has an important consequence: $(L(n) + R(n))/2$, the average of the left and right sums, is a better approximation than either of them. Its error will be the average of $LE(n)$ and $RE(n)$, and this will go to 0 faster than any $C/n$ because

$$\lim_{n\to\infty} n\frac{LE(n) + RE(n)}{2} = \frac{1}{2}\left(\lim_{n\to\infty} nLE(n) + \lim_{n\to\infty} nRE(n)\right) = 0$$

In fact, the average of the left and right sums has a name: the Trapezoid Rule. The "`student`" package calculates it using "`trapezoid`":

```
> trapezoid(g(x), x=a..b, n);
```

$$\frac{1}{2}\frac{(b-a)\left(g(a) + 2\left(\sum_{i=1}^{n-1} g\left(a + \frac{i(b-a)}{n}\right)\right) + g(b)\right)}{n}$$

The fact that the Trapezoid Rule approximation $T(n)$ is the average of $L(n)$ and $R(n)$ is fairly easy to see by hand calculation. Maple won't do it for symbolic $n$, but will for any particular value of $n$. For example, with $n = 5$:

```
> V:= value(trapezoid(g(x), x=a..b, 5)
>   - (leftsum(g(x), x=a..b, 5) + rightsum(g(x), x=a..b, 5))/2);
```

$$V := \frac{1}{2}\left(\frac{1}{5}b - \frac{1}{5}a\right)\left(g(a) + 2g\left(\frac{4}{5}a + \frac{1}{5}b\right) + 2g\left(\frac{3}{5}a + \frac{2}{5}b\right) + 2g\left(\frac{2}{5}a + \frac{3}{5}b\right) + 2g\left(\frac{1}{5}a + \frac{4}{5}b\right) + g(b)\right) - \frac{1}{2}\left(\frac{1}{5}b - \frac{1}{5}a\right)\left(g(a) + g\left(\frac{4}{5}a + \frac{1}{5}b\right) + g\left(\frac{3}{5}a + \frac{2}{5}b\right) + g\left(\frac{2}{5}a + \frac{3}{5}b\right) + g\left(\frac{1}{5}a + \frac{4}{5}b\right)\right) - \frac{1}{2}\left(\frac{1}{5}b - \frac{1}{5}a\right)\left(g\left(\frac{4}{5}a + \frac{1}{5}b\right) + g\left(\frac{3}{5}a + \frac{2}{5}b\right) + g\left(\frac{2}{5}a + \frac{3}{5}b\right) + g\left(\frac{1}{5}a + \frac{4}{5}b\right) + g(b)\right)$$

```
> simplify(V);
```

$$0$$

Now let's define $T(n)$ and its error $TE(n)$ analogously to $L(n)$, $LE(n)$ and the others.

```
> T:= unapply(value(trapezoid(f(x), x=a..b, n)), n);
> TE:= unapply(A - T(n), n);
```

Does this error go to 0 like $C/n^2$ as $n \to \infty$?

```
> limit(n^2 * TE(n), n=infinity);
```

$$-\frac{5}{6}a^3 + \frac{5}{12}b^2a^4 + \frac{5}{12}a^6 - \frac{5}{2}b^2a - \frac{5}{12}b^6 + \frac{5}{6}b^3 - \frac{5}{6}ba^5 + \frac{5}{6}b^5a - \frac{5}{12}b^4a^2 + \frac{5}{2}ba^2$$

Yes, it does go like $C/n^2$. Moreover, notice that the constant $C$ here is exactly $-2$ times the corresponding constant for $ME(n)$.

To see how close the actual errors $ME(n)$ and $TE(n)$ are to these forms $C/n^2$, let's plot $n^2ME(n)$ and $n^2TE(n)$ as functions of $n$, say for $n$ from 1 to 40. We'll need to select particular values of a and b, let's say $-1$ and 2. We'll use "seq" to form sequences of points $[n, n^2ME(n)]$ or $[n, n^2TE(n)]$ and then plot them with style=POINT.

```
> a:= -1; b:= 2;
> plot( [ seq([n, n^2 * ME(n)], n=1..40),
>       seq([n, n^2 * TE(n)], n=1..40) ], style=POINT);
```
(Figure 4.3)

After about $n = 7$, the two series of points are practically horizontal. We conclude that the errors $ME(n)$ and $TE(n)$ are quite well approximated by constant multiples of $n^{-2}$.

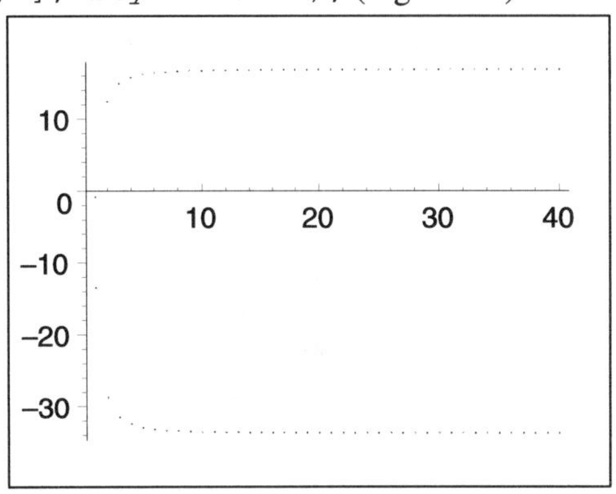


Figure 4.3. $n^2ME(n)$ and $n^2TE(n)$ vs $n$

Now let's take a different point of view: what $n$ would be required for each method to achieve a given degree of accuracy? Suppose (still with $a = -1$ and $b = 2$) we want the absolute value of the error to be at most $10^{-9}$ (approximately the accuracy Maple attains when "Digits" is set to 10). For the left or right sum:

```
> limit(n*LE(n), n=infinity);
```

$$27$$

Since $LE(n) \approx 27/n$, for this to be $10^{-9}$ we'd need $n \approx 2.7 \times 10^{10}$.

```
> LE(2.7 * 10^10);
```

$$-.3\ 10^{-8}$$

The reason this isn't $10^{-9}$ is roundoff error in evaluating the function $LE$. The value you get may vary from session to session (it will be affected by Maple's choice of which order to use for the terms). If we increase "Digits", we should get a more accurate value.

```
> Digits:= 20;
> LE(2.7 * 10^10);
```

$$0.9999999997\,10^{-9}$$

Of course, Maple does not actually add $2.7 \times 10^{10}$ terms in calculating this value! That would take a very long time even on a fast computer, and would also be subject to serious problems with roundoff error. So the left or right sum is completely impractical as a numerical method if we want an accurate approximation. The Midpoint Rule or Trapezoid Rule, by comparison, is not so bad.

```
> limit(n^2*TE(n), n=infinity);
```

$$\frac{-135}{4}$$

For $135/(4n^2) = 10^{-9}$ we would need $n \approx \sqrt{135/4 \times 10^9} \approx 183\,711.7$ (which we would have to round up to 183 712).

```
> evalf(TE(183712));
```

$$-.99999706831674294006\,10^{-9}$$

This is still a rather large number of steps, but not astronomically large: the computation should be feasible on a fast computer (don't try it if you're in a hurry and your computer is slow). The "`add`" function is similar to "`sum`", except that it always adds the individual terms rather than looking for a closed-form expression. I'll use a floating-point value for the step size "`dx`" so that Maple will add floating-point numbers rather than rational fractions.

```
> N:= 183712; dx:= evalf((b-a)/N);
> dx/2 * (f(a) + 2 * add(f(a+i*dx), i=0 .. N-1) + f(b));
```

$$-4.4999999990000029449$$

This took about 49 seconds on my computer, and the result agrees quite well with the closed-form value:

```
> evalf(T(N));
```

$$-4.4999999990000029317$$

Similarly, for the Midpoint Rule we get $\lim_{n\to\infty} n^2 ME(n) = \dfrac{135}{8}$, so we would need $n$ to be at least 129 904.

```
> N:= 129904; dx:= evalf((b-a)/N);
> dx * add(f(a+(i+.5)*dx), i = 0 .. N-1);
```

$$-4.5000000009999970868$$

This one took my computer about 36 seconds (at the same rate it would take over 86 days for $n = 2.7 \times 10^{10}$). And still better methods are ahead. ■

## Simpson's Rule

We have noticed (for our particular $f$, although it is true in general for sufficiently differentiable functions $f$) the $1/n^2$ term in the error for the Trapezoid Rule is $-2$ times the corresponding term for the Midpoint Rule. Just as taking the average of the left and right sums canceled the $1/n$ term (and produced the Trapezoid Rule), we might try to take a suitable combination of the Trapezoid Rule and the Midpoint Rule. In this case, the appropriate combination would be $(2/3)M(n) + (1/3)T(n)$. This is the Simpson's Rule approximation to our integral. We write it as $S_{2n}$ rather than $S_n$ because it uses the same $2n + 1$ equally spaced points as $T_{2n}$. The "`student`" package has a "`simpson`" command.

```
> a:= 'a'; b:= 'b';
> simpson(g(x), x=a..b, n);
```

$$\frac{1}{3}(b-a)\left(g(a)+g(b)+4\left(\sum_{i=1}^{\frac{1}{2}n} g\left(a+\frac{(2i-1)(b-a)}{n}\right)\right)\right.$$
$$\left.+2\left(\sum_{i=1}^{\frac{1}{2}n-1} g\left(a+2\frac{i(b-a)}{n}\right)\right)\right)/n$$

Note that I had to unassign "a" and "b", which still had values from the last section. Let's try it for our "typical" function $f(x) = x^5 - 5x^2$. We need to use "`value`" because "`simpson`" uses a "`Sum`" rather than a "`sum`".

```
> v:= value(simpson(f(x), x=a..b, n));
```

The result looks very complicated, and I won't bother reproducing it here. Let's use "`expand`" on it.

```
> v:= expand(v);
```

$$v := \frac{5}{3}\frac{a^2b^4}{n^4} + \frac{4}{3}\frac{ba^5}{n^4} - \frac{5}{3}b^3 + \frac{1}{6}b^6 + \frac{1}{3}\frac{b^6}{n^4} + \frac{5}{3}a^3 - \frac{1}{6}a^6 - \frac{1}{3}\frac{a^6}{n^4} - \frac{5}{3}\frac{b^2a^4}{n^4} - \frac{4}{3}\frac{ab^5}{n^4}$$

This is not so bad. Note that there are terms with no $n$ in them, and terms proportional to $1/n^4$. In the limit as $n \to \infty$, the terms with $1/n^4$ would go to 0, leaving the terms with no $n$. Thus the terms with no $n$ must give us $\int_a^b f(x)\,dx$, while those with $1/n^4$ contribute to the error.

```
> e:= int(f(x), x=a..b) - v;
```

$$e := -\frac{5}{3}\frac{a^2b^4}{n^4} - \frac{4}{3}\frac{ba^5}{n^4} - \frac{1}{3}\frac{b^6}{n^4} + \frac{1}{3}\frac{a^6}{n^4} + \frac{5}{3}\frac{b^2a^4}{n^4} + \frac{4}{3}\frac{ab^5}{n^4}$$

This is the error in Simpson's Rule for our function. It is of the form $C/n^4$, where the constant $C$ turns out to be $(f'''(a) - f'''(b))(b-a)^4/180$.

```
> simplify(e * n^4 - ((D@@3)(f)(a)-(D@@3)(f)(b)) * (b-a)^4 /180);
```

$$0$$

In general for a smooth function the error $SE(n)$ will be approximately $C/n^4$ (but not always exactly $C/n^4$), where again $C = (f'''(a) - f'''(b))(b-a)^4/180$. This means that Simpson's Rule should have much smaller error than the Trapezoid or Midpoint Rule when $n$ is large; in fact, it is often much better even when $n$ is not very large. For our example with $a = -1$ and $b = 2$:

```
> subs(a=-1, b=2, e);
```

$$-81\frac{1}{n^4}$$

What $n$ would we need for the absolute value of the error to be at most $10^{-9}$?

```
> evalf((81 * 10^9)^(1/4));
```

$$533.48382301167684037$$

So $n = 534$ should do it. This is quite reasonable, compared to 129 904 for the Midpoint Rule. Even a very slow computer should be able to do this computation quickly.

```
> N:= 534; dx:= evalf((b-a)/N);
> dx/3*(f(a)+f(b)+4*add(f(a+(2*i-1)*dx), i = 1 .. N/2)
>        + 2*add(f(a+2*i*dx), i = 1 .. N/2 - 1));
```

$$-4.4999999990038608933$$

In this example, we were able to use exact formulas for the errors in order to determine which $n$ to use. For most functions, Maple won't be able to find exact formulas for the sums, or even for the integral (after all, the point of numerical integration is to find a numerical value for the integral when no antiderivative can be found). How could we determine $n$ in this case?

One way is to use the following error estimates, which are found in many calculus texts:

If $f''$ is continuous on $[a, b]$ and $|f''(x)| \le K$ there, then

$$\left| \int_a^b f(x)\,dx - T_n \right| \le \frac{K(b-a)^3}{12n^2}$$

$$\left| \int_a^b f(x)\,dx - M_n \right| \le \frac{K(b-a)^3}{24n^2}$$

If $f^{(4)}$ is continuous on $[a, b]$ and $|f^{(4)}(x)| \le K$ there, then

$$\left| \int_a^b f(x)\,dx - S_n \right| \le \frac{K(b-a)^3}{180n^4}$$

These estimates are reliable (assuming that your $K$ is correct). However, they have two disadvantages.

(1) They require bounds on $f''$ (for Midpoint and Trapezoid Rules) or $f^{(4)}$ (for Simpson's Rule). It may require considerable effort to obtain these bounds.

(2) The actual errors may be much less than the estimates.

Another way is to use our knowledge of the limiting form of the error to guide us. This method is not foolproof, but when used cautiously it will usually give good results.

**EXAMPLE 2** Find $\displaystyle\int_1^3 \sin(x^2)\,dx$ to within an accuracy of $10^{-5}$, using Simpson's Rule.

**SOLUTION** How can we determine an $n$ that will provide sufficient accuracy? First we'll try using the error estimates.

```
> f:= x -> sin(x^2);
> f4:= (D@@4)(f)(x);
```

$$f4 := 16\sin\left(x^2\right)x^4 - 48\cos\left(x^2\right)x^2 - 12\sin\left(x^2\right)$$

```
> plot(abs(f4), x=1..3);
```
(Figure 4.4)

The graph shows three peaks. The maximum value of $|f^{(4)}(x)|$ on our interval appears to be about 1170, occurring somewhere between $x = 2.8$ and $x = 3$. We could obtain the maximum more precisely by using "fsolve" to find the critical point, but such precision is not really needed here. It looks safe to say that $|f^{(4)}(x)| < 1200$. Thus the error estimate for Simpson's Rule is

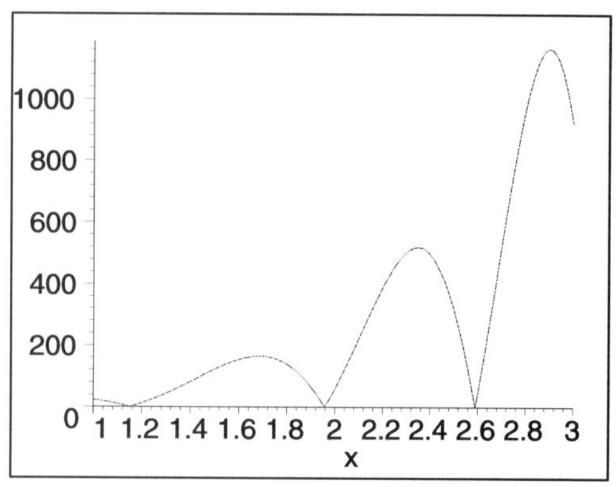


Figure 4.4. $|f^{(4)}(x)|$

$$\left| \int_1^3 f(x)\,dx - S_n \right| \le \frac{1200(3-1)^5}{180n^4}$$

For what $n$ would the error estimate be $10^{-5}$?

```
> Digits:= 10;
>  evalf((1200 * 2^5/(180 * 10^(-5)))^(1/4));
```

$$67.96176070$$

So $n = 68$ should do it.

```
> evalf(simpson(f(x), x=1..3, 68));
```

$$.4632950437$$

Now let's try it using the limiting form of the error: $(f'''(a) - f'''(b))(b-a)^4/(180n^4)$. When is that less than $10^{-5}$ in absolute value?

```
> C:= evalf(((D@@3)(f)(1)-(D@@3)(f)(3))*2^4/180);
```

$$C := -17.45670590$$

```
> evalf((-C/10^(-5))^(1/4));
```

$$36.34884151$$

So perhaps $n = 38$ will do it (we must use an even $n$ for Simpson's Rule).

```
> s38:= evalf(simpson(f(x), x=1..3, 38));
```

$$s38 := .4633026685$$

I said "perhaps" because although we know $C/n^4$ is the limiting form of the error, it isn't the exact error. We can't check the actual error without knowing the true value of the integral. But we can get some confirmation that the error is nearly $C/n^4$ for $n = 38$ by comparing $S_{38}$ to some nearby $S_n$, say $S_{40}$. If the error was exactly $C/n^4$, we would have

$$S_{40} - S_{38} = \frac{C}{38^4} - \frac{C}{40^4}$$

```
> evalf(C/38^4 - C/40^4);
```

$$-.1552948988\,10^{-5}$$

What do we actually have?

```
> s40:= evalf(simpson(f(x), x=1..3, 40));
```

$$s40 := .4633010965$$

```
> s40 - s38;
```

$$-.15720\,10^{-5}$$

This is quite close to what it should be, but slightly larger in absolute value. If it had been smaller, it would tend to indicate that the actual errors are smaller than the limiting form, and we might have been confident that $n = 38$ was adequate. As it is, since we calculated 38 as the least possible $n$ for which $|C/n^4| < 10^{-5}$, we might worry that the actual error is a bit larger in absolute value than $10^{-5}$. We might conjecture that for $n \geq 38$ the absolute value of the error is at most $C_1/n^4$, where $C_1 = \dfrac{.15720}{.1552948988}C$. What would this be for $n = 38$?

```
> .15720/.1552948988 * C/38^4;
```

$$-.8474679060\,10^{-5}$$

So we should still be OK.

Just for comparison, let's see Maple's value for the integral (which should be accurate to about 10 digits), and the actual error in $S_{38}$:

```
> s:= evalf(Int(f(x), x=1..3));
```

$$s := .4632942252$$

```
> s - v;
```

$$-.84433\,10^{-5}$$

■

## Romberg Integration

The Romberg method uses a sequence of Trapezoid Rule approximations, together with theoretical information on the limiting form of the errors, to obtain very accurate approximations. We let $T_n^0$ be the Trapezoid Rule approximation $T_{2^n}$ with $2^n$ intervals, and

$$T_n^j = \frac{4^j T_n^{j-1} - T_{n-1}^{j-1}}{4^j - 1} \text{ for } 1 \le j \le n$$

Then the Romberg approximation $R_n$ is $T_n^n$. Note that this requires evaluating $f$ at the same points as does $T_{2^n}$. It uses all the values $T_{2^j}$ for $1 \le j \le n$, but the points at which these evaluate $f$ are all points at which $T_{2^n}$ evaluates $f$.

Let's try it first with a general function $f(x)$ on the interval $[a, b]$, using a table `T[j,n]` for $T_n^j$, and going up to $n = 5$. This is the first time we've used a doubly indexed table, but this is really no different from one with a single index. We first use a "`for`" loop to calculate `T[0,n]` using the Trapezoid Rule, then a "`for`" loop inside another "`for`" loop to calculate the other `T[j,n]` values. This double loop will calculate `T[1,1]` (for $n = 1$), then `T[1,2]` and `T[2,2]` (for $n = 2$), etc. Note that `T[j-1,n]` and `T[j-1,n-1]` are calculated before they are used in calculating `T[j,n]`, so this method should work.

```
> f:= 'f';
> for nn from 0 to 8 do
>     T[0,nn]:= value(trapezoid(f(x), x=a..b, 2^nn))
> od:
> for nn from 1 to 8 do
>    for jj from 1 to nn do
>       T[jj,nn]:= (4^jj*T[jj-1,nn] - T[jj-1,nn-1])/(4^jj-1)
>    od
> od;
```

I used "`:`" on the first "`for`" loop because the last few "`T[0, nn]`" values fill several screens and are not particularly interesting. I didn't need to do this for the second "`for`" loop: you don't see the values of $T[j, n]$ as they are assigned, because this takes place inside a double loop. You can look at the values individually. To make the formulas simpler, I'll substitute $a = 0$ and $b = 1$ in these.

```
> subs(a=0, b=1, T[1,1]);
```

$$\frac{1}{6}f(0) + \frac{2}{3}f\left(\frac{1}{2}\right) + \frac{1}{6}f(1)$$

This is Simpson's Rule $S_2$. In fact, $T_n^1$ is $S_{2^n}$.

```
> subs(a=0, b=1, T[3,3]);
```

$$\frac{31}{810}f(0) + \frac{512}{2835}f\left(\frac{1}{8}\right) + \frac{176}{2835}f\left(\frac{1}{4}\right) + \frac{512}{2835}f\left(\frac{3}{8}\right) + \frac{218}{2835}f\left(\frac{1}{2}\right)$$
$$+\frac{512}{2835}f\left(\frac{5}{8}\right) + \frac{176}{2835}f\left(\frac{3}{4}\right) + \frac{512}{2835}f\left(\frac{7}{8}\right) + \frac{31}{810}f(1)$$

Part of the motivation for the Romberg method is that the limiting form of the error in $T_n^j$ as $n \to \infty$ (for a smooth $f$) is $Cn^{-4j}$. Moreover, $T_n^j$ gives the exact answer if $f$ is a polynomial of degree at most $2n + 1$. Let's try $R_n$ when $f$ is a polynomial of degree 10, with $a = 0$ and $b = 1$.

```
> f:= unapply(sum(c[n]*x^n, n=0..10), x);
> a:= 0; b:= 1;
> T[3, 3];
```

$$\frac{1}{4}c_3 + \frac{1}{5}c_4 + \frac{1}{6}c_5 + \frac{1}{7}c_6 + \frac{1}{8}c_7 + \frac{40963}{368640}c_8 + \frac{1639}{16384}c_9 + \frac{143141}{1572864}c_{10} + c_0 + \frac{1}{2}c_1 + \frac{1}{3}c_2$$

The coefficients of $c_0$ to $c_7$ are all correct. Moreover, even the coefficients that are not correct are not very far from the correct ones.

```
> T[3,3] - int(f(x), x=0..1);
```

$$\frac{1}{122880}c_8 + \frac{3}{81920}c_9 + \frac{1687}{17301504}c_{10}$$

**■ EXAMPLE 3** Find $\int_1^3 \sin(x^2)\,dx$ to within an accuracy of $10^{-12}$, using the Romberg method.

**SOLUTION** Note that these are the $f$, $a$, and $b$ from Example 2. We need to increase "`Digits`" in order to attain the desired accuracy (even to represent a number to this precision). To ensure that roundoff error isn't too much of a problem, I'll use 16 digits, and place the Romberg approximations $R_n$ for $n$ from 1 to 8 in a list[(7)].

```
> f:= x -> sin(x^2);
> a:= 1; b:= 3; Digits:= 16;
> R:= [seq(evalf(T[n,n]), n=1..8)];
```

$$R := [-.5912068370606863, .5228911937232021, .4684532863109115, .4631415662697726,$$
$$.4632948081679371, .4632942249810508, .4632942251702527, .4632942251703877]$$

How accurate are these?[(8)] We can use the difference between $R_{n-1}$ and $R_n$ as an estimate of the error in $R_n$. This is usually very conservative.

```
> seq(R[n] - R[n-1], n=2..8);
```

$$1.114098030783888, -.0544379074122906, -.0053117200411389,$$
$$.0001532418981645, -.5831868863\,10^{-6}, -.1892019\,10^{-9}, -.1350\,10^{-12}$$

We can be fairly confident that $R_8 = .4632942251703877$ is sufficiently accurate. To check this, let's see Maple's value for the integral, using "`evalf(Int(...`":

```
> J:= evalf(Int(f(x), x=a..b));
```

$$J := .4632942251703879$$

This is probably accurate to its full 16 digits; at least, that's what Maple tries to achieve. There may be some uncertainty about the last digit (you can check it by using "`evalf`" with a larger number of digits: this confirms that `J` is accurate). Now let's see the actual errors in the approximations $R_n$.

```
seq(J - R[n]), n=1..8);
```

$$1.054501062231074, -.0595969685528142, -.0051590611405236, .0001526589006153,$$
$$-.5829975492\,10^{-6}, .1893371\,10^{-9}, .1352\,10^{-12}, .2\,10^{-15}$$

Indeed, it seems that we were being rather conservative: $R_n$ being a much better approximation than $R_{n-1}$, $R_n - R_{n-1}$ is actually close to the error in $R_{n-1}$ and much larger in absolute value than the error in $R_n$. But to get a better estimate of the error in $R_n$, you'd want to calculate $R_{n+1} - R_n$ (and if you were calculating $R_{n+1}$ anyway, you'd use $R_{n+1}$ rather than $R_n$ as your best approximation to the integral).

It's also instructive to look at the error in $T_8^1 = S_{256}$, the Simpson's Rule approximation that evaluates $f$ at the same points as does $R_8$.

```
> evalf(J - T[1,8]);
```

$$-.40652177\,10^{-8}$$

Thus the Romberg approximation is a significant improvement on Simpson's Rule in this case. ■

---

(7) Since a list "`R`" starts with "`R[1]`", I'll leave out $R_0$.

(8) Note that your results might not be exactly the same as mine in the last digit or two. Maple sometimes produces slightly different results when these calculations are repeated, because they may be done in a different order; the order can make a difference in floating-point computation.

## Improper Integrals and Non-Smooth Functions

The various numerical integration methods we have considered all depend on the integrand being a "smooth" function. The error estimates for the Trapezoid and Midpoint Rules, for example, assume that $f''$ is continuous, while that for Simpson's Rule assumes $f^{(4)}$ is continuous. Although these rules will "work" for functions that do not have so many continuous derivatives, in the sense that the limit as $n \to \infty$ is correct, they may approach that limit quite slowly (e.g., see Exercise 3). We must therefore modify our approach in dealing with such functions.

There are even more serious problems in dealing with improper integrals. Here we often can't even get started (e.g., for an integral over an infinite interval, or an integrand that is undefined at an endpoint). Of course, some improper integrals diverge, and these are outside the reach of any numerical integration method (but we want to be sure to detect this divergence before trying numerical methods).

Our main tool for dealing with such integrals is a change of variables, by which we try to transform the integral into a proper integral with a smooth integrand. In some cases, we break up the interval of integration into several pieces to make this simpler.

■ **EXAMPLE 4** Evaluate $\displaystyle\int_0^\infty \frac{dx}{\sqrt{x^5+1}}$.

**SOLUTION** This is a "type I" improper integral, i.e., the integral of a continuous function on an infinite interval. We know it converges since the integrand is less than $x^{-5/2}$ and $\displaystyle\int_1^\infty x^{-5/2}\,dx$ converges. The simplest way to transform an infinite interval to a finite one is the change of variable $t = 1/x$, which takes $[a, \infty)$ to $(0, 1/a]$ for $a > 0$. Here we want to use $a = 0$, but that's only a minor annoyance: we can use $t = 1/(x+1)$, which takes $[0, \infty)$ to $(0, 1]$.

```
> A:= Int(1/sqrt(x^5+1), x=0..infinity);
> B:= changevar(1/(x+1)=t, A, t);
```

$$B := \int_0^1 \frac{1}{\sqrt{\frac{(-t+1)^5}{t^5}+1}\,t^2}\,dt$$

Maple should be able to simplify this. It is appropriate to use the "`symbolic`" option in "`simplify`" here, since we know that the integrand should be positive.

```
> B:= simplify(B, symbolic);
```

$$B := \int_0^1 t/\left(5\,t^5 - 10\,t^4 + 10\,t^3 - 5\,t^2 + t\right)^{1/2}\,dt$$

This is now a proper integral, but the integrand still isn't smooth: near $t = 0$ the "$t$" term in the square root is dominant, so the integrand behaves like $\sqrt{t}$ which is non-differentiable at $t = 0$. To smooth it out, we use another change of variables.

```
> B:= changevar(t=s^2, B, s);
```

$$B := \int_0^1 2s^3/\left(s^2 - 5s^4 + 10s^6 - 10s^8 + 5s^{10}\right)^{1/2}\,ds$$

We can take out a factor of $s$ from the numerator and denominator, again using the "`symbolic`" option in "`simplify`".

```
> B:= simplify(B, symbolic);
```

$$B := 2\int_0^1 s^2/\left(1 - 5s^2 + 10s^4 - 10s^6 + 5s^8\right)^{1/2}\,ds$$

This time there's no problem at $s = 0$. The integral can now be done using any of our numerical methods. Let's use Simpson's Rule with $n = 40$. Note the "2" in the definition of `f` below is from the "2" that appears in front of the last integral.

```
> f:= unapply(2*integrand(B), s);
```

$$f := s- > 2s^2/\left(1 - 5s^2 + 10s^4 - 10s^6 + 5s^8\right)^{1/2}$$

```
> evalf(simpson(f(x), x=0..1, 40));
```

$$1.549696108$$

For comparison, let's see Maple's value for the integral. We'll set "`infolevel`" to have Maple show us how it arrives at its answer.

```
> infolevel[all]:= 1;
> evalf(A);
```

The result is rather long, and I'll only go over the highlights.

```
evalf/int/improper: applying transformation x =  1/x
evalf/int/improper: interval is  0 .. 1 for the two integrands:
```

$$\frac{1}{\sqrt{x^5+1}}, \frac{1}{\sqrt{\frac{1}{x^5}+1\,x^2}}$$

Maple splits up the original interval $[0, \infty)$ into $[0, 1]$ and $[1, \infty]$, and applies the change of variable $t = 1/x$ (as we would write it) to the integral on $[1, \infty)$. For the original integrand on $[0, 1]$, it applies Curtis-Clenshaw quadrature ("`ccquad`").

```
evalf/int/control: from ccquad, error =    4.003155e-12
integrand evals = 19              result = .9388514394380
```

This procedure evaluates the integrand 19 times and comes up with a result, .9388514394380, which Maple believes accurate to within about $4 \times 10^{-12}$. Then Maple looks at the second, transformed, integral, concludes (just as we did) that it will have a singular point at 0, and applies another change of variable $t = s^2$. Now that the singularity has been removed, it applies "`ccquad`" again.

```
evalf/int/control: from ccquad, error =    6.108448e-14
integrand evals =   55            result =   .6108448383092
```

This one required evaluating the integrand 55 times, with a result of .6108448383092. Adding this to the previous result, it comes up with the answer:

$$1.549696278$$

■

There are various other tricks that Maple uses when it can't come up with suitable transformations. Sometimes it even looks for an antiderivative!

Maple can often tell when an improper integral diverges, and in that case "`evalf`" will produce an error message.

```
> evalf(Int(1/(x^2 - 3*x), x=1..infinity));
Error (in evalf/int) integrand has a pole in the interval
```

Sometimes, however, Maple will not detect that an integral diverges. It may try applying numerical methods to such an integral. Since the desired numerical accuracy can never be attained, Maple can take a **very** long time attempting such a calculation, and you may be forced to interrupt the computation before Maple reaches a conclusion. This happens, for example, in Release 4 with the following integral:

```
> evalf(Int(1/(tan(x) - 2*x), x=1..2));
```

## EXERCISES

**1.** Let $f(x) = 1/(1+x)$.

**(a)** Approximate $\int_0^1 f(x)\,dx$ using the Midpoint, Trapezoid, and Simpson's Rules for $n = 10$. Find the error in each approximation.

**(b)** Calculate the limits as $n \to \infty$ of $n^2$ times the error in the Midpoint Rule and the Trapezoid Rule, and $n^4$ times the error in Simpson's Rule.

**(c)** How large must $n$ be so that the absolute value of the error in the Midpoint Rule is less than .0001? Do the same for the Trapezoid Rule and Simpson's Rule.

**2.** Choose an interval $[a, b]$ and a function $f(x)$ so that Maple can evaluate the Midpoint, Trapezoid, and Simpson's Rule sums explicitly as functions of $n$, but so that these approximations are not exact (i.e., don't use a polynomial of degree $\leq 3$). You can try a polynomial of degree more than 3, an exponential, sine or cosine, a rational function with a denominator that factors (but its roots are outside the interval $[a, b]$), or a sum of these. Let $ME(n)$, $TE(n)$, and $SE(n)$ be the errors in the Midpoint, Trapezoid, and Simpson's Rules for your function on your interval.

**(a)** Find $ME(n), TE(n)$, and $SE(n)$ explicitly as functions of $n$.

**(b)** Plot $n^2ME(n)$ and $n^2TE(n)$ (on the same graph) and $n^4SE(n)$ (on a different graph) as functions of $n$. What can you conclude?

**(c)** Find the limits of $n^2ME(n)$, $n^2TE(n)$ and $n^4SE(n)$ as $n \to \infty$.

**(d)** What $n$'s would be needed for the errors in the Midpoint, Trapezoid, and Simpson's Rules to be less than $10^{-9}$?

**3.** Consider $\displaystyle\int_0^1 \sqrt{x}\,dx$. Let $ME(n)$, $TE(n)$ and $SE(n)$ be the errors in the Midpoint, Trapezoid, and Simpson's Rules for this integral.

**(a)** Plot $n^{3/2}ME(n)$, $n^{3/2}TE(n)$, and (for even $n$) $n^{3/2}SE(n)$ (on the same graph) as functions of $n$. What can you conclude?

**(b)** Approximately what $n$ would be needed for the Simpson's Rule error to be less than $10^{-9}$?

**4. (a)** Let $f(x) = x^{11}$. Find the error $TE(n)$ in the Trapezoid Rule for the interval $[a, b]$. Show that this is of the form

$$\sum_{j=1}^{5} c_j \frac{(b-a)^{2j}\left(f^{(2j-1)}(b) - f^{(2j-1)}(a)\right)}{n^{2j}}$$

where $c_j$ are constants (not involving $a$ or $b$). *Hint:* to find the coefficient of $n^{-2j}$, you can use "`coeff(TE(n),n,-2*j)`".

**(b)** Try it now for some other power $x^m$ (where $m$ is an integer $\geq 5$). Show that $TE(n)$ has the same form (with the same $c_j$'s for $j \leq 5$, but perhaps more terms). It turns out that for any function with at least $2m+2$ derivatives on $[a, b]$,

$$TE(n) = \sum_{j=1}^{m} c_j \frac{(b-a)^{2j}\left(f^{(2j-1)}(b) - f^{(2j-1)}(a)\right)}{n^{2j}} + O\left(n^{-2m-2}\right)$$

where $c_j$ are constants that don't depend on $n$, $f$, $a$, or $b$ (and $c_1$ to $c_5$ are as in part (a)).

**(c)** Compare the actual $TE(n)$ to the sum from $j = 1$ to 5 in the formula above, using $f(x) = \exp(x)$, $a = 0$, and $b = 1$. Calculate $n^{12}$ times the difference, for $n = 1$ to 10. Does this appear to approach a limit? You should use "`Digits:=30`" since the differences get very small.

**5. (a)** What would the error formula from Exercise 4 say about the error in using the Trapezoid Rule to approximate $\displaystyle\int_0^{2\pi} \frac{dx}{2+\cos x}$?

**(b)** How accurate are the Midpoint, Trapezoid, and Simpson's Rule approximations for $n = 20$? (make "`Digits`" at least 20). Do you find the result surprising?

**6.** **(a)** Find the Romberg approximations $R_n$, $n = 1,\ 2,\ \ldots 5$, for $\displaystyle\int_1^3 \frac{dx}{x^2 - 2x + 5}$.

**(b)** Calculate the error in $R_n$ and compare to $T_n^{n-1} - R_n$, for $\displaystyle\int_0^1 x^{11}\,dx$ and for $\displaystyle\int_0^1 x^{41}\,dx$ with $n = 1,\ 2,\ \ldots 5$.

**(c)** Is $T_n^{n-1} - R_n$ a good estimate of the error in $R_n$ for $\displaystyle\int_0^1 x^k\,dx$ when $k$ is large?

**(d)** What $n$ would be needed for $R_n$ to be exactly equal to $\displaystyle\int_0^1 x^{41}\,dx$? Would it be practical to calculate this $R_n$?

**7.** Transform $\displaystyle\int_0^\infty \frac{dx}{x + \sqrt{x^3 + 1}}$ into a proper integral, and use Simpson's Rule to find an approximation with error less than 0.00001.

**8.** Maple sometimes has trouble with integrals involving absolute values. For example,

```
int(x * abs(sin(x)), x=0 .. 10*Pi);
```

$$\int_0^{10\pi} x|\sin(x)|\,dx$$

Break up the interval $[0,\ 10\pi]$ into subintervals to find the exact value of this integral.

# LAB 14 PARAMETRIC AND POLAR CURVES

**Goals:** *To learn to use Maple to plot polar and parametric curves, and to study their properties. To learn how Maple deals with vectors, and how these can be useful in computer graphics.*

## Plotting Parametric Curves

A parametric curve in the plane is described by a pair of equations $x = f(t)$, $y = g(t)$, for $t$ in some interval of real numbers. Maple's "`plot`" command can draw the graph of such a curve.

■ **EXAMPLE 1** Plot the parametric curve $x = t\sin(t)$, $y = t - 2\cos(t)$ for $-5 \le t \le 5$. "Zoom in" on the part of the curve in the fourth quadrant.

**SOLUTION**

```
> f:= t -> t * sin(t);   g:= t -> t - 2 * cos(t);
> plot([ f(t), g(t), t=-5..5]); (Figure 4.5)
```

Inside square brackets we have first the expression $f(t)$ describing the $x$ coordinate, then $g(t)$ describing the $y$ coordinate, and finally the parameter interval[9]. Maple will show a region of the $xy$ plane just large enough to contain the curve. If you wish you can specify $x$ and $y$ coordinate intervals for the plot (after the square brackets). Thus to see an enlargement of the interesting-looking loop in the fourth quadrant of this plot, you could try

```
> plot([ f(t), g(t), t=-5..5],
>     0..2, -3..0);
```

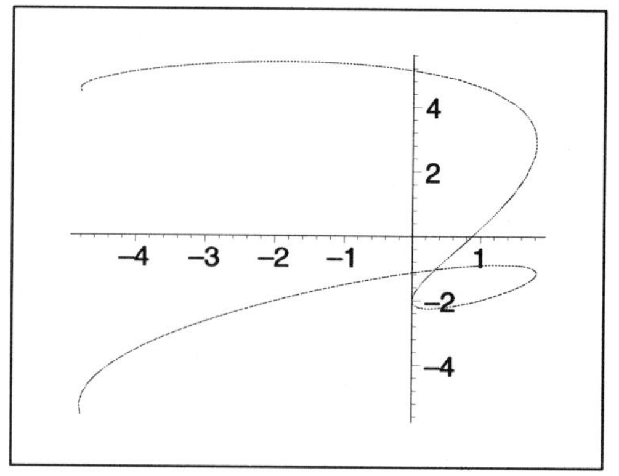


Figure 4.5. $x = t\ \sin t$, $y = t - 2\ \cos t$

Although this enlarges part of the graph, it doesn't really increase the level of detail, because the same number of points $(f(t), g(t))$ will be calculated. We could increase this number by specifying the "numpoints" option.

```
> plot([ f(t), g(t), t=-5 .. 5], 0 .. 2, -3 .. 0, numpoints=200);
```

This method is rather inefficient, because most of the points calculated are outside of the plotting region. It would be better to use only those $t$ values corresponding to the part of the curve in our region. What are those $t$ values? It may help to plot $f(t)$ and $g(t)$ separately as functions of $t$. To be able to tell which is which in one graph, we can plot the list `[f(t), g(t)]` and specify two different colours or linestyles as a list.

```
> plot([f(t), g(t)], t = -5 .. 5, linestyle = [1, 3]);
```
(Figure 4.6)

The solid curve is $f(t)$ and the dashed curve is $g(t)$. We want the interval where $f(t) \geq 0$ and $g(t) \leq 0$. This happens from the leftmost point where $f(t)$ crosses the $t$ axis (near $t = -3.1$) to the point where $g(t)$ crosses the axis (near $t = 1$). Looking at the formula for $f(t)$, it is easy to see that the first point is actually $t = -\pi$. To find the second point, we can use "fsolve":

```
> b:= fsolve(g(t)=0, t=.5 .. 1.5);
```

$$b := 1.029866529$$

We can now use this information to get the detail of the region we want. After producing the plot, I'll save it for future use.

```
> plot([ f(t), g(t), t=-Pi .. b ]);
> pplot:= %:
```
(Release 5) or `> pplot:= ":` (Release 4)

It's often useful to be able to locate points corresponding to particular $t$ values on the curve. This can be done by using "display" to combine the parametric curve with a plot of one or more points. Here, for example, I will take the last plot and draw a circle at each of the points where $t$ is an integer multiple of $1/10$. This is done using the options "style=POINT" and "symbol=CIRCLE". I will assign the list of points $(f(t), g(t))$ for these values of $t$ to the variable "points".

One circle looks much like another circle, so it may also be useful to have Maple print the value of $t$ at several of these points, say at integer multiples of $1/2$. This can be done using the "textplot" command from the "plots" package. The input to "textplot" is a list, each item of which is a list of three items: an $x$ coordinate, a $y$ coordinate, and the label to be printed at the point $(x, y)$. Thus:

```
> plots[textplot]([ [1,2,A], [3,4,B] ]);
```

will place the value of "A" (which can be a number, an expression, or just the name "A" itself) at $x = 1, y = 2$, and the value of "B" at $x = 3, y = 4$. I will assign the list of points and $t$ values for our plot to the variable "labels". An option to the "textplot" command is to specify the alignment

(9) Note the difference between plotting a list of functions and plotting a parametric curve: for the parametric curve the parameter interval is included within the brackets.

of the labels. The default is to have it centred on the specified point. In this case, it is better not to have the label right on top of the point, so I will choose the option "`align=BELOW`" (I could have used "`ABOVE`", "`LEFT`", "`RIGHT`", or even a combination such as "`{ABOVE,LEFT}`"). In fact, I'll allow a bit more room between the point and its label by subtracting .05 from the $y$ value.

```
> with(plots):
> points:= [ seq( [f(n/10), g(n/10)], n = -31 .. 10) ]:
> labels:= [ seq( [f(n/2), g(n/2)-.05, n/2], n=-6..2) ]:
> display({ pplot,
>          plot(points, style=POINT, symbol=CIRCLE),
>          textplot(labels, align=BELOW) }); (Figure 4.7)
```

■

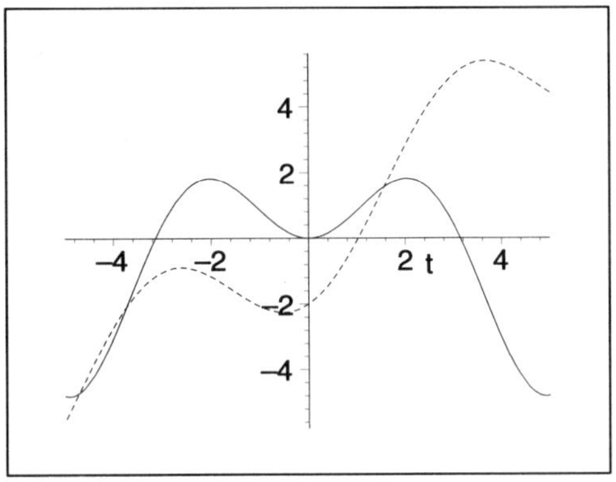


Figure 4.6. $f(t)$ and $g(t)$

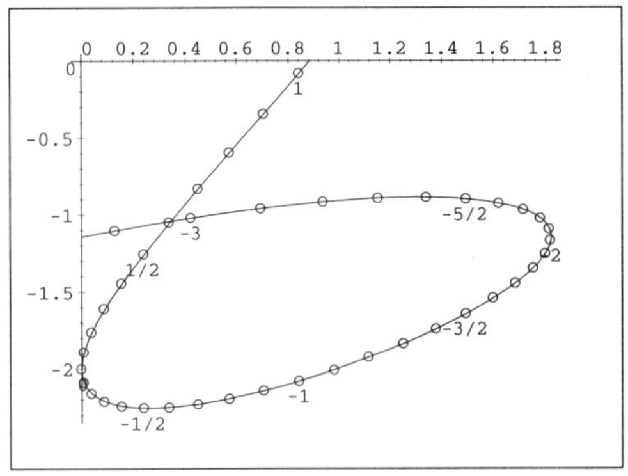


Figure 4.7. The lower right quadrant

We can also produce an animation of the curve with the symbol moving along it. Each frame will be a "`display`" with the curve "`pplot`" and a plot of a single point[10].

```
> display([ seq(display({ pplot,
>        plot([[f(n/10),g(n/10)]], style=POINT,symbol=CIRCLE) }),
>               n=-31..10) ], insequence=true);
```

## Some Features of the Curve

One of the conspicuous features of our curve is its self-intersection point. On the graph this appears to be somewhere near $x = .3$, $y = -1$. We can use "`fsolve`" to find it more exactly[11]. We want to find two distinct values $r$ and $s$ of the parameter which yield the same point, i.e., $f(r) = f(s)$ and $g(r) = g(s)$. To make sure that $r$ and $s$ are distinct, we can divide one or both of these equations by $r - s$.

```
>sol:=  fsolve({ f(r)=f(s), (g(r)-g(s))/(r-s)=0 }, {r, s});
```

$$sol := \{r = -3.027824597,\ s = .6046492583\}$$

We were rather lucky here: "`fsolve`" might well have found a solution with $r$ or $s$ outside our parameter interval. If that had happened, we could have specified intervals for $r$ and $s$.

```
> fsolve({ f(r)=f(s), (g(r)-g(s))/(r-s)=0 }, {r, s},
>          r = -5 .. 5, s = -5 .. 5);
```

The point of self-intersection is thus:

```
> eval(subs(sol, [f(r), g(r)]));
```

$$[.3437271125, -1.040753813]$$

---

(10) Actually, a list containing a single point.

(11) For some curves you might be able to use "`solve`", but it doesn't work in this case.

The slope of the curve at the point $(f(t), g(t))$ is $g'(t)/f'(t)$ (assuming $f'(t) \neq 0$).

```
> m:= unapply(D(g)(t)/D(f)(t), t);
```

$$m := t \rightarrow \frac{1 + 2\sin(t)}{\sin(t) + t\cos(t)}$$

It might also be of interest to see the points where the tangent line passes through the origin. Looking at the graph, there appear to be three of these points. One of these is at $t = 0$ where the curve is tangent to the $y$ axis. The other two can be obtained by solving $m(t) = g(t)/f(t)$.

```
> fsolve(m(t)=g(t)/f(t), t, t=-5 .. -Pi);
```

$$-4.148314918$$

```
> fsolve(m(t)=g(t)/f(t), t, t=-Pi .. 0);
```

$$-2.315596337$$

## Arc Length and Area

The length of the parametric curve $x = f(t)$, $y = g(t)$ for $a \leq t \leq b$ is $\int_a^b \sqrt{f'(t)^2 + g'(t)^2}\, dt$. Because of the square root, it is often impossible to find a formula for the antiderivative, even when $f$ and $g$ are quite simple. However, the numerical integration techniques should generally have no problem with this integral. In our example:

```
> evalf(Int(sqrt((D(f)(t))^2 + (D(g)(t))^2), t=-5 .. 5));
```

$$24.92593913$$

The area bounded by a non-self-intersecting closed curve is $\pm \int_a^b g(t) f'(t)\, dt$, $+$ if the curve is traversed clockwise as $t$ increases and $-$ if it is counterclockwise. In our curve, as we have seen, we have a closed loop from $t = -3.027824597$ to $t = .6046492583$. This curve is traversed clockwise. To find the area of the loop:

```
> int(g(t)*D(f)(t), t= -3.027824597 .. .6046492583);
```

$$1.677411651$$

## Vectors

Maple contains many commands for manipulating vectors. We will use only a few of them here, and meet a few more in Lab 17. Although there are other objects that Maple calls "vectors", for our purposes it will be best to consider a vector as a list.

```
> P:= [f(t), g(t)];
```

$$P := [t\sin(t), t - 2\cos(t)]$$

Since neither arrows above letters nor boldface type are available in Maple, I will use capital letters for the names of vectors and lower case for scalars (in any situation where both vectors and scalars are present). This vector "`P`" is the "position vector" for our parametric curve. Note that each component of this vector is actually an expression depending on $t$ rather than just a number. We could also define a vector whose components are functions instead of expressions, but there are some technical complications doing it that way.

The most basic operations on vectors are addition and scalar multiplication. You can write a normal-looking expression involving vectors and scalars, such as "`P - t * V`" (note the use of "*" for scalar multiplication just as for ordinary multiplication of numbers). Maple will perform these operations automatically (if the scalar is an explicit number):

```
> 2*P - [3,2*t];
```

$$[-3 + 2t\sin(t), -4\cos(t)]$$

For scalar multiplication by some other type of expression, you need the "evalm" command to perform the operations.

```
> Q:= t * P - [t, t^2];
```

$$Q := t[t\ \sin(t),\ t - 2\cos(t)] - [t,\ t^2]$$

```
> evalm(Q);
```

$$[t^2\ \sin(t) - t,\ t(t - 2\cos(t)) - t^2]$$

Unfortunately, Maple is not very consistent in its use of vectors, and the result of "evalm" is an "array" structure, not a list. We can convert it to a list using "convert(..., list)". It will be convenient to define a single command to combine "evalm" and "convert(..., list)".

```
> evl:= V -> convert(evalm(V), list);
> Q:= evl(Q);
```

$$Q := [t^2\ \sin(t) - t,\ t(t - 2\cos(t)) - t^2]$$

Let's differentiate **P** with respect to $t$. If we think of our curve as the path of a moving particle with the parameter $t$ representing time, the result will be the velocity vector **V**.

```
> V:= diff(P,t);
```

$$V := [\sin(t) + t\ \cos(t),\ 1 + 2\ \sin(t)]$$

The components of a vector "A" are called "A[1]" and "A[2]". We could define a "len" command to get the length of a vector:

```
> len:= A -> sqrt(A[1]^2 + A[2]^2);
> len(P);
```

$$\sqrt{t^2\ \sin(t)^2 + (t - 2\ \cos(t))^2}$$

Note that I'm not using "length", which is already a standard Maple command with a quite different meaning. It's generally a bad idea to redefine standard Maple commands, and Maple will return an error message if you try to do so.

The dot product of two vectors is performed by the "dotprod" command in the linalg package. Actually, there's a bit of a complication because the ordinary dotprod is set up for use with complex numbers: we really should use dotprod with a third input, the word "orthogonal", to get our version of the dot product.

```
> linalg[dotprod](V,P,orthogonal);
```

$$(\sin(t) + t\ \cos(t))t\ \sin(t) + (1 + 2\sin(t))(t - 2\cos(t))$$

This doesn't look much like the usual mathematical notation **V** • **P**. Fortunately, we can define an operation, which I'll call "&.", that can be used in the same way as "•", between two vectors. Note the back-quotes.

```
> `&.`:= (V,W) -> linalg[dotprod](V,W,orthogonal);
> V &. P;
```

$$(\sin(t) + t\ \cos(t))t\ \sin(t) + (1 + 2\sin(t))(t - 2\cos(t))$$

While I'm at it, I'll redefine "len" in terms of the dot product. This means the command will be useful in later labs as well, when we deal with vectors in three or more dimensions.

```
> len:= V -> sqrt(V &. V);
```

We will often want to produce a **unit vector** by dividing a vector by its length. Since division by the length is really a scalar multiplication, we use "evl" to perform it and convert the result back into a list if necessary.

```
> unit:= V -> evl(V/len(V));
```

The definitions of "`evl`", "`&.`", "`len`", and "`unit`" (plus one more we will meet in Lab 17) are in the file "`dot.def`", which should be available to you. If not, you should save them in a file of that name for future reference.

One of the uses of the dot product is to calculate the angle between two vectors, using the formula $|\mathbf{U}||\mathbf{V}|\cos\theta = \mathbf{U} \bullet \mathbf{V}$. Actually, the `linalg` package contains a command "`angle`" that does this for us.

```
> a:= linalg[angle]([1,2],[3,1]);
```

$$a := \arccos\left(\frac{1}{10}\sqrt{5}\sqrt{10}\right)$$

```
> simplify(a);
```

$$\frac{1}{4}\pi$$

## An Arrowhead for the Curve

■ **EXAMPLE 2** In Figure 4.7, draw an arrowhead showing the direction of the parametric curve[12].

**SOLUTION** Of course, it's not hard to draw a simple arrowhead (two straight lines will do), but the trick here is to aim it in the right direction.

The tip of the arrowhead will be at a point on the curve (the position vector $\mathbf{P}$ for some value of the parameter $t$). Suppose the other two vertices (the barbs) are at $\mathbf{Q}$ and $\mathbf{R}$. The midpoint of these barbs is $(\mathbf{Q}+\mathbf{R})/2$, which is on the central axis of the arrow. Thus $\mathbf{P}-(\mathbf{Q}+\mathbf{R})/2$ is in the direction of the arrow, which should be the direction of the forward tangent to the curve at $\mathbf{P}$. On the other hand, $\mathbf{Q}-\mathbf{R}$ should be perpendicular to this direction.

The velocity vector $\mathbf{V}$ is in the same direction as our arrow, but its length (the speed of a particle tracing out the curve with the parameter $t$ as time) will vary. The tangent vector $\mathbf{T}$ is a unit vector in the same direction.

```
> T:= unit(V);
```

$$T := \left[\frac{\sin(t)+t\cos(t)}{\sqrt{(\sin(t)+t\cos(t))^2+(1+2\sin(t))^2}}, \frac{1+2\sin(t)}{\sqrt{(\sin(t)+t\cos(t))^2+(1+2\sin(t))^2}}\right]$$

Now here's another vector, perpendicular to $\mathbf{T}$ (you can check this using "`&.`"), and of length $1/2$.

```
> U:= [-T[2]/2, T[1]/2];
```

I will take the barbs $\mathbf{Q}$ and $\mathbf{R}$ of the arrowhead to be $\mathbf{P}-h(\mathbf{T}-\mathbf{U})$ and $\mathbf{P}-h(\mathbf{T}+\mathbf{U})$, for then $\mathbf{P}-(\mathbf{Q}+\mathbf{R})/2 = h\mathbf{T}$ and $\mathbf{Q}-\mathbf{R} = 2h\mathbf{U}$. The constant $h$ will be the length of the arrowhead, and also the distance between the barbs. I made $\mathbf{U}$ have length $1/2$ rather than 1 so that, with this choice, the arrowhead has about the right shape. You can, if you wish, alter the shape to your taste by changing the length of $\mathbf{U}$.

```
> Q:= evl(P - h * (T - U));
> R:= evl(P - h * (T + U));
```

Now "`P`", "`Q`", and "`R`" are expressions involving `h` and the parameter `t`. I would like to define a command "`ahead`" so that "`ahead(s,r)`" plots an arrowhead of length $r$ at the point on the curve for $t = s$. To do this we will need to use "`subs`" to substitute the values of "`s`" and "`r`" for "`t`" and "`h`" in these expressions. Then we want to use "`plot`" to plot our points, joining them by straight lines.

---

(12) You may think this is unnecessary, since you can tell the direction from the labels, but there is a purpose to this example: it is intended to show how vectors can be useful for drawing an object of a certain shape pointed in a certain direction.

```
> ahead:= (s,r) -> plot(subs(t=s, h=r, [Q,P,R]));
```

And now to try it out. Remember to use "`constrained`" scaling so the arrowhead is not distorted.

```
> display({pplot, ahead(-1, 0.1)}, scaling=constrained);
```

■

We could also try an animation with a moving arrowhead:

```
> display([ seq(display({ pplot, ahead(n/10, 0.1) }),
>            n=-31..10) ], scaling=constrained, insequence=true);
```

Similar techniques can be used to animate an object of any shape moving along a parametric curve and pointed in varying directions.

## Polar Coordinates

Maple can plot curves in polar coordinates. The form is similar to the ordinary (Cartesian) parametric plot command, with the "`coords=polar`" option at the end.

```
> plot([ sin(t), t, t=0 .. Pi ], coords=polar);
```

This plots the curve $r = \sin(\theta)$ for $0 \leq \theta \leq \pi$. It is really a special case of a "polar parametric plot". Inside the list which forms the first input for "`plot`" we have first the definition of $r$ as a function of parameter $t$, then $\theta$ as a function of $t$, and then the parameter interval. In our case we have $\theta = t$, but in general both $r$ and $\theta$ could depend on the parameter $t$ in an arbitrary way.

After the "`]`" we can also specify $x$ and $y$ intervals for the plot, "constrained" scaling (which is usually a good idea for polar plots), or other options.

Note that "`coords=polar`" won't work in Release 4 for plotting a list of points. You would have to translate them from polar to rectangular coordinates and then plot. It doesn't work for an implicit plot in early versions of Release 5 (this is a `bug` which is corrected in the `EC1 patch` for Windows, and in Release 5.1 for all platforms).

Maple doesn't have built-in commands to transform from polar to rectangular coordinates or vice versa, but it is not hard to do, using the formulas. In finding $\theta$ for a point given in rectangular coordinates, we can use the two-variable version `theta = arctan(y,x)`, which works in all quadrants of the plane.

■ **EXAMPLE 3** The limaçon of Pascal is given in polar coordinates by $r = 1 + 2\cos\theta$. Plot it, and find an equation for it in rectangular coordinates.

**SOLUTION** Since $1 + 2\cos\theta$ is periodic in $\theta$ with period $2\pi$, we need only plot this over an interval of length $2\pi$.

```
> plot([1 + 2*cos(t), t, t=0..2*Pi],
>        coords=polar, scaling=constrained);
```
(Figure 4.8)

To find the equation of this curve in rectangular coordinates, we could do a substitution into the polar equation: $r = \sqrt{x^2 + y^2}$ and $\cos(\theta) = x/r = x/\sqrt{x^2 + y^2}$.

```
> eq:= sqrt(x^2 + y^2) = 1 + 2 * x/sqrt(x^2 + y^2);
```

There's one thing wrong with this equation: $r$ is not always positive in the limaçon's polar equation, so we won't necessarily want the positive square root of $x^2 + y^2$. Let's try it for a typical $\theta$ for which $1 + 2\cos\theta$ is negative: $\theta = 3.0$.

```
> r3:= 1 + 2*cos(3.0);
```

$$r3 := -.979984993$$

```
> subs( x=r3*cos(3.0),
>    y=r3*sin(3.0), eq);
```

$$.9799849930 = 2.979984993$$

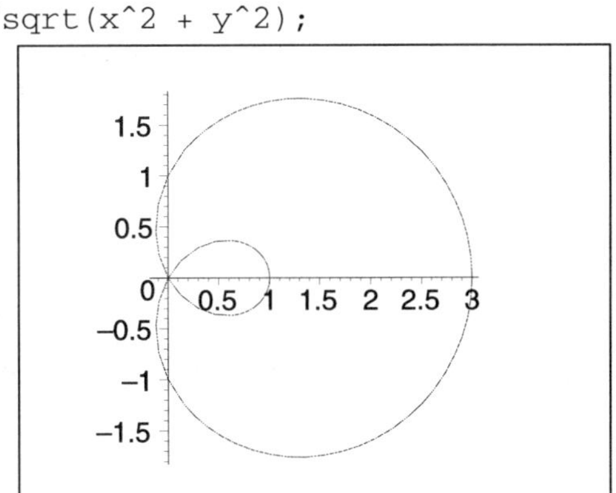


Figure 4.8. The limaçon

How can we remedy the situation? Let's isolate

the square root on one side of the equation and then square both sides, getting an equation that doesn't involve square roots. Then the choice of square root won't matter. We first multiply by the square root and simplify.

```
> eq:= simplify(eq * sqrt(x^2 + y^2));
```

$$eq := x^2 + y^2 = \sqrt{x^2 + y^2} + 2x$$

Now we want to subtract $2x$ from both sides and square them. The simplest way of doing this is with "map", which applies a function to all operands of an expression (in the case of an equation, it applies the function to both sides of the equation).

```
> eq:= map(t -> (t-2*x)^2, eq);
```

$$eq := \left(x^2 + y^2 - 2x\right)^2 = x^2 + y^2$$

Let's see how this one works with $\theta = 3.0$.

```
> subs( x=r3*cos(3.0), y=r3*sin(3.0), eq);
```

$$.9603705873 = .9603705866$$

It's close enough, I think (some roundoff error has affected the last two decimal places). We can also use "implicitplot" to check that "eq" defines the same curve as the polar equation.

Can we get $y$ explicitly as a function of $x$, rather than an implicit equation?

```
> solve(eq, y);
```

$$\frac{1}{2}\sqrt{-4x^2 + 8x + 2 + 2\sqrt{8x+1}},\ -\frac{1}{2}\sqrt{-4x^2 + 8x + 2 + 2\sqrt{8x+1}},$$
$$\frac{1}{2}\sqrt{-4x^2 + 8x + 2 - 2\sqrt{8x+1}},\ -\frac{1}{2}\sqrt{-4x^2 + 8x + 2 - 2\sqrt{8x+1}},$$

Looking at Figure 4.8, we should expect four different solutions, because there are vertical lines (for $x$ in the intervals $(-1/8, 0)$ and $(0, 1)$) that intersect the curve in four points. For other values of $x$ there are fewer real solutions, because some of these involve the square root of a negative number. ■

## EXERCISES

**1. (a)** Plot the parametric curves $x = a\cos t$, $y = b\sin t$ for $0 \le t \le 2\pi$, with various positive values of $a$ and $b$.

**(b)** Show that these curves are ellipses. What is the significance of $a$ and $b$?

**(c)** Does allowing negative values of $a$ or $b$ produce any new curves?

**2. (a)** Plot the parametric curves $x = \cos t$, $y = \sin(kt)$ for $k = 1, 2, 3$.

**(b)** What do you expect the curve $x = \cos t$, $y = \sin(6t)$ to look like? Try it and see.

**(c)** Now try $x = \cos(2t)$, $y = \sin(5t)$. Why doesn't this look like a closed curve? *Hint:* compare what you get at $t$ and $\pi - t$.

**3. (a)** Find the parametric equations of the ellipse $x^2/a^2 + y^2/b^2 = 1$ using as parameter $m$, the slope of the line from $(-a, 0)$ to $(x, y)$. *Hint:* find the equation for $y$ in terms of $m$ and $x$, substitute into the equation of the ellipse, and solve for $x$.

**(b)** Write an expression for the arc length of the ellipse as an integral involving $m$.

**(c)** The more usual parametrization of the ellipse is $x = a\cos t$, $y = b\sin t$, $0 \le t \le 2\pi$. Write $t$ as a function of $m$. *Hint:* use the two-variable form of "arctan".

**4.** If a circle rolls without slipping on the outside of a second, stationary circle, the path of a point on the circumference of the rolling circle is called an **epicycloid**. Suppose the stationary circle, centred at the origin, has radius $a$, the rolling circle has radius $b$, and the point whose motion we follow is initially at $(a, 0)$. A convenient parameter is the angle $\theta$ so that the centre of the rolling circle is at $C = ((a+b)\cos\theta,\ (a+b)\sin\theta)$. The parametric equations of the epicycloid are then

$$x = (a+b)\cos\theta - b\cos((1 + a/b)\theta), \qquad y = (a+b)\sin\theta - b\sin((1 + a/b)\theta)$$

**(a)** Plot an epicycloid with $a = 3$ and $b = 2$ for $0 \leq \theta \leq 2\pi$. On the same graph, plot the stationary circle. Use "constrained" scaling so that the circles look circular. *Hint:* use $\theta$ as a parameter for the circle as well as the epicycloid.

**(b)** Choose some $t$ with $0 < t < \pi/2$. Add to the graph from (a) the rolling circle and the straight line from its centre to the point $P$ for $\theta = t$.

**(c)** For $0 \leq \theta \leq 2\pi$, this epicycloid is not a closed curve. What should $T$ be so that the epicycloid for $0 \leq \theta \leq T$ is a closed curve?

**(d)** Produce an animation of the picture from (c) (i.e., the stationary circle, the epicycloid, the rolling circle at $\theta = t$, and the straight line from the centre of the rolling circle to the point that traverses the epicycloid) for $0 \leq t \leq T$. Thus in viewing the animation you will see the rolling circle roll and the point on its circumference trace out the epicycloid. You may wish to use a single plot of the whole epicycloid in this (which saves some computer time), or else in the frame for $t$ just draw the part of the epicycloid for $0 \leq \theta \leq t$ (this part of the exercise can require a fairly large amount of computer time if done on a slow machine).

**5. (a)** For what values of $a$ and $b$ is the epicycloid for $0 \leq \theta \leq 2\pi$ a closed curve?

**(b)** Find the length of the epicycloid from one cusp to the next, when $a = 2$, $b = 1$.

**(c)** Find the area inside the epicycloid, if $a$ and $b$ are as in part (a).

**6. (a)** What are the velocity and acceleration vectors for the epicycloid, if $\theta$ is taken as time?

**(b)** When do the velocity and acceleration vectors have their maximum magnitudes? Their minimum magnitudes?

**7. (a)** Substitute $x = r\cos\theta$ and $y = r\sin\theta$ into the implicit equation for the limaçon (the last "eq"), and solve for $r$. Do you get the polar equation of the limaçon?

**(b)** Show that $r = -1 + 2\cos(\theta)$ produces the same curve as $r = 1 + 2\cos(\theta)$. How can this be?

**(c)** A line moves in the plane, one point $A$ on it travelling in a circle of radius 1, and the line always passing through a stationary point $B$ on the circumference of that circle. Take $B$ to be the origin, with the centre $C$ of the circle at $(1, 0)$. Consider two points $P$ and $Q$ on the line, on opposite sides of $A$ and at distance 1 from it. Show that the path of $P$ and $Q$ is the limaçon.

**(d)** Produce an animation of the limaçon with the circle and the moving line segment $PQ$.

**8. (a)** Express the length of the limaçon as an integral, simplified as much as possible. Evaluate this approximately.

**(b)** Find the area between the two loops of the limaçon.

**9. (a)** Let $C$ be the parabola $y = x^2$. Suppose a point $P$ moves so that the angle between the two tangents to $C$ that pass through $P$ is a constant $\alpha$. Find parametric equations for the path of the point, using as parameter the $x$ coordinate of one of the points of tangency.

**(b)** What happens when $\alpha = \pi/2$?

**(c)** Plot the curve for $\alpha = \pi/3$.

**10.** There are many interesting-looking, or even beautiful, polar curves. Create one that strikes your fancy and plot it. Suggestion: try making $r$ a periodic function of $\theta$, the period being some multiple of $2\pi$. This can produce a curve that winds around the origin several times.

*CHAPTER 5*

# Series

## LAB 15 INFINITE SERIES

**Goals:** *To use Maple in determining whether an infinite series converges. To learn some methods of finding good approximations to the sum of a convergent series, with Maple's help.*

### Sums in Maple

We begin by looking at Maple's own facilities for handling series. In practice, we would rely on these most of the time. However, in most of this lab we want to learn how to do the analysis ourselves, using Maple as an aid in the calculations.

Maple's "sum" function can be used for both partial sums and infinite series.

```
> sum(1/k^2, k=1..6);
```

$$\frac{5369}{3600}$$

```
> sum(1/k^2, k=1..infinity);
```

$$\frac{1}{6}\pi^2$$

Maple can find the sums of some infinite series in "closed form", i.e., as an exact expression. If it can't do this, "sum" just returns the series.

```
> S:= sum(k^2/(2^k  + 1), k=1 .. infinity);
```

$$S := \sum_{k=1}^{\infty} \frac{k^2}{2^k + 1}$$

Even if there is no exact expression for the sum, "evalf" may be able to approximate the sum.

```
> evalf(S);
```

$$5.412796795$$

On the other hand, there are series for which "evalf" doesn't come up with an answer either.

```
> evalf(sum(2^k/(2^k + k + 1), k=1..infinity));
```

$$\sum_{k=1}^{\infty} \frac{2^k}{2^k + k + 1}$$

Perhaps we were being a bit unfair to Maple here: this series diverges[(1)]. We should have determined that before trying "evalf". But there are also convergent series for which "evalf" fails:

---

(1) We might have hoped that Maple would detect this divergence: sometimes it does detect that a series diverges to $\infty$, but not in this case.

```
> evalf(sum(ln(k+1)/k^3, k=1..infinity));
```

$$\sum_{k=1}^{\infty} \frac{\ln(k+1)}{k^3}$$

There are also divergent series for which Maple returns an answer.

```
> sum((-2)^n, n = 1 .. infinity);
```

$$\frac{-2}{3}$$

This is not really a bug: there is a theory of summation methods for divergent series, according to which this answer is correct. But in the context of elementary calculus, it is wrong. Thus we can't trust "`sum`" without first satisfying ourselves that the series converges.

Note: as we saw in Lab 11, the "`sum`" command has an "inert" form "`Sum`" which returns the sum unevaluated. One way this is useful is that `evalf(Sum(...))` will evaluate the sum numerically without first trying to find an exact answer for it.

## Convergence Tests

Maple does not have a specific command to determine whether a series converges or diverges[(2)]. However, it can be used to apply the standard convergence tests for a series, such as the integral test, comparison test, ratio test, root test, and alternating series test.

■ **EXAMPLE 1** Does $\displaystyle\sum_{k=1}^{\infty} \frac{k^3}{k^4+k+1}$ converge?

**SOLUTION** Will the integral test work? It is easy to see that $f(x)$ is positive and continuous on $(0, \infty)$. Is it non-increasing? Certainly not for small $x$, but perhaps for large $x$. Let's look at the derivative.

```
> f:= k -> k^3/(k^4 + k + 1);
> normal(D(f)(x));
```

$$-\frac{x^2(x^4-2x-3)}{(x^4+x+1)^2}$$

For large $x$, the $x^4$ terms will dominate and so $f'(x)$ will be negative. Certainly this will be true if $x \geq 2$, as then $x^4 \geq 8x \geq 2x+12$ so $x^4 - 2x - 3 \geq 9$. Now, does the improper integral converge?

```
> int(f(x), x=2..infinity);
```

$$\infty$$

Apparently it diverges, and by the integral test so does the sum.

Integration of a complicated function is often time-consuming and not always successful, so the integral test is probably not the best choice. Let's try the comparison test. The usual choice for comparison is a "$p$-series" $\sum k^{-p}$. For a rational function such as $f$ it is easy to see which $p$ to use.

```
> limit(f(k)/k^(-1), k=infinity);
```

$$1$$

Since the limit exists and is not 0, and $\displaystyle\sum_{k=1}^{\infty} k^{-1}$ diverges, our series diverges as well.

This was a "limit comparison test". Could we have used the ordinary comparison test instead? We would want to show that $f(k) \geq A/k$ for all sufficiently large $k$, where $A$ is a positive constant. According to our limit result, we need $A < 1$. Let's try it with $A = 1/2$.

---

(2) Sometimes "`sum`" returns $\infty$ for a divergent series: this is the case in Release 5 for Example 1. But we can't always rely on "`sum`" to do this.

```
> d:=  f(k) - 1/(2*k);
```

$$d := \frac{k^3}{k^4 + k + 1} - \frac{1}{2}\frac{1}{k}$$

```
> normal(d);
```

$$\frac{1}{2}\frac{k^4 - k - 1}{(k^4 + k + 1)\,k}$$

Thus all we need is to ensure $k^4 - k - 1 \geq 0$. This is clearly true if $k \geq 2$. ■

■ **EXAMPLE 2** Does $\displaystyle\sum_{k=1}^{\infty} \frac{k\ln(k^2+1)}{k^3 - k + 1}$ converge?

**SOLUTION** Without the factor $\ln(k^2 + 1)$, we could use $1/k^2$ for a comparison test. That won't work here, though.

```
> f:= k -> k * ln(k^2+1)/(k^3-k+1);
> limit(f(k)/ k^(-2), k=infinity);
```

$$\infty$$

Fortunately, we have some "room" between $1/k$ and $1/k^2$. We could try $k^{-3/2}$.

```
> limit(f(k)/k^(-3/2), k=infinity);
```

$$0$$

Therefore this one converges. ■

■ **EXAMPLE 3** For what values of the constant $a$ does $\displaystyle\sum_{k=1}^{\infty} \frac{a^k(k+1)^{2k}}{(2k+1)!}$ converge?

**SOLUTION** Series involving $k$'th powers or factorials are usually good candidates for the ratio test or the root test. We'll try the ratio test (the root test would give the same answer). Let $b = |a|$ (the test is actually for absolute convergence).

```
> f:= k -> b^k * (k+1)^(2*k)/(2*k+1)!;
> limit(f(k+1)/f(k), k=infinity);
```

$$\frac{1}{4} b\,\mathrm{e}^2$$

We conclude that the series converges absolutely when $|a| < 4/\mathrm{e}^2$, and diverges when $|a| > 4/\mathrm{e}^2$. The ratio test is inconclusive when $|a| = 4/\mathrm{e}^2$. Evidently for this value of $|a|$, the convergence is a more delicate matter.

To deal with the case $|a| = 4/\mathrm{e}^2$, we could try a comparison test with a $p$-series, but it's not clear which $p$ to use. Let's see: if $f(k)$ was $ck^{-p}$, we would have $\ln f(k) = \ln c - p\ln k$, so $\ln(f(k))/\ln(k) \to -p$ as $k \to \infty$.

```
> b:= 4*exp(-2);
> limit(ln(f(k))/ln(k), k=infinity);
> simplify(%); (needed in Release 5 only)
```

$$-\frac{3}{2}$$

From this result, it seems most likely that $f(k)$ behaves like $k^{-3/2}$ as $k \to \infty$. It's not conclusive, but suggestive.

```
> limit(f(k)*k^(3/2), k=infinity);
```

```
> simplify(%); (needed in Release 5 only)
```

$$\frac{1}{4}\frac{(e)^2}{\sqrt{\pi}}$$

Since the limit is finite, we conclude that for this value of $|a|$ the series converges absolutely.

An alternative method would have been to use some $p$ between 1 and $3/2$. We weren't sure (before trying it) whether the comparison with $p = 3/2$ would work. But a comparison with, say, $p = 5/4$ would definitely work: if $\lim\limits_{k\to\infty} \frac{\ln f(k)}{\ln k} < -p$ then $\lim\limits_{k\to\infty} f(k)\, k^p = 0$[3]. ■

## Approximating the Sum

Once we know a series converges, we can try to find a good approximation for the sum $s$ of the series. Suppose we want the absolute value of the error in our approximation to be less than some positive number $\epsilon$. Let $s_n$ be the sum of the first $n$ terms of the series. We would like to find bounds for the "tails" $s - s_n$ of the series, say $A_n \le s - s_n \le B_n$. Then if we can find $n$ so that $B_n - A_n < 2\epsilon$, we can use $s_n^* = s_n + (A_n + B_n)/2$ as an approximation for $s$. This will ensure that $-\epsilon < s - s_n^* < \epsilon$.

■ **EXAMPLE 4** Find $\sum\limits_{k=1}^{\infty} \frac{k}{k^3 - k + 1}$ with an error of less than .00005 in absolute value.

**SOLUTION** We might try using a bound for the tail coming from the integral test. It is easy to check that $f$ is decreasing on $(1, \infty)$. We then have

$$\int_n^\infty f(x)\,dx \le s - s_n \le \int_{n+1}^\infty f(x)\,dx$$

Thus we want $\int_n^{n+1} f(x)\,dx < .0001$. The integral itself is rather complicated, but (again since $f$ is decreasing) $\int_n^{n+1} f(x)\,dx < f(n)$, so it suffices to have $f(n) < .0001$. The approximation will be

$$\begin{aligned} s_n^* &= s_n + \frac{1}{2}\left(\int_n^\infty f(x)\,dx + \int_{n+1}^\infty f(x)\,dx\right) \\ &= s_n + \frac{1}{2}\int_n^{n+1} f(x)\,dx + \int_{n+1}^\infty f(x)\,dx \end{aligned}$$

```
> f:= k -> k/(k^3 - k + 1);
> fsolve(f(n)=.0001, n, n=1..infinity);
```

100.0049499

Thus $n = 101$ should do it.

```
> evalf(Sum(f(k), k=1..101)
>   + 1/2 * Int(f(x), x=101 .. 102)
>             + Int(f(x), x=102 .. infinity));
```

1.695667180

For comparison, here is Maple's answer for the sum, which should be accurate to about 10 digits.

```
> evalf(Sum(f(k), k=1..infinity));
```

1.695666862

It seems that we've actually achieved a bit more accuracy than we expected: the actual error is about −.0000003. ■

---

[3] This is not very hard to prove, but we won't do it here.

**■ EXAMPLE 5** Find $\sum_{k=1}^{\infty} \frac{k^3}{2^k}$ with an error of less than .00005 in absolute value.

**SOLUTION** This is a series for which the ratio test works:

```
> f:= k -> k^3 / 2^k;
> limit(f(k+1)/f(k), k=infinity);
```

$$\frac{1}{2}$$

We conclude that for any $r > 1/2$ we will have $f(k+1) \le r f(k)$ when $k$ is sufficiently large. If this is true for $k \ge n$, we will have $f(k) \le f(n)\, r^{k-n}$ for $k > n$, and thus

$$s - s_n \le f(n) \sum_{k=n+1}^{\infty} r^{k-n} = \frac{f(n)\, r}{1-r}$$

(assuming $r < 1$ to make the series $\sum_{k=n+1}^{\infty} r^{k-n}$ converge). To keep things simple, I'll just use 0 as a lower bound for $s - s_n$. Thus we want to have $f(n)\, r/(1-r) \le .0001$, together with $f(k+1) \le r f(k)$ for $k \ge n$.

```
> expand(f(k+1)/f(k));
```

$$\frac{1}{2} + \frac{3}{2}\frac{1}{k} + \frac{3}{2}\frac{1}{k^2} + \frac{1}{2}\frac{1}{k^3}$$

This is obviously a decreasing function of $k$. Thus to have $f(k+1) \le r f(k)$ for $k \ge n$ it will suffice to have it true for $k = n$. We may as well take $r = f(n+1)/f(n)$. Then we need $r < 1$ and $f(n)r/(1-r) \le .0001$.

```
> r:= n -> f(n+1)/f(n);
> g:= unapply(f(n) * r(n)/(1 - r(n)), n);
```

Now where is $g(n) = .0001$? Let's try a few values.

```
> seq(evalf(g(n)), n=1..10);
```

$$-.6666666667,\ -4.909090909,\ -21.60000000,\ 166.6666667,\ 24.81617647,$$
$$13.00702247,\ 7.885057471,\ 4.942372881,\ 3.108795033,\ 1.942906857$$

It's nowhere near .0001 yet, but starting to decrease. Eventually it should decrease by about half from one $n$ to the next. We have a factor of about 1/20000 to go, which would mean about 15 halvings.

```
> seq(evalf(g(n)), n=25..30);
```

$$.0005985419618, .0003332492722,\ .0001848659656,\ .0001022020609,$$
$$.00005632093942,\ .00003094369169$$

So $n = 29$ will do it.

```
> evalf(Sum(f(n), n=1..29) + 1/2 * f(29) * r(29)/(1-r(29)));
```

$$25.99997231$$

Does that look suspiciously close to 26? Maple can actually sum this series exactly.

```
> sum(f(n), n=1..infinity);
```

$$26$$

■

**■ EXAMPLE 6** Find $\displaystyle\sum_{k=0}^{\infty} \frac{(-1)^k}{k^2+1}$ with an error of less than $10^{-10}$ in absolute value.

**SOLUTION** This is an alternating series, but all we can conclude from the alternating series test is that $s - s_n$ is between 0 and the next term $a_{n+1}$. What $n$ would we need for $|a_{n+1}| < 10^{-10}/2$?

```
> evalf(solve(1/((n+1)^2+1) = 1/2 * 10^(-10), n));
```

$$-141422.3562,\ 141420.3562$$

Calculating 141 421 terms of the series might not be very practical, so we look for alternatives. The first idea is to group the terms into pairs:

$$\left(\frac{1}{0^2+1} - \frac{1}{1^2+1}\right) + \left(\frac{1}{2^2+1} - \frac{1}{3^2+1}\right) + \left(\frac{1}{4^2+1} - \frac{1}{5^2+1}\right) + \dots$$
$$= \sum_{k=0}^{\infty} \left(\frac{1}{(2k)^2+1} - \frac{1}{(2k+1)^2+1}\right)$$

Now we have a sum of positive terms, to which we could apply the bounds coming from the integral test. Instead, we will try a different method: "improving the convergence" of the series. The idea is this: we write our series $\sum a_n$ as $\sum b_n + \sum(a_n - b_n)$, where $\sum b_n$ is a series whose partial sums we can calculate exactly (a telescoping series is ideal), and $|a_n - b_n|$ goes to 0 much faster than does $|a_n|$ as $n \to \infty$.

What should we take for $b_n$? The idea is to match as closely as possible the behaviour of $a_n$ as $n \to \infty$. Let's take a closer look at our $a_n$.

```
> a:= n -> 1/((2*n)^2+1) - 1/((2*n+1)^2+1);
> simplify(a(n));
```

$$\frac{1}{2}\frac{4n+1}{(4n^2+1)(2n^2+2n+1)}$$

For large $n$, the dominant terms are the $4n$ in the numerator and the $4n^2$ and $2n^2$ in the denominator, so this should be approximately $1/(4n^3)$. Actually, Maple's "`asympt`" command can do this calculation with even more details (we will see more about how this works in Lab 16).

```
> ap:= asympt(a(n), n);
```

$$ap := \frac{1}{4}\frac{1}{n^3} - \frac{3}{16}\frac{1}{n^4} + O\left(\frac{1}{n^6}\right)$$

So in addition to the main term $1/(4n^3)$ that we already know, there is another term $-3/(16n^4)$ (which will be much smaller in absolute value than the first one when $n$ is large), and the rest is bounded by some constant times $1/n^6$. We could actually get terms in more negative powers of $n$: "`asympt(a(n), n, p)`" would give us all terms up to the one in $1/n^{p-1}$. But this will be sufficient for our purposes.

One handy family of telescoping series is $\displaystyle\sum_n \left(n^{-p} - (n+1)^{-p}\right)$. We might try a combination of these. The main term in $n^{-p} - (n+1)^{-p}$ would be a constant times $n^{-p-1}$. In order to match our $a_n$ as well as possible, we could use $p = 2$, 3, and 4.

```
> b:= n -> A/n^2 + B/n^3 + C/n^4
>          - (A/(n+1)^2 + B/(n+1)^3 + C/(n+1)^4);
> bp:= asympt(b(n), n);
```

$$bp := 2\frac{A}{n^3} + \frac{-3A+3B}{n^4} + \frac{4A-6B+4C}{n^5} + O\left(\frac{1}{n^6}\right)$$

Now we want to choose $A$, $B$, and $C$ to match this to the terms for $a(n)$. We have three equations to solve, corresponding to the coefficients of $n^{-3}$, $n^{-4}$, and $n^{-5}$ respectively. We can use "`coeff`" to extract these coefficients, and then "`solve`" to find $A$, $B$, and $C$. Note that because it's a set (enclosed in "`{}`") rather than a list, the equations in "`eqs`" may be shown in any order. After finding the solution, we assign these values to `A`, `B`, and `C` using the "`assign`" command.

```
> eqs:= { seq(coeff(ap-bp, n, -p)=0, p=3..5) };
```

$$eqs := \left\{-4A + 6B - 4C = 0,\ -\frac{3}{16} + 3A - 3B = 0,\ \frac{1}{4} - 2A = 0\right\}$$

```
> sol:= solve(eqs, {A,B,C});
```

$$sol := \left\{B = \frac{1}{16},\ C = \frac{-1}{32},\ A = \frac{1}{8}\right\}$$

```
> assign(sol);
```

Let's look at the difference between $a(n)$ and $b(n)$. It should be very small, but how small?

```
> c:= unapply(simplify(a(n)-b(n)), n);
```

$$c := n \to -\frac{1}{32}\,\frac{-4\,n + 16\,n^3 + 70\,n^4 + 60\,n^6 + 108\,n^5 - 1 - 4\,n^2}{(4\,n^2+1)\,(2\,n^2+2\,n+1)\,n^4\,(n+1)^4}$$

It is easy to see that $c(n) < 0$ for $n \geq 1$. Also, the numerator of $c(n)$ is a polynomial of degree 6 while the denominator has degree 12, so $c(n) = O\left(n^{-6}\right)$ as we wished. Moreover $n^6 c(n)$ should have a limit as $n \to \infty$.

```
> limit(n^6 * c(n), n=infinity);
```

$$\frac{-15}{64}$$

We'd prefer to have a bound, rather than just a limit. We could plot $n^6 c(n)$:

```
> plot(n^6 * c(n), n=1..100);
```

(Figure 5.1)

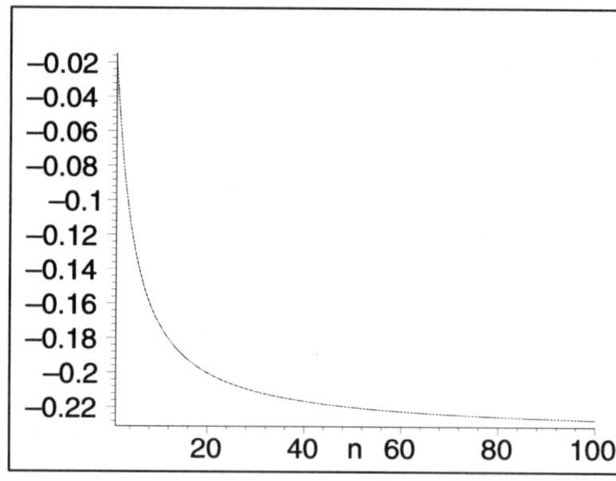


Figure 5.1. $n^6 c(n)$

It looks like $0 \geq n^6 c(n) \geq -15/64$ for all $n$ (although from the plot alone we couldn't be sure that this was true for large $n$). Alternatively, we can look at the derivative of $n^6 c(n)$.

```
> simplify(diff(n^6 * c(n), n));
```

$$-\frac{1}{16}n\left(-6\,n - 1 + 768\,n^{10} + 2296\,n^9 + 3400\,n^8 + 3216\,n^7 + 326\,n^4 + 1054\,n^5 + 40\,n^3\right.$$
$$\left.+2168\,n^6 - 11\,n^2\right) / \left((4\,n^2+1)^2\,(2\,n^2+2\,n+1)^2\,(n+1)^5\right)$$

It is easy to see that this is negative (i.e., $n^6 c(n)$ is decreasing) for $n \geq 1$. Thus for any $n > 1$ we have

$$0 > \sum_{k=n+1}^{\infty} c(k) \geq -\frac{15}{64}\sum_{k=n+1}^{\infty} k^{-6} \geq -\frac{15}{64}\int_n^{\infty}\frac{dx}{x^6} = -\frac{3}{64n^5}$$

To ensure the absolute value of the error is less than $10^{-10}$, we want $(1/2)(3/64)n^{-5} < 10^{-10}$.

```
> evalf((1/2 * 3/64 * 10^10)^(1/5));
```

$$47.20437556$$

Thus it should work for $n = 48$. In order to avoid problems with roundoff error, we should use more than 10 digits when calculating the final result. This will consist of the sum of the following terms: $a(0)$ (needed because $b(0)$ is not defined), $\sum_{k=1}^{\infty} b(k) = A + B + C$, $\sum_{k=1}^{48} c(k)$, and the average of our bounds for $\sum_{k=49}^{\infty} c(k)$, namely $-(1/2)\,(3/64)\,48^{-5}$.

```
> Digits:= 20;
> ans:= evalf(a(0) + A + B + C + sum(c(k), k=1..48)
>         - 1/2 * 3/64 * 48^(-5));
```

$$ans := .63601452756440015465$$

To check our result, here are Maple's answer, and the difference between the two.

```
> mapleans:= evalf(Sum((-1)^k/(k^2+1), k=0..infinity));
```

$$mapleans := .63601452749106658147$$

```
> ans - mapleans ;
```

$$.7333357318\ 10^{-10}$$

■

## EXERCISES

**1.** Find the partial sums $s_n$ of the series $\sum_{k=1}^{\infty} \frac{1}{k^3}$ for $n = 10, 20, 40$, and $80$. Examine the differences between these. What trends do you notice? Assuming these trends continue, what can you say about the convergence of the series and about its sum?

**2.** Test the following series for convergence.

**(a)** $\sum_{n=1}^{\infty} \frac{1}{\sqrt{2n^2 - 1}}$  **(b)** $\sum_{n=1}^{\infty} \frac{n}{2^n - n}$

**(c)** $\sum_{n=1}^{\infty} \frac{n^n}{(n!)^2}$  **(d)** $\sum_{n=1}^{\infty} \frac{(-1)^n}{\ln(n+1)}$

**3. (a)** Suppose $a_k > 0$ for all $k$ and $\lim_{k\to\infty} \frac{\ln(k a_k)}{\ln(\ln k)} = p$ (possibly $+\infty$ or $-\infty$). Show that $\sum_{k=1}^{\infty} a_k$ diverges if $p > -1$ and converges if $p < -1$.

**(b)** Use this "loglog test" to determine whether $\sum_{k=1}^{\infty} \frac{k^{2k}(k^2-1)!}{(k^2+k)!}$ converges.

**(c)** Do the same for $\sum_{k=1}^{\infty} \frac{k^{2k}\sqrt{(2k^2)!}}{2^{k^2}(k^2+k)!}$

**(d)** Construct, using powers, logarithms and/or factorials, a series $\sum a_k$ with $a_k > 0$ for which this test is inconclusive.

**4.** For what values of the constant $p$ does $\sum_{n=1}^{\infty} \frac{(2n)!\,n^p}{4^n (n!)^2}$ converge? *Hint:* you can use "`asympt`" on $a_n/n^p$ if it doesn't work on $a_n$ directly.

**5. (a)** Does the series $\sum_{n=2}^{\infty} \frac{1}{(-1)^n \sqrt{n} + 1}$ converge? *Hint:* this is not an alternating series. Group the terms into pairs.

**(b)** Now try it for $\sum_{n=2}^{\infty} \frac{1}{(-1)^n\, n + 1}$

**6. (a)** Let $a_0 = 2$ and $a_{n+1} = 2 - \sqrt{4 - a_n}$ for $n \geq 0$. Show that $\sum_{n=0}^{\infty} a_n$ converges. *Hint:* first show that $a_n$ is decreasing. The ratio test works well if you express $a_n$ in terms of $a_{n+1}$ rather than the reverse.

**(b)** Find the sum to within .0001.

**7.** Show that $\sum_{n=1}^{\infty} \frac{3n^2 - 2n - 11}{n^4 + 4n^3 + 6n^2 + 4n + 5}$ is a telescoping series, and find its sum. *Hint:* use partial fractions.

**8.** Use the method of improving convergence to find the sum of the series $\sum_{n=1}^{\infty} \ln(n)/n^3$ to within .0001. You will need to use the telescoping series $\sum \left(\ln(n)\, n^{-p} - \ln(n+1)\,(n+1)^{-p}\right)$ as well as $\sum \left(n^{-p} - (n+1)^{-p}\right)$, and match the coefficients of $\ln(n)\, n^{-p}$ and of $n^{-p}$ separately. Note: if `S` is a sum of terms, some containing $\ln(n)$ and some not, a quick way to extract only the terms that don't contain $\ln(n)$ is `subs(ln(n)=0,S)`.

**9.** The harmonic series $\sum_{n=1}^{\infty} \frac{1}{n}$ diverges. This exercise develops a way of estimating its partial sums $s_n$.

**(a)** Show that $g(x) = 1/x - \ln((x+1)/x)$ is a decreasing function of $x$ for $x > 0$. Conclude that for integers $0 < A < B$, $\int_A^B g(x)\, dx > \sum_{n=A+1}^{B} g(n) > \int_{A+1}^{B+1} g(x)\, dx$.

**(b)** Using the fact that $\sum \ln((n+1)/n)$ is a telescoping series, find upper and lower bounds for $\sum_{n=A+1}^{B} \frac{1}{n}$ where $1 < A < B$.

**(c)** Find the least positive integer $N$ so that the partial sum $s_N > 7$. *Hint:* take $A = 10$, and calculate $s_{10}$ explicitly. Use "`fsolve`" to find $B$ so the upper bound of (b) is $7 - s_{10}$.

# LAB 16 POWER SERIES

**Goals:** *To learn how to obtain, use, and manipulate Taylor series in Maple. To use graphical methods to examine the convergence of Taylor series.*

## Taylor Series

Maple has the "`taylor`" command for finding a Taylor series of an expression.

```
> S:= taylor(exp(x), x=c, 5);
```

$$S := e^c + e^c(x-c) + \frac{1}{2}e^c(x-c)^2 + \frac{1}{6}e^c(x-c)^3 + \frac{1}{24}e^c(x-c)^4 + O\left((x-c)^5\right)$$

The first input to "`taylor`" is the expression for which you want a series, the second tells Maple that $x$ is the variable and you want a series about $x = c$, and the third tells it to do its series calculations using powers up to, but not including, 5. The $O\left((x-c)^5\right)$ is a reminder of the terms not shown, constituting the rest of the series. Indeed, according to Taylor's Theorem, this remainder is $O\left((x-c)^5\right)$, i.e., its absolute value is less than some constant times $|x-c|^5$ for $x$ in some interval around $c$.

The third input was actually optional. If you don't include it, Maple will use as a default the value of a variable called "`Order`". This is initially set to 6, but you may change it by assigning some other number to "`Order`".

Although the result of "`taylor`" looks like an ordinary polynomial with "$O\left((x-c)^5\right)$" tacked on, it's really something a bit different: a special "series" data structure. For some purposes, such as computing an approximation to the expression at a particular value of $x$, we need the Taylor polynomial. You can't just subtract the "$O(\ldots)$" term, you must use "`convert`".

```
> convert(S, polynom);
```

$$e^c + e^c(x-c) + \frac{1}{2}e^c(x-c)^2 + \frac{1}{6}e^c(x-c)^3 + \frac{1}{24}e^c(x-c)^4$$

Not all expressions have Taylor series. Maple will return an error message if you ask it for a Taylor series that doesn't exist.

```
> taylor(sqrt(x), x=0);
Error, does not have a taylor expansion, try series()
```

## Convergence of Maclaurin Series

In this section we explore graphically the way the Maclaurin series of a function converges to the function. Previously we have studied the convergence of a series of numbers to its limit, but here we have a series of functions converging to a function. We first set up some machinery for conveniently finding Maclaurin polynomials and evaluating them at a point.

```
> P:= (n,t) ->
>          subs(x=t, convert(taylor(f(x), x=0, n+1), polynom));
> R:= (n,t) -> f(t) - P(n,t);
```

What will this do? First "`taylor`" finds the Maclaurin series to $O\left(x^{n+1}\right)$ for the function $f$ (whatever `f` will be defined to be at the time), using the variable `x`. Be sure that no value is assigned to `x` when you use `P`. Then "`convert(..., polynom)`" removes the $O\left(x^{n+1}\right)$ so that we have a Maclaurin polynomial of degree $n$. Finally, "`subs(x=t, ...)`" substitutes `t` for `x` in that polynomial. This allows us to evaluate the Maclaurin polynomial at any point. So "`P(n,t)`" will be the Maclaurin polynomial $P_n(t)$ and "`R(n,t)`" will be the error $R_n(t)$. Note that "n" must be a number, not a symbolic variable.

```
> P(3,x);
```

$$f(0) + D(f)(0)\,x + \frac{1}{2}D^{(2)}(f)(0)\,x^2 + \frac{1}{6}D^{(3)}(f)(0)\,x^3$$

■ **EXAMPLE 1** Study the convergence of the Maclaurin series for the function exp.

**SOLUTION** We plot $\exp(x)$ and its Maclaurin polynomials $P_n(x)$ of degrees 1 through 10.

```
> f:= exp;
> plot({ f(x), seq(P(n,x), n=1..10) }, x=-6..2, y=-2..8);
```

(Figure 5.2)

It's rather hard to see what's happening for $x > 0$ because $\exp(x)$ rises so quickly, but I think you can see some of what happens, and for $x < 0$ the behaviour is easy to see. Each Maclaurin polynomial is quite close to $f(x)$ on some interval around 0, so close that in many cases you can't tell the difference between them, but eventually moves away from $f(x)$ as $x$ gets farther away from 0. For any given $x$, $P_n(x)$ to $f(x)$ as $n \to \infty$. The larger $n$ is, the larger is the interval on which $P_n(x)$ is close to $f(x)$.

We can produce an animation of the Maclaurin polynomials approaching $f(x)$. It's simpler to use "display" here rather than "animate".

```
> with(plots, display);
> p1:= plot(f(x), x=-6..2, y=-2..8, colour=blue):
> display([ seq(display({ p1, plot(P(n,x), x=-6..2, y=-2..8,
>                colour=red)}), n=1..12)], insequence=true);
```

A better picture of what is happening, especially for $x > 0$, can be obtained by plotting the remainders $R_n(x)$.

```
> plot({ seq(R(n,x), n=1..10) }, x=-6..6, y=-3..3);
```

(Figure 5.3)

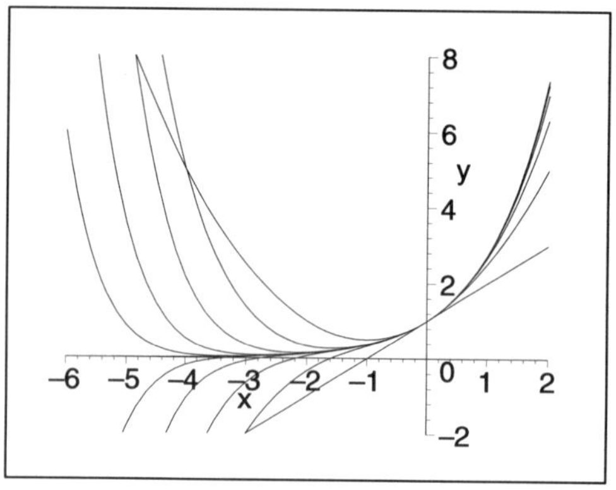


Figure 5.2. Convergence to exp

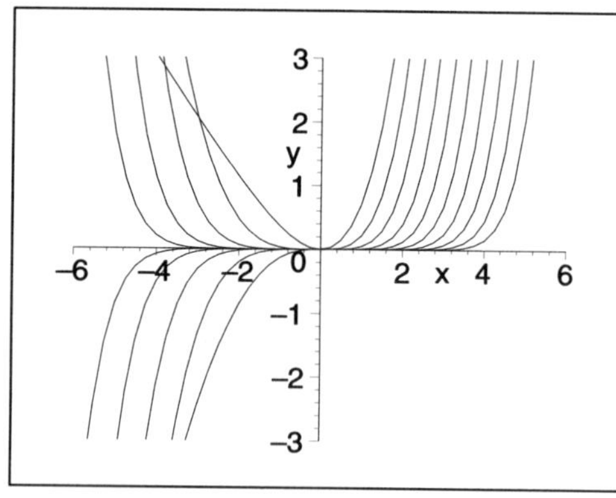


Figure 5.3. Remainders for exp

The result is rather striking. On the right we have a family of curves that appear to be approximately parallel and separated from each other horizontally by equal distances. On the left, except for the curve for $n = 1$ which is rather different, we have a similar situation with curves going up alternating with curves going down. Let's look at the points on the right where the curves intersect a horizontal line, say $y = 1$. The $x$ values will be $c_n$ where $0 < c_n < \infty$ and $R_n(c_n) = 1$. We'll find them using "fsolve" and put them into a table. If the curves are separated horizontally by (approximately) equal distances, the differences between successive $c_n$ should be approximately constant.

```
> for nn from 1 to 10 do
>       c[nn]:= fsolve(R(nn,x) = 1, x, x=0..infinity) od;
```

$$c_1 = 1.146193221$$

...

$$c_{10} = 4.703272391$$

Are the differences between these approximately constant?

```
> seq(c[n+1] - c[n], n = 1 .. 9);
```

$$.421926771, .408039750, .400018633, .394750158, .391007726,$$
$$.388204333, .386021917, .384272467, .382837415$$

Yes, the differences do seem to be rapidly approaching a limit. A full analysis of the reasons for this limit would be difficult. However, we can get a rough estimate of $c_n$ from the Lagrange formula for the remainder:

$$R_n(x) = \frac{f^{(n+1)}(X)}{(n+1)!} x^{n+1}$$

for some $X$ between 0 and $x$. In our case we have

$$1 = R_n(c_n) = \frac{\exp(X_n)}{(n+1)!} c_n^{n+1}$$

Since $X_n > 0$ we obtain $1 \geq c_n^{n+1}/(n+1)!$, so $c_n \leq ((n+1)!)^{1/(n+1)}$. Let's define the latter as "`cp(n)`".

```
> cp:= n -> ((n+1)!)^(1/(n+1));
```

Now we compare these bounds to the actual $c_n$ values we calculated.

```
> seq(evalf(cp(n) - c[n]), n=1..10);
```

$$.268020341, .249000601, .237204097, .228992710, .222866633,$$
$$.218078900, .214210008, .211003765, .208293712, .205966389$$

It appears that the difference between $c_n$ and its upper bound approaches a constant. For the bound, we can actually find a limit for the difference from one value to the next.

```
> L := limit(cp(n+1) - cp(n), n = infinity);
```

$$L := e^{(-1)}$$

```
> evalf(L);
```

$$.3678794412$$

Indeed, from Maple's calculations above it is plausible that $c_{n+1} - c_n$ approaches this number. ■

■ **EXAMPLE 2** Study the convergence of the Maclaurin series for the function arctan.

**SOLUTION** We don't need the $R_n(x)$ for even $n$: arctan is an odd function, so its Maclaurin series only involves odd powers of $x$.

```
> f:= arctan;
> plot({ seq(R(2*n-1,x), n=1..6) }, x=-2..2, y=-1..1); (Figure 5.4)
```

The picture is much different from the previous example. For $x$ between $-1$ and 1, the remainders converge to 0, quickly when $x$ is well inside the interval, and slowly when $x$ is near $\pm 1$. In particular the remainders converge to 0 at $x = \pm 1$ (but very slowly). Outside the interval $[-1, 1]$, the remainders do not converge to 0, and in fact their absolute values go to $\infty$. This is because the radius of convergence of the series is 1.

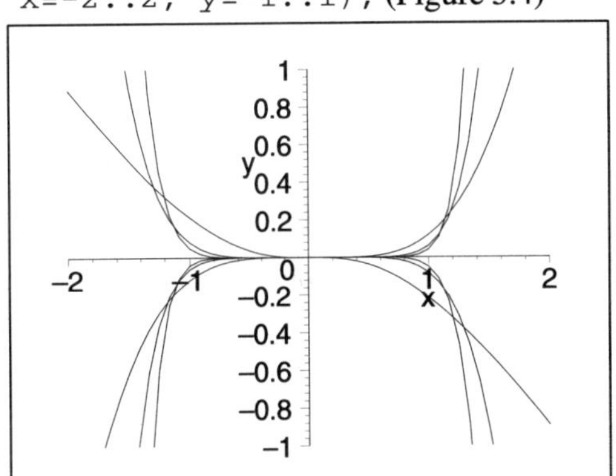


Figure 5.4. Remainders for arctan

We might take a closer look, using an animation. We'll use the absolute value of the remainder, rather than the remainder itself (which alternates in sign).

```
> display([ seq( plot(abs(R(2*n-1,x)), x=.8..1.2, y=0..1),
>                 n=1..10) ], insequence=true);
```

■

## Manipulating Power Series

Various operations can be done to obtain new series from old series: the basic operations of arithmetic, as well as substitution, differentiation, and integration. In this section we will look at some examples of these operations. Most of the operations are actually done with the Taylor polynomials rather than Maple's "series" structures.

■ **EXAMPLE 3** Starting with "well-known" series, obtain the degree 10 Taylor polynomial for $\ln(1+x^2)\cos(\cos(x)))$ in powers of $x$.

**SOLUTION** We'll begin with the $\ln(1+x^2)$. This can be obtained from $\ln(1+t)$ by substituting $t = x^2$. You may already know the series for $\ln(1+t)$, but we'll derive it from the series for its derivative, $1/(1+t)$.

```
> S:= taylor(1/(1+t), t=0);
```

$$S := 1 - t + t^2 - t^3 + t^4 - t^5 + O\left(t^6\right)$$

Now we ask Maple for an antiderivative. The one we want will have no constant term, because $\ln(1+0) = 0$.

```
> S:= int(S, t);
```

$$S := t - \frac{1}{2}t^2 + \frac{1}{3}t^3 - \frac{1}{4}t^4 + \frac{1}{5}t^5 - \frac{1}{6}t^6 + O\left(t^7\right)$$

We can't just substitute $t = x^2$ in the series as it is. What we must do is convert it to a polynomial, then substitute. We'll save the result for future use, calling it "`s1`".

```
> s1:= subs(t=x^2, convert(S, polynom));
```

$$s1 := x^2 - \frac{1}{2}x^4 + \frac{1}{3}x^6 - \frac{1}{4}x^8 + \frac{1}{5}x^{10} - \frac{1}{6}x^{12}$$

Note that these terms of the series are more than enough: the remainder will be $O\left(x^{14}\right)$.

Now for the series for $\cos(\cos(x))$. We know the series for $\cos(x)$. We want this up to the $x^{10}$ term. We'll save it as "`s2`".

```
> s2:= taylor(cos(x), x=0, 11);
```

$$s2 := 1 - \frac{1}{2}x^2 + \frac{1}{24}x^4 - \frac{1}{720}x^6 + \frac{1}{40320}x^8 - \frac{1}{3628800}x^{10} + O\left(x^{11}\right)$$

Now to get $\cos(\cos(x))$, we want to substitute this into a series for $\cos(t)$. But it shouldn't be the series in powers of $t$: because $\cos(x)$ is not near 0 when $x$ is near 0, $O\left((\cos(x))^n\right)$ would not be small. Instead, we need to use the series for $\cos(t)$ in powers of $t-1$, because $\cos(0) = 1$.

```
> s3:= taylor(cos(t), t=1);
```

$$s3 := \cos(1) - \sin(1)(t-1) - \frac{1}{2}\cos(1)(t-1)^2 + \frac{1}{6}\sin(1)(t-1)^3 + \frac{1}{24}\cos(1)(t-1)^4 - \frac{1}{120}\sin(1)(t-1)^5 + O\left((t-1)^6\right)$$

And now we substitute $t = s2$ in this series (after converting both to polynomials). Note that $t-1$ will then be $-x^2/2 + \ldots$, so that $O\left((t-1)^6\right) = O\left(x^{12}\right)$. Now `s2` is missing terms of order $x^{11}$ and up, which will also contribute to the error term. But again all these contributions will be $O(x^{11})$ or more, so the result below is $\cos(\cos x)$ correct to $O\left(x^{11}\right)$.

```
> s3:= subs(t=convert(s2, polynom), convert(s3, polynom));
```

$$s3 := \cos(1) - \sin(1)\%1 - \frac{1}{2}\cos(1)\%1^2 + \frac{1}{6}\sin(1)\%1^3 + \frac{1}{24}\cos(1)\%1^4 - \frac{1}{120}\sin(1)\%1^5$$

$$\%1 := -\frac{1}{2}x^2 + \frac{1}{24}x^4 - \frac{1}{720}x^6 + \frac{1}{40320}x^8 - \frac{1}{3628800}x^{10}$$

We could use "expand" on this, but the result will be a mess, with many terms in high powers of $x$. Instead, we simply use "taylor" to extract the terms we're interested in.

```
s3:= taylor(s3, x=0, 11);
```

$$s3 := \cos(1) + \frac{1}{2}\sin(1)\,x^2 + \left(-\frac{1}{24}\sin(1) - \frac{1}{8}\cos(1)\right)x^4 + \left(-\frac{7}{360}\sin(1) + \frac{1}{48}\cos(1)\right)x^6 + \left(\frac{209}{40320}\sin(1) + \frac{1}{960}\cos(1)\right)x^8 + \left(-\frac{1259}{3628800}\sin(1) - \frac{193}{241920}\cos(1)\right)x^{10} + O\left(x^{12}\right)$$

After converting this to a polynomial, we multiply it by s1, which was the Taylor polynomial for $\ln(1+x^2)$. Then we extract the degree 10 Taylor polynomial of the product.

```
> taylor(convert(s3, polynom) * s1, x=0, 11);
```

$$\cos(1)\,x^2 + \left(-\frac{1}{2}\cos(1) + \frac{1}{2}\sin(1)\right)x^4 + \left(-\frac{7}{24}\sin(1) + \frac{5}{24}\cos(1)\right)x^6 + \left(\frac{121}{720}\sin(1) - \frac{1}{6}\cos(1)\right)x^8 + \left(-\frac{4999}{40320}\sin(1) + \frac{143}{960}\cos(1)\right)x^{10} + O\left(x^{12}\right)$$ ■

■ **EXAMPLE 4** Find the first six terms of the Taylor series for $f(x)$ in powers of $x-1$, if $y = f(x)$ satisfies the equation $x^4 + y^4 = 2xy$ with $y(1) = 1$.

**SOLUTION** This is perhaps a bit more interesting than finding the series for an explicitly given function, which Maple can do very well by itself. Here we only know the function by an implicit equation. How do we know it even has a Taylor series? There is a theoretical justification, which we will omit: we simply assume the function is **analytic** at 1, i.e., in some interval around 1 it is the sum of a Taylor series in powers of $x-1$, and calculate what that series must be. Note, by the way, that $x = 1, y = 1$ does satisfy the equation.

```
> ys:= 1 + sum(a[n] * (x-1)^n, n=1..5);
```

$$ys := 1 + a_1(x-1) + a_2(x-1)^2 + a_3(x-1)^3 + a_4(x-1)^4 + a_5(x-1)^5$$

The only coefficient we know so far in the Taylor series for $y$ is the constant term, $y(1) = 1$. To find the other coefficients, we substitute "ys" for $y$ in the equation for $x$ and $y$.

```
> eq:= x^4 + ys^4 = 2 * x * ys;
```

$$eq := x^4 + \left(1 + a_1(x-1) + a_2(x-1)^2 + a_3(x-1)^3 + +a_4(x-1)^4 + a_5(x-1)^5\right)^4 = 2x\left(1 + a_1(x-1) + a_2(x-1)^2 + a_3(x-1)^3 + +a_4(x-1)^4 + a_5(x-1)^5\right)$$

Because ys is missing the terms in $(x-1)^6$ and higher, this equation is not really correct. Terms in $(x-1)^6$ and higher may be different on one side than on the other. But the terms in lower powers should be the same. We can therefore use "taylor" to order 6 on the difference of the two sides, and get a series whose terms (except for the $O(\ldots)$) should all be 0 if the $a_n$ are correct.

```
> g:= taylor(op(1,eq) - op(2,eq), x=1, 6);
```

$$g := (2a_1 + 2)(x-1) + \left(6a_1^2 + 2a_2 + 6 - 2a_1\right)(x-1)^2 + \left(4 + 4\left(a_1^2 + 2a_2\right)a_1 + 4a_2a_1 + 2a_3 - 2a_2\right)(x-1)^3 + \left(1 + 4a_3a_1 + 2a_2^2 + 2a_4 + 4\left(2a_2a_1 + 2a_3\right)a_1 + \left(a_1^2 + 2a_2\right)^2 - 2a_3\right)(x-1)^4 + \left(2a_5 + 4a_1a_4 + 4a_2a_3 + 4a_1\left(2a_3a_1 + a_2^2 + 2a_4\right) + 2\left(a_1^2 + 2a_2\right)\left(2a_2a_1 + 2a_3\right) - 2a_4\right)(x-1)^5 + O\left((x-1)^6\right)$$

Each coefficient here gives us an equation in the $a_n$'s, since each coefficient of $g$ must be 0. We can use "coeff" to extract the coefficients. Here, for example, is the coefficient of $(x-1)^2$:

```
> coeff(g, x-1, 2);
```

$$6a_1^2 + 2a_2 + 6 - 2a_1$$

And here is the whole set of equations:

```
> eqs:= { seq( coeff(g, x-1, n) = 0, n=1..5) };
```

$$eqs := \{2\,a_1 + 2 = 0,\ 6\,a_1^2 + 2\,a_2 + 6 - 2\,a_1 = 0,$$
$$4 + 4\,(a_1^2 + 2\,a_2)\,a_1 + 4\,a_2 a_1 + 2\,a_3 - 2\,a_2 = 0,$$
$$1 + 4\,a_3 a_1 + 2\,a_2^2 + 2\,a_4 + 4\,(2\,a_2 a_1 + 2\,a_3)\,a_1 + (a_1^2 + 2\,a_2)^2 - 2\,a_3 = 0,$$
$$2\,a_5 + 4\,a_1 a_4 + 4\,a_2 a_3 + 4\,a_1\,(2\,a_3 a_1 + a_2^2 + 2\,a_4) + 2\,(a_1^2 + 2\,a_2)\,(2\,a_2 a_1 + 2\,a_3) - 2\,a_4 = 0\}$$

Can we solve the equations? Yes: we can get $a_1$ from the first, then $a_2$ from the second, $a_3$ from the third, etc. Of course we only need one "solve" command to do it.

```
> S:= solve(eqs, {a[1], a[2], a[3], a[4], a[5]});
```

$$S := \{a_4 = -449,\ a_3 = -49,\ a_1 = -1,\ a_5 = -4627,\ a_2 = -7\}$$

Here, then, is our Taylor series solution:

```
> subs(S, ys);
```

$$2 - x - 7(x-1)^2 - 49(x-1)^3 - 449(x-1)^4 - 4627(x-1)^5$$

(where the "$2 - x$" should really be $1 - (x - 1)$).

The coefficients seem to be growing rather rapidly. This should lead us to believe that the series has a rather small radius of convergence. As a matter of fact, it turns out that the function $y(x)$ is not defined for $x > 1.067592398$ (see Exercise 5), so the radius of convergence can't be more than .067592398. ■

## Asymptotics

In Example 6 of Lab 15 we wanted to see how a sequence $a_n$ (really a function $a(n)$) behaved as $n \to \infty$, and used the "asympt" command. Now we can see how this worked. The idea is this: substitute $n = 1/t$, so $n \to \infty$ becomes $t \to 0$, use a Maclaurin series in the variable $t$, and then substitute back $t = 1/n$. Here is how it works in that example.

```
> a:= n -> 1/((2*n)^2+1) - 1/((2*n+1)^2+1);
> at:=subs(n=1/t, a(n));
```

$$at := \frac{1}{4\frac{1}{t^2} + 1} - \frac{1}{\left(2\frac{1}{t} + 1\right)^2 + 1}$$

```
> at:= simplify(at);
```

$$at := \frac{1}{2}\frac{t^3(4+t)}{(4+t^2)(2+2t+t^2)}$$

While the initial result of the substitution might have appeared to have a singularity at $t = 0$ (because of the $1/t^2$ and $1/t$), after simplification this apparent singularity has gone away and we have a function that should be analytic (at least for small $t$). Here are the first few terms of its Maclaurin series.

```
> s:= taylor(at, t=0);
```

$$s := \frac{1}{4}t^3 - \frac{3}{16}t^4 + O\left(t^6\right)$$

Finally, we substitute back to get this in terms of $n$. The result is the same as we obtained from "asympt".

```
> subs(t=1/n, s);
```

$$\frac{1}{4}\frac{1}{n^3} - \frac{3}{16}\frac{1}{n^4} + O\left(\frac{1}{n^6}\right)$$

Not all functions can be dealt with in this way, and "asympt" actually uses more general types of series when necessary. Sometimes there are positive or fractional powers of $n$ or terms involving logarithms, which we are not yet able to handle.

## EXERCISES

1. For each of the following functions $f(x)$, plot the function and several of its Maclaurin polynomials $P_n(x)$ on the same graph, using appropriate intervals, to see how $P_n(x)$ converges to $f(x)$. Then plot the remainders $R_n(x)$. As $n$ increases, what happens to the interval around 0 on which $|R_n(x)| < 0.1$? *Hint:* if the function is even or odd, you need only consider even or odd partial sums, respectively.

   **(a)** $\sin x$ **(b)** $\tan x$ **(c)** $\ln(1+x)$ **(d)** $\sin x^2$ **(e)** $x/(1+x^2)$

2. Consider the power series $\sum_{n=1}^{\infty} x^n/(n+1)$.

   **(a)** What, according to Maple, is the sum $f$ of this series? What is the radius of convergence?

   **(b)** Plot the remainders $R_n(x)$ for several different $n$. For what $x$ values do they appear to be converging to 0?

   **(c)** How many terms of the series would be needed to approximate the sum to an accuracy of .001 for $x = .5$? For $x = .9$?

3. Starting with "well-known" series, obtain the degree 9 Maclaurin polynomial for $f(x) = \cos(\sin(x)\ln(1+x))$.

4. **(a)** Let $f(x) = \dfrac{1}{2\pi}\displaystyle\int_0^{2\pi} \exp(x\cos(t))\,dt$. Maple can't evaluate this integral, but in fact it turns out to be a Maple function: `BesselI(0,x)`. Evaluate $f$ and `BesselI(0,...)` at several points to convince yourself that they are the same.

   **(b)** Find the Maclaurin series of $f(x)$ by starting with the Maclaurin series of exp. *Hint:* $\displaystyle\int_0^{2\pi} \cos^n(t)\,dt = 0$ if $n$ is odd and $2^{1-n}\pi n!/((n/2)!)^2$ if $n$ is even.

   **(c)** Check your answer for (b) by using "`taylor`" on $f(x)$ and `BesselI(0,x)`.

   **(d)** Plot `BesselI(0,x)` and its Maclaurin polynomials $P_n(x)$ for even $n$ from 2 to 10 and $x$ from 0 to 5.

   **(e)** The function $f(x)$ satisfies the differential equation $xf''(x) + f'(x) - xf(x) = 0$. What is $xP''_{10}(x) + P'_{10}(x) - xP_{10}(x)$?

5. **(a)** Convert the equation $x^4 + y^4 = 2xy$ to polar coordinates and plot it. What happens near $x = 1.07$?

   **(b)** Using implicit differentiation, find the largest possible $x$ value in this equation. *Hint:* you can use "`solve`" or "`fsolve`" on two equations in two unknowns.

   **(c)** Suppose $y = y(x)$ is an analytic solution of $x^4 + y^4 = 2xy$ with $y(0) = 0$. Find the Maclaurin polynomial $P_{11}(x)$ of degree 11.

   **(d)** Plot $x^4 + P_{11}(x)^4 - 2xP_{11}(x)$ for $-1.1 \le x \le 1.1$.

   **(e)** How close is $P_{11}(0.5)$ to the true $y(0.5)$ (obtainable using "`fsolve`")? *Hint:* be sure to specify an interval in the "`fsolve`" command, so that it makes the proper choice of solution.

6. For every positive integer $k$ there is $x(k)$ with $k\pi < x(k) < (k+1/2)\pi$ and $\tan x(k) = x(k)$. We are going to find a series for $x(k) - (k+1/2)\pi$ in powers of $1/k$.

   **(a)** Substitute $x(k) = (k+1/2)\pi - w(1/k)$ into the equation $\tan(x(k)) - x(k) = 0$, and simplify. You will need to substitute $\sin(k\pi) = 0$ manually, since Maple doesn't know $k$ is supposed to be an integer. Then substitute $k = 1/t$ and simplify again. Let "`eq`" be the numerator of the left side of the resulting expression.

   **(b)** Suppose $w(t)$ is an analytic solution of the equation `eq=0` with $w(0) = 0$. Find the Maclaurin polynomial $P_8(t)$ of degree 8.

   **(c)** Compare $(k+1/2)\pi - P_8(1/k)$ to $\tan((k+1/2)\pi - P_8(1/k))$ for $k = 1, 2, \ldots 5$.

**7.** Suppose a function $f(x)$ analytic at 0 satisfies the differential equation

$$4(x - x^2)f''(x) + (6 - 8x)f'(x) - f(x) = 0$$

and the initial condition $f(0) = 1$. What is its degree 5 Maclaurin polynomial $P_5(x)$? *Hint:* the coefficients of $P_5(x)$ determine the degree 4 Maclaurin polynomial of $4(x - x^2)f''(x) + (6 - 8x)f'(x) - f(x)$.

*CHAPTER 6*

# *Vectors and Geometry*

## LAB 17 THREE DIMENSIONS

**Goals:** *To learn some more of Maple's facilities for dealing with vectors and matrices, and how to use these for solving three-dimensional geometric problems.*

We have already met vectors in two dimensions (i.e., in a plane) in Lab 14. Now we wish to deal with vectors in three-dimensional space. Most of what we did in two dimensions carries over directly into three dimensions, the only difference being that vectors now have three components instead of two.

We begin by reading in the definitions in the file "`dot.def`" which we first saw in Lab 14. If this file is not available to you, you can type in the definitions now from Appendix I.

```
> read `dot.def`;
```

The standard basis vectors for 3-space are usually denoted by **i**, **j**, and **k**. I generally use upper-case letters for vectors, but Maple reserves "`I`" for $\sqrt{-1}$. Therefore I'll call them "`II`", "`JJ`", and "`KK`".

```
> II:= [1,0,0]; JJ:= [0,1,0]; KK:= [0,0,1];
```

■ **EXAMPLE 1** A dodecahedron is a solid with 12 faces, each a regular pentagon of the same size. A sphere can be circumscribed around it (with the 20 vertices on the surface of the sphere), and another, concentric sphere can be inscribed in it (touching the centre of each face). If the edges of the dodecahedron have length 1, what are the radii of these spheres?

**SOLUTION** Figure 6.1 is a picture of a dodecahedron, produced as follows:

```
> plots[polyhedraplot]([1,1,1], polytype=dodecahedron,
>    style=PATCH, scaling=constrained, orientation=[50,60]);
```

The picture can be rotated to look at the dodecahedron from different angles. Click and drag the mouse on the plot. In Release 5 under Windows 95, 98, or NT, the object itself rotates as you drag the mouse. In earlier releases, a box appears, which encloses the plotted object: you rotate the box to the desired orientation by dragging it with the mouse, and then click the **R** button or press `Enter` to redraw the picture in the new orientation.

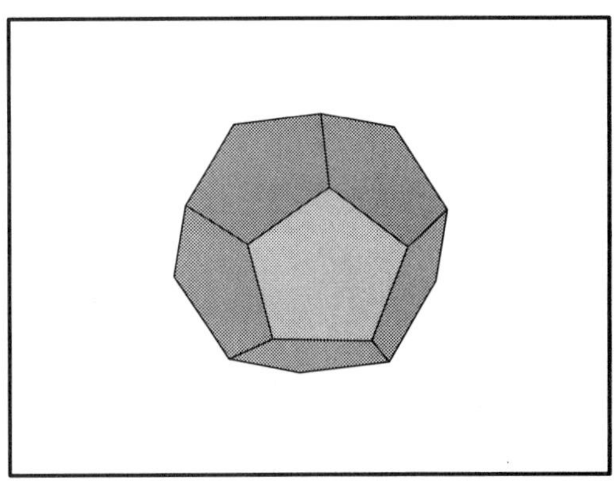

Figure 6.1. The dodecahedron

We begin our analysis by placing one face on the $xy$ plane, centred at the origin. One vertex $A$ will be on the positive $x$ axis, at $(d, 0, 0)$ (where $d$ will have to be determined). The others are at the same distance $d$ from the origin on lines at angles $2\pi/5$,

$4\pi/5$, $6\pi/5$, and $8\pi/5$ (counterclockwise) from this axis in the $xy$ plane. We'll let $B$ and $C$ be the vertices adjacent to $A$. Our strategy will be as follows: first we'll find the coordinates of $A$, $B$, and $C$, determining $d$ in the process. Then we'll find the coordinates of the third vertex $E$(1) joined to $A$. This will let us find the centre $F$ of the spheres. Finally, the distances from $F$ to $A$ and from $F$ to the $xy$ plane are the radii of the circumscribed and inscribed spheres respectively.

```
> A:= evl(d * II);
> B:= evl(d * (cos(2*Pi/5)*II + sin(2*Pi/5)*JJ));
> C:= evl(d * (cos(2*Pi/5)*II - sin(2*Pi/5)*JJ));
```

We could try to work with exact expressions, but these expressions would become rather complicated. It is simpler to use "`evalf`" and work with numerical approximations.

```
> B:= evalf(B);
```

$$B := [.3090169945d,\ .9510565160d,\ 0]$$

```
> C:= evalf(C);
```

$$C := [.3090169945d,\ -.9510565160d,\ 0]$$

Now $d$ can be determined from the fact that the length of $AB$, which is an edge of the dodecahedron, is supposed to be 1.

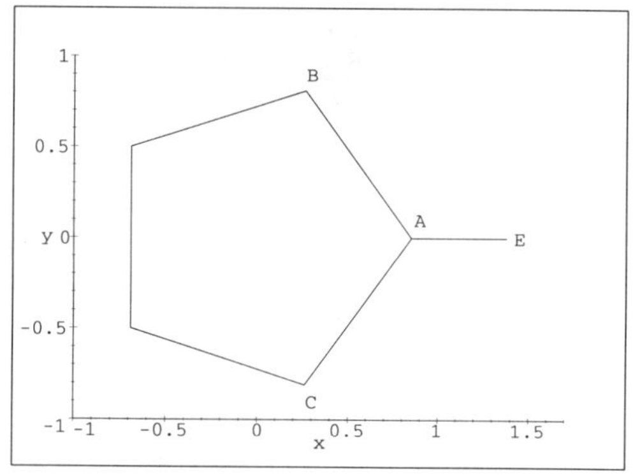


Figure 6.2. $A$, $B$, $C$, and $E$

```
> L:= len(A-B);
```

$$L := 1.175570504\sqrt{d^2}$$

Maple uses $\sqrt{d^2}$ instead of $d$, because it doesn't know that $d > 0$. In any case, the number 1.175570504 should be the first operand of this expression.

```
> d:= 1/op(1,L);
```

$$d := .8506508088$$

Now we find $E$, the vertex joined to $A$ by an edge not in the $xy$ plane. Look at Figure 6.2, which is a view from the positive $z$ direction. By symmetry, $E$ lies above the $x$ axis. Let $t$ be the $x$ component of $E - A$. Since the length of an edge is 1, the $z$ component is $\sqrt{1-t^2}$.

```
> E:= evl(A + t * II + sqrt(1-t^2) * KK);
```

$$E := \left[.8506508088 + t,\ 0,\ \sqrt{1-t^2}\right]$$

Now the angle between the edges $AE$ and $AB$ is the angle between two adjacent edges of a regular pentagon, namely $3\pi/5$. Actually it's simpler to deal directly with the dot products.

```
> eq:= (E-A) &. (B-A) = (C-A) &. (B-A);
```

$$eq := -.5877852525t = -.3090169943$$

```
> t:= solve(eq, t);
```

$$t := .5257311118$$

We can now find the centre $F$ of the spheres. Again using symmetry, it must be on the $z$ axis. Let $ri$ be its $z$ coordinate. This is the distance from $F$ to the centre of the face that is in the $xy$ plane, so it is the radius of the inscribed sphere. The vertices $A$ and $E$ should be the same distance from $F$, and that should be enough to determine $ri$. It's usually easier to work with squares of distances rather than the distances themselves, to avoid square roots.

---

(1) I'm avoiding "`D`" which is used by Maple for differentiation.

```
> F:= evl(ri * KK);
```

$$F := [0, 0, ri]$$

```
> eq:= len(F-E)^2 =
>    len(F-A)^2;
```

$$eq := 1.894427192 + (ri - .8506508086)^2 = .7236067985 + ri^2$$

At first sight this is a quadratic equation, but on expanding the square on the left the two $ri^2$ terms will cancel.

```
> ri:= solve(eq, ri);
```

$$ri := 1.113516364$$

That's the radius of the inscribed sphere. The radius of the circumscribed sphere is the distance from $F$ to $A$.

```
> rc:= len(F-A);
```

$$rc := 1.401258539$$

■

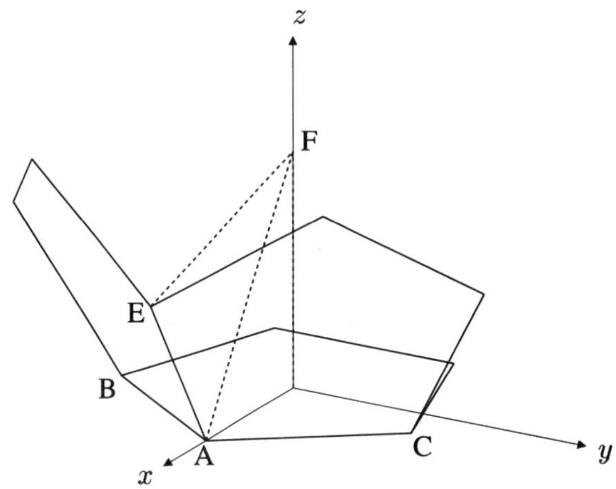


Figure 6.3. Finding $F$

## The Cross Product

Perhaps the most unique feature of three-dimensional space is the cross product. In Maple this operation is performed by the "`crossprod`" command in the "`linalg`" package. To obtain the cross product of vectors "`U`" and "`V`" you would write "`crossprod(U,V)`". However, we would prefer something that looks more like the standard mathematical notation $\mathbf{U} \times \mathbf{V}$. Moreover, the result of "`crossprod`" is an array, which we want to convert to a list. Therefore, just as I did for the dot product with "`&.`" in Lab 14, I defined a cross product operation "`&x`". This definition is part of "`dot.def`".

```
> `&x`:= (U, V) -> convert(linalg[crossprod](U, V), list);
```

Let's create three arbitrary vectors (as lists) **U**, **V**, and **W** and investigate some of the properties of the cross product.

```
> U:= [u1, u2, u3]; V:= [v1, v2, v3]; W:= [w1, w2, w3];
> U &x V;
```

$$[u2\, v3 - u3\, v2,\ u3\, v1 - u1\, v3,\ u1\, v2 - u2\, v1]$$

```
> evl(U &x V + V &x U);
```

$$[0, 0, 0]$$

The scalar triple product of **U**, **V**, and **W** is $\mathbf{U} \bullet (\mathbf{V} \times \mathbf{W})$. Note that the result is a scalar.

```
> normal(U &. (U &x V));
```

$$0$$

```
> normal(U &. (V &x W) - (U &x V) &. W);
```

$$0$$

All three of these results of 0 verify important facts about the cross product:

(i) $\mathbf{U} \times \mathbf{V} = -\mathbf{V} \times \mathbf{U}$

(ii) $\mathbf{U} \times \mathbf{V}$ is perpendicular to $\mathbf{U}$

(iii) $\mathbf{U} \bullet (\mathbf{V} \times \mathbf{W}) = (\mathbf{U} \times \mathbf{V}) \bullet \mathbf{W}$.

## Planes and Lines

**EXAMPLE 2** Find the equation of the plane through points $A$, $B$, and $E$ of the dodecahedron in Example 1.

**SOLUTION** The normal to the plane is perpendicular to $A - B$ and $A - E$, so we can take it to be their cross product.

```
> N:= (A-B) &x (A-E);
```

$$N := [.6881909605, .5000000003, -.4253254043]$$

The second component should really be exactly .5 (the extra .0000000003 being due to roundoff error). Let's correct this. It's possible to assign values to individual entries in a list, so you can simply write:

```
> N[2]:= .5;
```

$$N_2 := .5$$

The equation of the plane can be written in the form $\mathbf{N} \bullet \mathbf{R} = \mathbf{N} \bullet \mathbf{R}_0$, where $\mathbf{R} = x\mathbf{i} + y\mathbf{j} + z\mathbf{k}$ and $\mathbf{R}_0$ is the position vector of any point in the plane. We may as well use $A$ for this point.

```
> R:= [x,y,z]:
> eq:= N &. R = N &. A;
```

$$eq := .6881909605x + .5y - .4253254043z = .5854101972$$

■

**EXAMPLE 3** Find another vertex of the dodecahedron on the face containing $A$, $B$, and $E$.

**SOLUTION** One of these vertices, call it $G$, is on a line through $E$ parallel to $BA$. We can take $B - A$ as a direction vector for the line.

I'd like to use "`t`" as a parameter for the parametric equation of the line, but "`t`" has already been used in defining "`E`". If we undefine `t`, that will affect `E` as well. We can prevent this by assigning the **value** of `E` to "`E`".

```
> E:= E;
```

$$E := [1.376381921, 0, .8506508086]$$

```
> t:= 't';
> G:= evl(E + t * (B-A));
```

$$G := [1.376381921 - .5877852525t, .8090169945t, .8506508086]$$

Now how shall we determine $t$? One way is that the angle between $G - B$ and $A - B$ is $3\pi/5$, the same as the angle between $B - A$ and $C - A$. Since these are unit vectors, the dot products are also the same.

```
> t:= solve((G-B) &. (A-B) = (B-A) &. (C-A), t);
```

$$t := 1.618033988$$

Finally, we look at the value of `G` with this $t$.

```
> G;
```

$$[.4253254048, 1.309016994, .8506508086]$$

■

## Distances, Projections, and Reflections

Two basic tools in dealing with points, lines, and planes are the scalar and vector projections of a vector on another vector. It is convenient to define these as functions.

```
> sproj:= (U,V) -> (U &.V)/len(V);
> vproj:= (U,V) -> evl((U &. V) * V/(V &. V));
```

Thus "`sproj(U,V)`" is the scalar projcction of **U** on **V**, and "`vproj(U,V)`" is the vector projection of **U** on **V**. Of course, **V** must not be the zero vector.

**EXAMPLE 4** Find the distance from $C$ to the plane we found in Example 2.

**SOLUTION** For the distance from $C$ to the plane, we want the scalar projection of $C - P$ on the normal **N** to the plane where $P$ is any point on the plane. We may as well use $A$.

```
> sproj(C-A, N);
```

$$-.8506508086$$

The distance from $C$ to the plane is the absolute value of this scalar projection, or .8506508086. ■

**EXAMPLE 5** Find the reflection of point $E$ across the plane through $F$, $B$ and $G$.

**SOLUTION** A normal to the plane is $(F - B) \times (F - G)$.

```
> N:= (F-B) &x (F-G);
```

$$N := \left[1.244949142, -.4045084979, -.4\,10^{-9}\right]$$

If $E'$ is the reflection of $E$ in the plane, then $E - E'$ is a vector that is normal to the plane and whose length is twice the distance from the plane to $E$. Thus $(E - E')/2$ is the vector projection of $E - P$ on **N**, where $P$ is any point in the plane. I'll take $P = B$.

```
> Ep:= evl(E - 2*vproj(E-B, N));
```

$$Ep := [-1.113516365, .8090169966, .8506508094]$$

Note that $E'$ should be another vertex of the dodecahedron, because the dodecahedron is symmetric under this reflection. Using more reflections, we can actually find all the vertices. ■

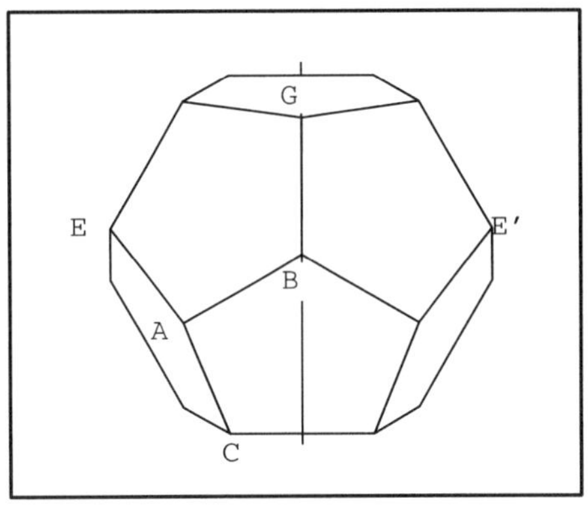


Figure 6.4. **E** and its reflection

## Matrices

A matrix can be produced with the "`matrix`" command in the "`linalg`" package. The input to "`matrix`" is a list of the rows of the matrix, each row being itself a list of numbers.

```
> M:= linalg[matrix]([[1,2,3], [4,5,6], [7,8,9]]);
```

$$M := \begin{bmatrix} 1 & 2 & 3 \\ 4 & 5 & 6 \\ 7 & 8 & 9 \end{bmatrix}$$

The matrix element in row $i$ and column $j$ of matrix $M$ is "`M[i,j]`".

```
> M[2,3];
```

$$6$$

A matrix can be transposed using the "`transpose`" command in "`linalg`".

```
> L:= linalg[transpose](M);
```

$$L := \begin{bmatrix} 1 & 4 & 7 \\ 2 & 5 & 8 \\ 3 & 6 & 9 \end{bmatrix}$$

Matrix multiplication uses "`&*`" rather than "`*`". This is not performed automatically, however: you need to use "`evalm`".

```
> evalm(L &* M);
```

$$\begin{bmatrix} 66 & 78 & 90 \\ 78 & 93 & 108 \\ 90 & 108 & 126 \end{bmatrix}$$

For the purposes of matrix multiplication, a list on the right of "`&*`" is considered as a column vector (even though Maple shows it as a row).

```
> V:= evl(M &* [x,y,z]);
```

$$V := [x + 2y + 3z,\ 4x + 5y + 6z,\ 7x + 8y + 9z]$$

The determinant of an $n \times n$ matrix is obtained using "`det`".

```
> det(M);
```

$$0$$

An identity matrix in an "`evalm`" expression is written using the rather strange notation "`&*()`".

```
> N:= evalm(M + &*() );
```

$$N := \begin{bmatrix} 2 & 2 & 3 \\ 4 & 6 & 6 \\ 7 & 8 & 10 \end{bmatrix}$$

The inverse of a matrix $M$ is "`M^(-1)`". Again, this must be performed using "`evalm`". Of course the matrix must be nonsingular, otherwise there will be an error message.

```
> evalm(N^(-1));
```

$$\begin{bmatrix} -6 & -2 & 3 \\ -1 & \frac{1}{2} & 0 \\ 5 & 1 & -2 \end{bmatrix}$$

A linear system $A\mathbf{X} = \mathbf{B}$ (where $A$ is a matrix, $\mathbf{B}$ a vector, and $\mathbf{X}$ an unknown vector) can be solved by "`linsolve(A,B)`". The "`linsolve`" command is in "`linalg`". The result is a vector, and we need to convert it to a list with "`evl`".

```
> linalg[linsolve](N, [14,32,50]);
```

$$[2, 2, 2]$$

This example could also have been done as

```
> evl(N^(-1) &* [14,32,50]);
```

If we used the matrix $M$ instead of $N$, we couldn't use the latter form because $M$ is singular, but "`linsolve`" would still work (if any solutions exist).

```
> linalg[linsolve](M, [14,32,50]);
```

$$[\_t_1 - 2, -2\_t_1 + 8, \_t_1]$$

Here there are many solutions, one for any real value of the parameter "$\_t_1$".

## EXERCISES

1. Find the following using Maple, where $\mathbf{A} = [1, 2, 3]$, $\mathbf{B} = [3, 2, 1]$, and $\mathbf{C} = [4, 1, 3]$.
   **(a)** $(\mathbf{A} \times \mathbf{B}) \times \mathbf{C}$ **(b)** $\mathbf{A} \times (\mathbf{B} \times \mathbf{C})$
   (did you expect these to give the same result?)
   **(c)** The scalar triple product of **A**, **B**, and **C**.
   **(d)** The angle between $\mathbf{A} \times \mathbf{B}$ and $\mathbf{B} \times \mathbf{C}$.
2. Find the following using Maple, where $A$, $B$, and $C$ are the points $(1, 2, 3)$, $(3, 2, 1)$, and $(4, 1, 3)$ respectively.
   **(a)** The distance from $A$ to the line through $B$ and $C$.
   **(b)** An equation for the plane through $A$, $B$ and $C$.
   **(c)** The distance from the plane through $A$, $B$, and $C$ to the origin.

**3.** In a methane molecule, the nuclei of four hydrogen atoms form the vertices of a regular tetrahedron, with the nucleus of a carbon atom at the centre. Find the angle (well-known to organic chemists!) between the vectors from the carbon nucleus to two hydrogen nuclei. *Hint:* start with two hydrogen nuclei at given points $H_1$ and $H_2$ a distance $L$ apart. Place the third hydrogen nucleus at a point $H_3$ at a distance $L$ from both $H_1$ and $H_2$, and the fourth at $H_4$, the same distance $L$ from the first three. The carbon atom's position vector is $\mathbf{C} = (\mathbf{H}_1 + \mathbf{H}_2 + \mathbf{H}_3 + \mathbf{H}_4)/4$, where $\mathbf{H}_i$ are the position vectors corresponding to $H_i$.

**4.** Show using vectors that the midpoints of the sides of a rhombus are the vertices of a rectangle. *Hint:* a rhombus is a parallelogram whose sides are equal. Let one vertex be at the origin, and let $\mathbf{U}$ and $\mathbf{V}$ be the position vectors for the adjacent vertices. What vectors go from one midpoint to an adjacent one? Use the dot product.

**5.** **(a)** Where does the line $\mathcal{L}$ through $(1, 2, 3)$ and $(2, 0, 4)$ cut the plane $\mathcal{P}$ through $(1, 5, 0)$, $(3, 4, 5)$, and $(3, 2, 1)$?

**(b)** What is the angle between $\mathcal{L}$ and $\mathcal{P}$? *Hint:* if the angle between the line and a normal to the plane is $\theta$, $0 \le \theta \le \pi/2$, then the angle between the line and the plane is $\pi/2 - \theta$.

**(c)** Find a point on $\mathcal{P}$ whose distance from $\mathcal{L}$ is 6 and whose distance from the plane $x+y+z=3$ is 7.

**6.** What is the angle between two adjacent faces of the dodecahedron? *Hint:* this is the same as the angle between the normals to the faces.

**7.** Redo Example 1 using exact expressions. Make frequent use of "`simplify`".

**8.** A body of mass 10 kg is attached by springs to the three points $(1, 1, 2)$, $(1, 2, 1)$, and $(3, 0, 1)$ (with distances measured in metres). Each spring has a natural length of 2 metres and a spring constant of 50 newtons per metre. Thus if the body is at $P$, the spring attaching it to $R$ exerts a force on the body of $50(|P - R| - 2)$ newtons in the direction of $R - P$ (assuming that $|P - R| > 2$). Gravity also acts on the body, exerting a downward force of 98 newtons. Find a position where the body will rest at equilibrium, with all forces in balance.

**9.** Find the equation of a sphere tangent to the lines with parametric equations

$$R = [2,\ 0,\ -7] + t[3,\ 2,\ -2]$$
$$R = [-6,\ 6,\ 2] + t[-4,\ -1,\ 3]$$
$$R = [6,\ 4,\ -4] + t[-1,\ 0,\ -3]$$
$$R = [3,\ 7,\ -2] + t[6,\ -1,\ 2]$$

*Hint:* the centre of the sphere will be the same distance from each of the lines. Use "`fsolve`" on a set of three equations, with the coordinates of the centre of the sphere as unknowns. It's probably better to use the squares of the distances rather than the distances themselves.

**10.** Verify the vector identity $(\mathbf{A} \times \mathbf{B}) \times (\mathbf{A} \times \mathbf{C}) = (\mathbf{A} \bullet (\mathbf{B} \times \mathbf{C}))\mathbf{A}$.

**11.** Consider the matrix

$$M = \begin{bmatrix} 9/11 & -6/11 & 2/11 \\ -6/11 & -7/11 & 6/11 \\ 2/11 & 6/11 & 9/11 \end{bmatrix}$$

Show that this has the following properties:

**(a)** $M = M^{-1}$

**(b)** $\det(M) = -1$

**(c)** For any vector $\mathbf{X}$, $M\mathbf{X}$ is the reflection of $\mathbf{X}$ across the plane $x + 3y - z = 0$.

# LAB 18 GRAPHING FUNCTIONS OF SEVERAL VARIABLES

**Goals:** *To learn several methods of visualizing functions of two variables using Maple's three-dimensional plotting facilities.*

## Three-Dimensional Plotting

Except for "plot3d", all the plotting commands we consider in this lab are in the "plots" package, which you should load now.

```
> with(plots):
```

The graph of a function of two variables $f(x, y)$ is a surface in three-dimensional space, given by the equation $z = f(x, y)$. Of course, the computer screen is only two-dimensional, but we can draw a picture of how the surface would look when viewed from a certain position. Maple's "plot3d" command will do this for us.

```
> plot3d(x^2 + y^2, x=-1..1, y= -1..1);
```

This is the simplest form of the command. The first input to "plot3d" is the expression to be plotted, which depends on two variables. We usually use "x" and "y", but any names (not currently assigned values) could be used. Then come a range for the first variable and a range for the second variable. The result appears in a window or inline plot region, similar to those for two-dimensional plots that we have seen in earlier labs, but with a number of additional features. It shows the portion of the surface for $x$ and $y$ in the given intervals. The endpoints of the $x$ interval must be constants, but the endpoints of the $y$ interval may depend on $x$, as in the following example.

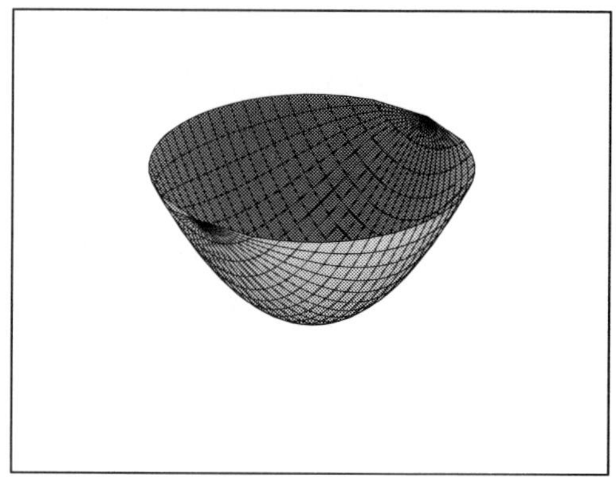

Figure 6.5. $z = x^2 + y^2$ for $|y| \le \sqrt{1 - x^2}$

```
> plot3d(x^2 + y^2, x=-1..1,
>           y= -sqrt(1-x^2) .. sqrt(1-x^2)); (Figure 6.5)
```

The "plot3d" command has many options, of which I'll mention only a few. In some cases, besides specifying them in the command itself, you can change them in the plot's menu or context bar, which you obtain by clicking the mouse on the plot. In Release 5 under Windows 95, 98, or NT the plot changes immediately when you change the option; in earlier releases you click **R** or press `Enter` to redraw the plot.

You can specify the option "view=$a$..$b$" where $a$ and $b$ are numbers, to show only the part of the graph for $a \le z \le b$. This is particularly useful if $f(x, y)$ can be very large for some values of $x$ and $y$. If you don't specify "view", Maple will show the entire surface.

As we saw with Figure 6.1 in Lab 17, three-dimensional plots can be rotated to look at the surface from different angles. Click and drag the mouse on the plot. In Release 5 under Windows 95, 98, or NT, the plot itself rotates as you drag the mouse. In earlier releases, a box appears, representing the solid region where $x$, $y$, and $z$ go from their minimum to their maximum values on the surface. Drag the mouse to rotate the box, then click **R** or press `Enter` to redraw the picture in the new orientation. You can also change the orientation by changing the numbers for $\vartheta$ ("theta") and $\varphi$ ("phi") in the plot's context bar. Here $\vartheta$ and $\varphi$ are angles, measured in degrees, specifying the direction of the viewpoint. $\vartheta$ is the "longitude", the counterclockwise angle between the positive

$x$ axis and the vertical plane containing the viewpoint: $\vartheta = 0$ for a viewpoint in the $xz$ plane for positive $x$, $\vartheta = 90$ for a viewpoint in the $yz$ plane for positive $y$, etc. $\varphi$ is the "colatitude", the angle between the positive $z$ axis and the direction of the viewpoint: $\varphi = 0$ for a viewpoint on the positive $z$ axis, $\varphi = 90$ for a viewpoint in the $xy$ plane, etc. The initial orientation can also be specified in the `plot3d` command as the option "`orientation=[`$\vartheta, \varphi$`]`".

As with two-dimensional plots, you can specify "`scaling=constrained`". With the default "unconstrained" scaling, the scales on the $x$, $y$, and $z$ axes are adjusted so that the box fits nicely in the plot, with corners near the top, bottom, left, and right edges. With "constrained" scaling, the scales on the $x$, $y$, and $z$ axes are equal, only the overall size of the box being adjusted to fit in the plot.

When `plot3d` plots a surface, it divides up the $x$ interval into a certain number of equal subintervals, say $x_i \leq x \leq x_{i+1}$ for $i = 1..m - 1$. Similarly, it divides up the $y$ interval into equal subintervals $y_j \leq y \leq y_{j+1}$, $j = 1..n - 1$. The function $f(x, y)$ will be evaluated at each of the $mn$ points $(x_i, y_j)$. You can control $m$ and $n$ with the option "`grid=[m,n]`". Larger values of $m$ and $n$ will usually result in a smoother-looking and more accurate surface. However, this will also require more computer time and memory, both proportional to $mn$. The default `grid` is `[25,25]`.

There are several different "styles" for plotting surfaces. "`style=POINT`" just plots the points $[x_i, y_j, f(x_i, y_j)]$ for $i = 1..m$ and $j = 1..n$. This is usually not very useful for our purposes. "`style=WIREFRAME`" joins the points with straight-line segments, so that you see a rectangular grid. In this style you can see all parts of the surface, even those that are behind other parts. On the other hand, with "`style=HIDDEN`" ("Hidden line" in the plot's Style menu), the surface becomes "opaque" and parts of it that are behind other parts will not be seen. However, in this style you still see only the grid lines. With "`style=PATCH`", in addition to the grid lines the surface itself is shown (coloured and/or shaded). Again, the surface is opaque and hides any other parts of the surface that are behind it. With "`style=PATCHNOGRID`" ("Patch w/o grid" in the menu), the surface itself is shown but there are no grid lines. This is especially useful in cases where there are very many grid intervals, and thus the grid lines are so close together in "`PATCH`" style that the surface patches themselves are hard to see. With "`style=CONTOUR`", you see curves $z = constant$ ("contour lines") on the surface. This style is like "`WIREFRAME`" in that the surface itself is transparent. Finally, "`style=PATCHCONTOUR`" ("Patch and Contour" in the menu) has both the surface elements of "`PATCH`" or "`PATCHNOGRID`" and the contour lines of "`CONTOUR`".

There are various ways of shading or colouring surfaces. With the option "`shading=XYZ`", red, green, and blue components of the colour vary with the $x, y$, and $z$ coordinates. With "`shading=XY`", the colours vary with $x$ and $y$ but don't depend on $z$. In the next three "`shading`" options, they only depend on $z$ and not on $x$ or $y$. With "`Z`" the colours go through various shades of red (at the highest $z$ values) to violet (at the lowest). With "`ZHUE`" ("Z (Hue)" in the menu), they also go from red to violet, but with all shades of the rainbow in between. With "`ZGREYSCALE`" ("Z (Greyscale)" in the menu), all colours are shades of gray, from very light at the highest $z$ values to very dark at the lowest. With "`NONE`" ("No Coloring" in the menu), the surface is white, with no shading or colouring, and the grid lines are black. However, shading and colour will be added when lighting is used. With a good choice of lighting, "`shading=NONE`" and "`style=PATCHNOGRID`" can produce a fairly realistic three-dimensional effect.

If none of these colour options suits you, you can define your own colour function. See the help page "`?plot3d[colorfunc]`" for details.

Lighting simulates the effect of having the surface lit by light coming from a certain direction, in addition to "ambient light" that comes from all directions. The parts of the surface where the light falls perpendicularly will be brightest, those parts where the light falls obliquely will be darker, and the darkest parts will be where the light comes from behind the surface. However, the surface does not actually cast a shadow. You can specify a light source with the option "`light=[phi,theta,r,g,b]`" where "`phi`" and "`theta`" are angles in degrees giving the direction of the light source, and "`r`", "`g`", and "`b`" are numbers from 0 to 1 giving the intensities of the red, green, and blue components of the light. Note that the order of the angles is reversed from that of the "`orientation`" option. Also, the direction of the light source is not relative to the $x$, $y$, and $z$ coordinate system but relative to the viewpoint. Thus "`light(0,...)`" makes a light that always comes vertically "down" in the plot, while "`light(90,0,...)`" comes from the same direction as the viewpoint, and "`light(90,90,...)`" comes from 90 degrees to the right of the viewpoint. The option "`ambientlight=[r,g,b]`" specifies ambient light, where again you specify the three colour components of the light. There are predefined lighting options available on the Color menu. You can also modes access these directly in the "`plot3d`" command with the option "`lightmodel=m`" where "`m`" is "`light1`", "`light2`", "`light3`", or "`light4`".

The choice of many of these options can involve aesthetic considerations. It is possible to make beautiful pictures in Maple. However, you should remember that our primary goal when plotting in these labs is not to produce art but to show as clearly as possible the mathematical features of the objects being plotted. Often some experimentation with various options is necessary before you obtain a good picture.

You can plot several surfaces in one picture. Instead of one expression $f(x, y)$, you use a set of expressions within braces "`{}`".

```
> plot3d({(x^2 + y^2)/4, (1 - x^2 - 2*y^2)/4},
>                x=-1 .. 1, y=-1 .. 1);
```

This method will use the same options for all surfaces that you plot. Sometimes you might want to use different options, e.g., to use different colours to help distinguish between the surfaces. Unfortunately you can't do this by the method that works for two-dimensional plots: plotting a list of expressions, and specifying a list of colours. This is because a three-dimensional plot of a list would be interpreted as a parametric surface. Instead, you can combine the results of different "`plot3d`" commands using the "`display`" command.

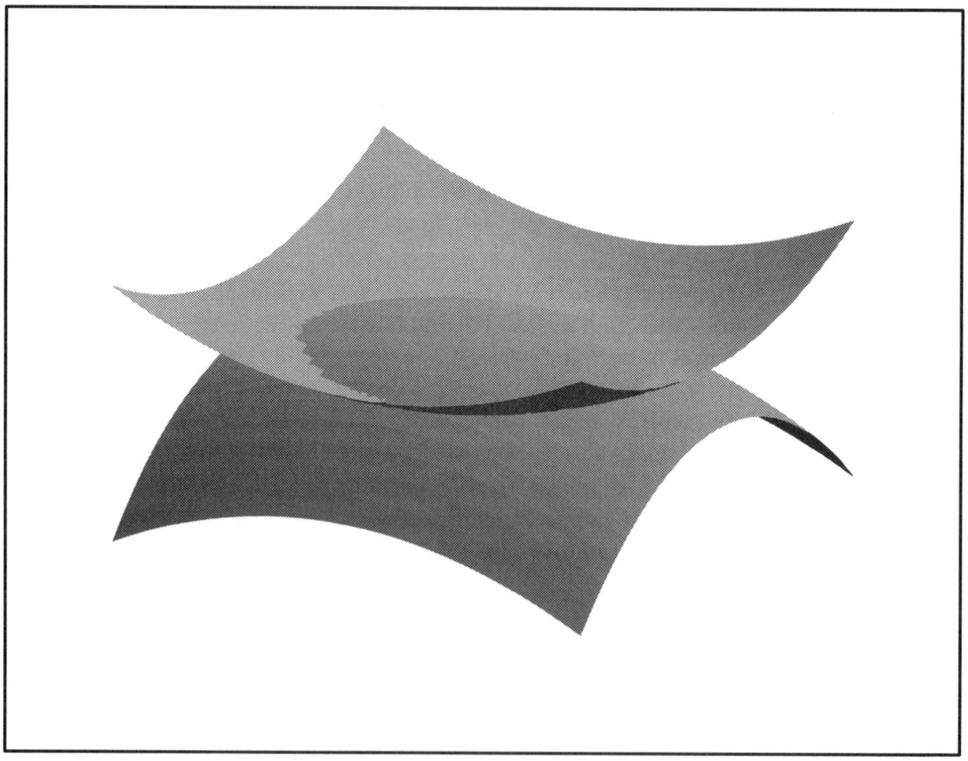

Figure 6.6. Intersecting surfaces

```
> p1:= plot3d((x^2 + y^2)/4, x=-1 .. 1, y=-1 .. 1,
>                          colour=cyan, style=PATCHNOGRID):
> p2:= plot3d((1 - x^2 - 2*y^2)/4, x=-1..1, y=-1 .. 1,
```

```
>                          colour=coral, style=PATCHNOGRID):
> display({p1,p2}, light=[30,75,1,1,1], ambientlight=[.3,.3,.3],
>      orientation = [120,70], scaling = constrained);
```
(Figure 6.6)

With "colour" specified in the "plot3d" commands, the shading options from the menu have no effect. If I hadn't specified lighting, there would have been flat colours with no shading. I chose a light source with white light (1,1,1), 75 degrees to the right of the viewpoint (Theta=75) and 60 degrees above it (Phi=30). I specified a small amount of ambient light, without which parts of the surface would be quite dark. Of course, in the printed image the colours show only as shades of gray.

In the printed image, you may notice a ragged appearance of the places where the two surfaces intersect. This ragged effect may or may not appear on the screen(2). This is due to the way the surfaces are rendered. In the printed image (which was produced using the PostScript graphics language) Maple draws the surface patches in a certain order, so that the patches that are closer to the viewer are drawn later and can hide parts of the surface that are farther away. However, when surface patches intersect parts of each should be behind parts of the other. Since Maple draws a whole patch at once, it can't draw this situation correctly. On the screen, however, it appears that a different method may be used, taking advantage of the three-dimensional capabilities of the video card, which allows intersecting patches to be drawn correctly.

One way to avoid the ragged appearance, if it occurs, is to make the intersection of the surfaces be on the grid lines, or on the boundaries of the patches of surface being plotted. We can do this because the $y$ interval can depend on $x$. In our case, $x^2 + y^2 = 1 - x^2 - 2y^2$ when $2x^2 + 3y^2 = 1$, i.e., $y = \pm\sqrt{(1-2x^2)/3}$ where $-1/\sqrt{2} \le x \le 1/\sqrt{2}$. I'll divide each surface into five parts. For $1 \le x \le -1/\sqrt{2}$ and for $1/\sqrt{2} \le x \le 1$ we have $-1 \le y \le 1$, but for $-1/\sqrt{2} \le x \le 1$ we separate the intervals $-1 \le y \le -\sqrt{(1-2x^2)/3}$, $-\sqrt{(1-2x^2)/3} \le y \le \sqrt{(1-2x^2)/3}$, and $\sqrt{(1-2x^2)/3} \le y \le 1$. The intersection of the surfaces thus occurs on the boundaries between the third and fourth parts and between the fourth and fifth parts. This makes 10 surfaces in all, which might take a lot of time and memory. You can reduce this by using smaller grid settings. To save a bit of typing I'll define a function "makep" to produce the plots. The inputs to the functions are the $x$ and $y$ ranges. Each "makep" should produce two plots, one for each surface. It can be done as follows:

```
> makep:= (rx, ry) -> (plot3d((x^2 + y^2)/4, x=rx, y=ry,
>                            colour=cyan, grid=[15,15]),
>                      plot3d((1 - x^2 - 2*y^2)/4, x=rx, y=ry,
>                            colour=coral, grid=[15,15]));
```

Note that I will change some of the $x$ intervals very slightly to avoid square roots of negative numbers in the $y$ intervals.

```
> eps:= 0.00001;
> p1:= makep(-1..-1/sqrt(2), -1 .. 1):
> p2:= makep(1/sqrt(2) .. 1, -1 .. 1):
> p3:= makep(-1/sqrt(2)+eps .. 1/sqrt(2)-eps,
>                -1 .. -sqrt((1-2*x^2)/3)):
> p4:= makep(-1/sqrt(2)+eps .. 1/sqrt(2)-eps,
>                -sqrt((1-2*x^2)/3) .. sqrt((1-2*x^2)/3)):
> p5:= makep(-1/sqrt(2)+eps .. 1/sqrt(2)-eps,
>                sqrt((1-2*x^2)/3) .. 1):
> display({p1,p2,p3,p4,p5}, light = [30,75,1,1,1],
>            ambientlight=[.3,.3,.3], orientation = [120,70],
>            scaling = constrained);
```

The result should show the surfaces intersecting on a smooth curve.

---

(2) In the cases I have tested, it does not appear in the versions that allow "real-time" rotation of 3-dimensional plots, i.e., Release 5 under Windows 95, 98, or NT.

## Contour Plots

Instead of drawing the three-dimensional surface $z = f(x, y)$, you can produce strictly two-dimensional representations of a function of two variables $f(x, y)$. One method is to draw a "contour map" showing the level curves $f(x, y) = C$ for various choices of $C$. Maple does this with the "`contourplot`" command.

```
> contourplot(x^2-y^3, x=-1..1, y=-1..1, axes=BOXED);
```
(Figure 6.7)

An important option for "`contourplot`" is "`contours`", which controls the number of values $C$ for which $f(x, y) = C$ is plotted. If you use "`contours=`$n$", there will actually be $n - 1$ such values, evenly spaced between the maximum and minimum $f$ values on the grid of points that Maple uses for the plot. The default, which is used in Figure 6.7, is "`contours=8`". Instead of a number, you can use a list of $C$ values:

```
> contourplot(x^2-y^3, x=-1..1, y=-1..1, axes=BOXED,
>             contours=[-1, 0, .5, 1]);
```

The "`filled=true`" option causes Maple to fill in the regions between the contours with various colours. You can also specify the colours (of contours, or of the regions between them in the case of "`filled=true`"): "`coloring=[`$a$`,`$b$`]`"(3) makes these colours go from "$a$" (at the lowest) to "$b$" (at the highest). The default is `[red,yellow]`.

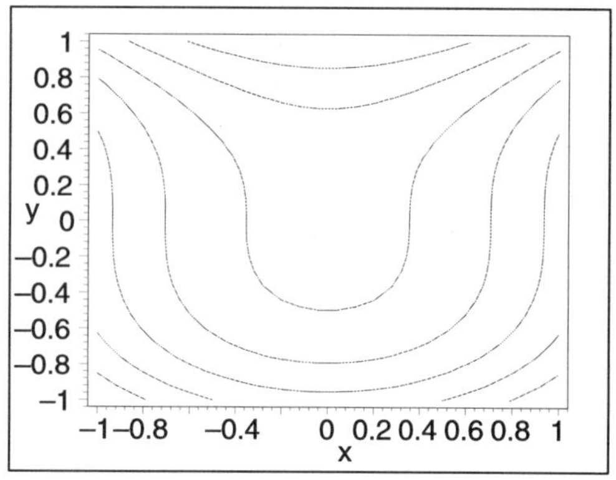


Figure 6.7. Contours: $x^2 - y^3$

Instead of the two-dimensional result of "`contourplot`", you can make a three-dimensional contour plot with the "`contourplot3d`" command. This is essentially the same as "`plot3d`" with the option "`style=contour`". You could produce a similar effect to the two-dimensional contour plot by using `orientation=[-90,0]`. The main advantage of this version is that it can be much faster.

## Density Plots

Another two-dimensional representation of a function of two variables $f(x, y)$ is a "density plot" where the shading or colour at a point $(x, y)$ depends on $f(x, y)$. Maple has the "`densityplot`" command for this purpose.

```
> densityplot(
>   2 * x^2 * exp(-sqrt(x^2+y^2)),
>   x = -5 .. 5, y = -3 .. 3,
>   axes = FRAMED);
```

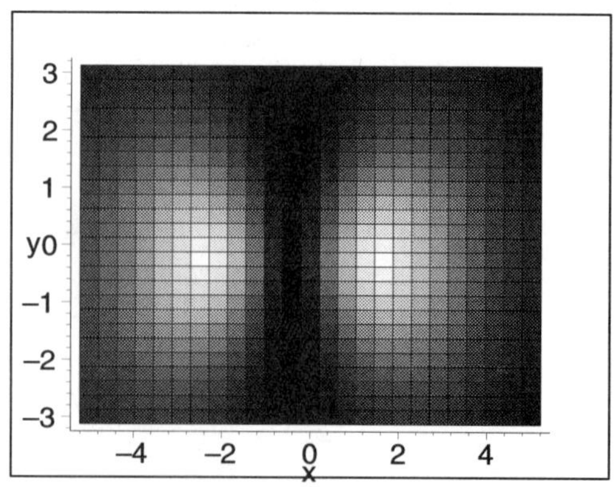


Figure 6.8. Density plot: $x^2 e^{-\sqrt{x^2+y^2}}$

This is a two-dimensional plot structure. Each rectangle in the grid is a shade of gray, depending on the value of $f(x, y)$ at the centre of the rectangle, darkest at the lowest $f(x, y)$ values and lightest at the highest values. I chose "`FRAMED`" for the axes rather than the default "`NORMAL`" axes, which in my opinion interfere too much with the shading. The other possible choices for axes are "`BOXED`" or "`NONE`". These choices can also be made from the plot's Axes menu, or by buttons on its context bar.

(3) Although generally "`color`" and "`colour`" are interchangeable in Maple, "`colouring`" is not allowed here.

Unfortunately Release 5 (including Release 5.1) has a bug, which causes the $x$ and $y$ directions in a density plot to be interchanged (thus Figure 6.8 would show one light-coloured region above the other, rather than the two side-by-side). A work-around is to plot $Re(f(x, y))$ instead of $f(x, y)$:

```
> densityplot(Re(2*x^2 * exp(-sqrt(x^2+y^2))),
>      x = -5 .. 5, y = -3 .. 3, axes = FRAMED);
```
(Figure 6.8)

Another bug affects the relation between the $f$ values and the shade of gray. The best results are obtained when the minimum $f$ value is 0 and the maximum is 1 or more: this makes the shades go from black at the lowest $f$ value to white at the highest.

It is possible to leave out the grid lines between the rectangles, using the option "`style = PATCHNOGRID`" (or choose "Patch w/o grid" from the plot's Style menu).

Another option is "`colourstyle=HUE`", which will use colour instead of shades of gray. The lowest $f(x, y)$ values are red, the highest are violet, and all the colours of the rainbow are in between. There is also "`colourstyle=RGB`", but this is not as useful: only the blue component of the colour will depend on $f(x, y)$, while the red and green depend on $x$ and $y$ respectively.

Density plots tend to be rather limited in the amount of information they can convey: they have uniformly shaded rectangles, often with only a relatively small number of different shades available, rather than smoothly varying shading. They should be used mainly for visualizing a function that really does represent some sort of "density". For example, the function in Figure 6.8 could represent the electron density in the $xy$ plane for a "p" state in a hydrogen atom.

## Three-Dimensional Implicit Plots

Just as "`implicitplot`" plots a curve defined by an equation in two variables, there is a command "`implicitplot3d`" to plot a surface defined by an equation in three variables. You specify the function, and ranges for the three variables. Most other options are similar to those of "`plot3d`".

```
> implicitplot3d(x^2*y - y^2*z + z^2*x=1,
>                   x=-2..2, y=-2..2, z=-2..2);
```

Here is essentially what "`implicitplot3d`" does. It evaluates the difference between the two sides of the equation at points $(x, y, z)$ in a three-dimensional rectangular grid. When the sign of this difference changes between grid points, Maple will draw a piece of the surface there. For example, suppose $(0, 0, 0)$, $(1, 0, 0)$, $(0, 1, 0)$, and $(0, 0, 1)$ are neighbouring grid points, and the differences between the two sides of the equation at these points are $1$, $-1$, $-1$, and $-2$ respectively. Then Maple will draw a triangular surface element with corners at $(1/2, 0, 0)$, $(0, 1/2, 0)$, and $(0, 0, 1/3)$. It chooses the point $(0, 0, 1/3)$, one third of the way from $(0, 0, 0)$ to $(0, 0, 1)$, because 0 is one third of the way from 1 (the value at $(0, 0, 0)$) to $-2$ (the value at $(0, 0, 1)$).

The "`grid`" option specifies the number of points in the $x$, $y$, and $z$ directions for the grid. Thus for an `implicitplot3d` with

```
x=0..1, y=0..1, z=0..1, grid=[3,4,5]
```

the equation will be evaluated at $x = 0$, $1/2$ and $1$, $y = 0$, $1/3$, $2/3$ and $1$, and $z = 0$, $1/4$, $1/2$, $3/4$ and $1$. If you don't specify "`grid`", the default `grid=[10,10,10]` is used. Larger numbers in the "`grid`" option will produce a more accurate and smoother-looking surface. However, this will also use more computer time and memory. With the default grid, Maple evaluates the equation $10 \times 10 \times 10 = 1000$ times. With `grid=[100, 100, 100]`, Maple would evaluate it one million times. On most computers, this would not be practical unless you are willing to wait many hours (and even then, Maple might run out of memory).

In the Student Edition, the "`implicitplot3d`" command is severely affected by the limitation on the maximum size of arrays. With any grid size larger than `[5,5,5]`, you will encounter the error message

```
Error, (in plots/iplot3d/implicit3d)
array size limited to 5120 elements in Student Edition
```

Maple draws lines on the edges of the surface elements in the "PATCH", "HIDDEN", and "WIREFRAME" styles. The result is usually not very informative or aesthetically pleasing. You will probably get better results with "PATCHNOGRID" or "PATCHCONTOUR" (the contours, as usual, are the curves $Z = constant$).

```
> implicitplot3d(x^2*y - y^2*z + z^2*x=1, x=-2..2, y=-2..2,
>   z=-2..2, grid=[15,15,15], style=PATCHCONTOUR, shading=XY,
>   orientation=[20,60], scaling=constrained, axes=BOXED);
```

(Figure 6.9)

I chose "BOXED" axes in this one because this is the clearest way to show how the edges of the plotted surface lie on the faces of the plotting region. Real-time rotation of the surface (in Release 5 under Windows 95, 98, or NT) is very useful here to help you visualize the actual shape of the surface.

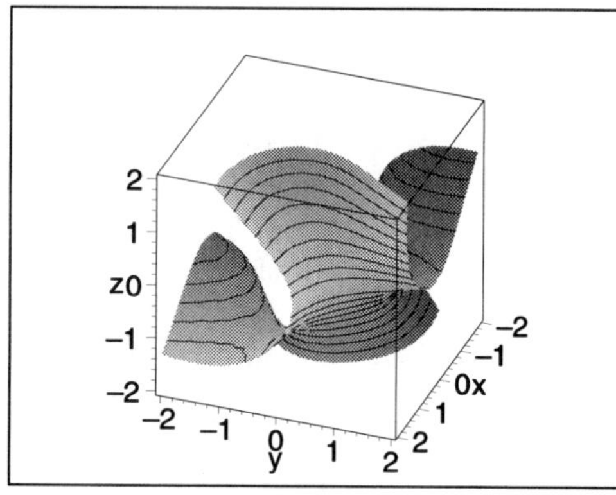


Figure 6.9. $x^2y - y^2z + z^2x = 1$

## Animation in Three Dimensions

Animation provides an opportunity to produce a "four-dimensional" graph, using time as the fourth dimension. Often we use it to show the effect of changing a parameter gradually. The three-dimensional animation command is "animate3d". It is similar to "plot3d", but with an extra range for the frame parameter (the variable that changes from frame to frame). As in "animate", there is an option "frames=..." to specify the number of frames.

Three-dimensional animations can take a very long time to produce, and use a lot of memory. The problem is worst if you use a large number of frames. Don't try to compete with Hollywood!

```
> animate3d({ x^2 + y^2, t - x^2 - 2*y^2 }, x=-1..1, y=-1..1,
>              t=0..1, style=PATCH);
```

This example shows two intersecting surfaces. As the frame parameter $t$ goes from 0 to 1, one surface moves upward.

Although in "plot3d" the $y$ range is allowed to depend on $x$, this is not true in "animate3d". The $x$ and $y$ ranges can't depend on the frame parameter $t$ either.

Another, somewhat more flexible, form of animation uses the "display" command with option "insequence=true". This allows you to animate any type of three-dimensional plot. The other options (shading, style, etc.) should be in the "display" command, and should come after "insequence=true".

```
> for kk from -5 to 5 do
>   p[kk]:= implicitplot3d(x^2 + y^2 - z^2 - kk/8 * (z-2) = 1,
>      x=-3..3, y=-3..3, z=-2..2, grid = [15,15,15]) od:
> display([ seq(p[k], k=-5..5)], insequence=true,
>            style=PATCHCONTOUR, shading=XY, axes=BOXED);
```

## EXERCISES

1. Plot in three dimensions the graphs of the following functions. Use option settings that show the shapes of the graphs as clearly as possible.
   **(a)** $\sin(xy)$. **(b)** $\sin(x)\sin(y)$ **(c)** $xy/(1 + x^2 + y^2)$
   **(d)** $x/(1 + y^2) - y/(1 + x^2)$
2. Plot the graphs of the following functions. Pay attention to the fact that $f(x, y)$ is not defined for all $(x, y)$. You may need to use the "view" option if $f(x, y)$ gets very large at some points.
   **(a)** $xy/(x^2 + y^2)$ **(b)** $x^y$ **(c)** $x \ln y + y \ln x$ **(d)** $\sqrt{\sin x \sin y}$

3. Produce graphs of the following objects using "`plot3d`". In most cases you will need to plot two different functions. For those objects that are unbounded, cut them off at a circular cross-section. Watch out for square roots of quantities that can become negative. You may need to adjust the endpoints of the $y$ intervals slightly (roundoff error can make a quantity negative when it should be 0).

   **(a)** A cone. **(b)** A sphere.

   **(c)** An ellipsoid with axes of lengths 1, 2, and 3. **(d)** A hyperboloid of one sheet.

4. The function $f(x, y) = xy/(x^2 - y^2)$ is not defined when $x = \pm y$. Plot it by dividing the plotting region into four pieces (avoiding $x = \pm y$) and combining them using "`display`". You should also use the "`view`" option in "`display`" to avoid plotting when $f(x, y)$ is large.

5. A "tent" has a circular base of radius $a$ centred at $(0, 0)$ and a horizontal ridge bar at height $b$ above the $x$ axis. Each cross-section perpendicular to the $x$ axis is an isosceles triangle with base a chord of the circle and vertex on the ridge bar. Represent the tent surface in the form $z = f(x, y)$. Plot it in three dimensions with $a = 1$ and $b = 1$.

6. Produce contour plots of the functions in Exercise 2. Observe how features of the three-dimensional graphs are related to features of the contour plots. Also produce density plots. Which type of plot would be more useful and informative?

7. In our example for "`display(..., insequence=true)`", the intersection of the surface with the plane $z = 2$ was constant in time. Why? Can you produce a similar example where the intersections with both $z = 2$ and $z = -2$ are both constant? Can you produce one where the intersections with $z = 2$ and $x = 0$ are constant?

8. If you used the options "`contours=12, grid=[25,25]`" in Figure 6.7, you would see a sharp "cusp" in one of the contours. With some other settings, the cusp disappears. Why? Should there really be a cusp in this plot?

9. This is not just a happy face, it's a contour plot of $((x - 1)^2 + y^2 - 1/2)((x + 1)^2 + y^2 - 1/2)(x^2/10 + (y + 1)^2 - 1/2)$. Using an appropriate setting for "`contours`", produce a plot where the corners of the "mouth" are included.

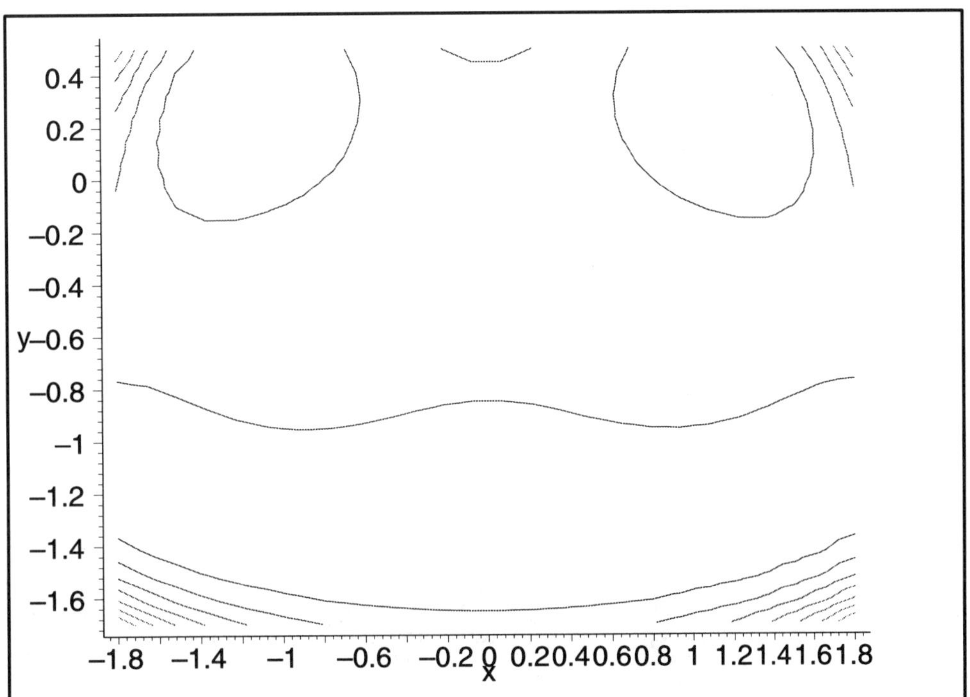


Figure 6.10. A face?

*CHAPTER 7*

# *Partial Differentiation*

## LAB 19 PARTIAL DERIVATIVES AND GRADIENTS

**Goals:** *To learn how to use Maple in calculations involving partial derivatives and the Chain Rule. To better understand and visualize the concepts of the gradient vector, directional derivatives, and tangent planes to surfaces.*

### Partial Derivatives

Maple's "`diff`" and "`D`" operators can be used to produce partial derivatives for functions of several variables. If "`expr`" is an expression involving a variable "`x`", then "`diff(expr,x)`" is the partial derivative of `expr` with respect to `x`.

```
> diff(x^2 * y + z, x);
```

$$2xy$$

```
> diff(f(x,y), y);
```

$$\frac{\partial}{\partial y} f(x,y)$$

You can obtain higher-order partial derivatives, pure or mixed, simply by placing additional variables as inputs to "`diff`".

```
> diff(f(x,y), x, y);
```

$$\frac{\partial^2}{\partial y\,\partial x} f(x,y)$$

You can also use "`$`" in repeated partial differentiation with respect to the same variable.

```
> diff(f(x,y), x$2, y);
```

$$\frac{\partial^3}{\partial y\,\partial x^2} f(x,y)$$

For a function "`f`" of several variables, "`D[k](f)`" is the partial derivative of `f` with respect to the $k$'th variable.

```
> f:= (x,y,z) -> x^2 * y + z;
> D[2](f);
```

$$(x,y,z) \to x^2$$

For higher partial derivatives, you specify several indices instead of one. Again you can use "`$`" to differentiate repeatedly with respect to one of the variables.

```
> D[1,2](f)(x,y,z);
```

$$2x$$

```
> D[1$2](f)(x,y,z);
```

$$2y$$

```
> D[1$2,2](f)(x,y,z);
```

$$2$$

## The Chain Rule

Calculations involving the Chain Rule for functions of several variables can be quite complicated to do by hand, especially for higher derivatives. With Maple they are easy to do (although the results may still be quite complicated).

■ **EXAMPLE 1** Let $u(s,t) = f(x(s,t),\, y(s,t),\, x(s,t)y(s,t))$ for some differentiable functions $f$, $x$ and $y$. Find $\partial u/\partial s$.

**SOLUTION**

```
>f:= 'f';  u:= f(x(s,t), y(s,t), x(s,t)*y(s,t));
> diff(u,s);
```

$$D_1(f)(x(s,t), y(s,t), x(s,t)y(s,t))\left(\frac{\partial}{\partial s}x(s,t)\right) + D_2(f)(x(s,t), y(s,t), x(s,t)y(s,t))\left(\frac{\partial}{\partial s}y(s,t)\right) + D_3(f)(x(s,t), y(s,t), x(s,t)y(s,t))\left(\left(\frac{\partial}{\partial s}x(s,t)\right)y(s,t) + x(s,t)\frac{\partial}{\partial s}y(s,t)\right)$$ ■

■ **EXAMPLE 2** Convert the expression $A = \dfrac{\partial^2 u}{\partial x^2} + \dfrac{\partial^2 u}{\partial y^2}$ (the left side of the Laplace equation) to polar coordinates.

**SOLUTION** Here $u$ is a function of two variables $x$ and $y$. We want to use polar coordinates $r$ and $\theta$, i.e., we write $u(x,y) = v(r,\theta)$. We can either consider $x$ and $y$ as functions of $r$ and $\theta$, or $r$ and $\theta$ as functions of $x$ and $y$. In this case, since we know the expression in terms of $x$ and $y$, it may be easier to do the latter. I'll write $r = \sqrt{x^2+y^2}$ and $\theta = \arctan(y,x)$ (using Maple's two-variable form of "arctan").

```
> u:= (x,y) -> v(sqrt(x^2 + y^2), arctan(y, x)):
> A:= D[1$2](u)(x,y) + D[2$2](u)(x,y);
```

The result is rather large and messy, and I won't reproduce it here. It is, of course, expressed in terms of $x$ and $y$. I want an expression in terms of $r$ and $\theta$, so I'll substitute in the expressions for $x$ and $y$ in terms of $r$ and $\theta$.

```
> A:= subs(x=r*cos(theta), y=r*sin(theta), A);
```

The "subs" command doesn't do any simplification, and our expression is certainly in need of some. I'll use "simplify".

```
> A:= simplify(A);
```

$$A := ((r^2\mathrm{csgn}(r)D_{1,1}(v)(\mathrm{csgn}(r)r,\ \arctan(r\ \sin(\theta),\ r\ \cos(\theta))) + D_1(v)(\mathrm{csgn}(r)r, \arctan(r\ \sin(\theta),\ r\ \cos(\theta)))r + \mathrm{csgn}(r)D_{2,2}(v)(\mathrm{csgn}(r)r, \arctan(r\ \sin(\theta),\ r\ \cos(\theta)))\mathrm{csgn}(r)))/r^2$$

The csgn($r$) terms arise from simplifying $\sqrt{r^2}$ (we didn't tell Maple that $r > 0$). Moreover, $\arctan(r\sin\theta, r\cos\theta)$ should be $\theta$, but again we didn't tell Maple that $-\pi < \theta \le \pi$. We can remove the csgn($r$), and also substitute $\theta$ for $\arctan(r\ \sin(\theta),\ r\ \cos(\theta))$, using "subs".

```
> A:= subs(csgn(r)=1,arctan(r*sin(theta),r*cos(theta))=theta,A);
```

$$A := \frac{r^2 D_{1,1}(v)(r,\theta) + D_1(v)(r,\theta)r + D_{2,2}(v)(r,\theta)}{r^2}$$

This is the desired formula. We usually write it in "$\partial$" notation.

```
> convert(A,diff);
```

$$\left(r^2\left(\frac{\partial^2}{\partial r^2}v(r,\theta)\right)+\left(\frac{\partial}{\partial r}v(r,\theta)\right)r+\left(\frac{\partial^2}{\partial \theta^2}v(r,\theta)\right)\right)/r^2$$ ■

## Gradients and Directional Derivatives

The "grad" command, which is in the "linalg" package, computes the gradient of an expression in several variables. In addition to the expression itself, you must provide the list of variables. This ensures that the components of the result are in the proper order. The result of "grad" is a vector.

```
> with(linalg, grad):
> grad( x^2 - y^2 + x * y * z, [x,y,z]);
```

$$[2x+yz,\ -2y+xz,\ xy]$$

We will also use the definitions in the file "dot.def"(1), which we first met in Lab 14. In particular, we will use "evl" to convert vectors to lists.

```
> read `dot.def`;
```

■ **EXAMPLE 3** Find the directional derivative of $f(x,y)=xy^3$ in the direction of $\mathbf{i}+2\mathbf{j}$ at the point $(2,3)$.

**SOLUTION** Maple doesn't have a specific command for directional derivatives, but we can use "grad" together with the dot product. We then define the vector $\mathbf{V}=\mathbf{i}+2\mathbf{j}$, and take the dot product of the gradient of $f$ with a unit vector in the direction of $\mathbf{V}$.

```
> f:= (x, y) -> x*y^3;
> V:= [1,2];
> G:= evl(grad(f(x,y), [x,y]));
```

$$G := \left[y^3,\ 3xy^2\right]$$

```
> dd:= unit(V) &. G;
```

$$dd := \frac{1}{5}\sqrt{5}y^3+\frac{6}{5}\sqrt{5}xy^2$$

Now to obtain our answer, we evaluate this at the point $(2,3)$, i.e., substitute $x=2$, $y=3$.

```
> subs(x=2, y=3, dd);
```

$$27\sqrt{5}$$ ■

Let's take another look at the directional derivatives, the gradient, and what they mean. We start with the parametric equations for the line through the point $(2,3)$ at angle $t$ counterclockwise from the positive $x$ direction. The parameter $s$ represents distance along the line.

```
> xp:= 2 + s * cos(t); yp:= 3 + s * sin(t);
```

Now let's look at our function $f$ along this line. We make this a function of $s$ and $t$, though we will really think of it as a function of $s$ for each value of $t$.

```
> fp:= unapply(f(xp, yp), (s,t));
```

$$fp := (s,t)\rightarrow(2+s\cos(t))\,(3+s\sin(t))^3$$

What is the derivative of this function (with respect to $s$)?

```
> D[1](fp);
```

$$(s,t)\rightarrow\cos(t)\,(3+s\sin(t))^3+3\,(2+s\cos(t))\,(3+s\sin(t))^2\sin(t)$$

We evaluate this at $s=0$ (which corresponds to our point $(2,3)$).

(1) If this file is not available to you, you can type in the definitions from Appendix I.

```
> D[1](fp)(0,t);
```

$$27\cos(t) + 54\sin(t)$$

This is the directional derivative of $f$ at $(2,3)$ in the direction of the unit vector $[\cos(t),\ \sin(t)]$ (i.e., the direction of our line, at angle $t$ counterclockwise from the positive $x$ direction). It should correspond to the dot product of this vector with the gradient of $f$ at $(2,3)$ (which I'll call $\mathbf{G}_0$).

```
> G0:= subs(x=2, y=3, G);
```

$$G0 := [27,\ 54]$$

```
> dd:= [cos(t), sin(t)] &. G0;
```

$$dd := 27\cos(t) + 54\sin(t)$$

Let's plot this as a function of $t$.

```
> plot(dd, t=0..2*Pi); (Figure 7.1)
```

The maximum value of the directional derivative appears to occur somewhere around $t = 1.1$. Let's find the maximum exactly, by finding where the derivative of "dd" is zero.

```
> tm:= solve(diff(dd, t)=0, t);
```

$$tm := \arctan(2)$$

Actually, as you see on the graph, there are two critical points in the interval $[0,\ 2\pi]$, one a maximum and the other a minimum. As it happens, this one is the maximum (recall that $\arctan(x)$ is between $-\pi/2$ and $\pi/2$, and positive when $x$ is positive). What is the unit vector giving the direction corresponding to this?

```
> V:= [cos(tm), sin(tm)];
```

$$V := \left[\frac{1}{5}\sqrt{5},\ \frac{2}{5}\sqrt{5}\right]$$

This should be the direction of the gradient vector $\mathbf{G}_0$.

```
> unit(G0);
```

$$\left[\frac{1}{5}\sqrt{5},\ \frac{2}{5}\sqrt{5}\right]$$

The length of the gradient vector should be the maximum value of the directional derivative. Let's check:

```
> len(G0) = simplify(subs(t=tm, dd));
```

$$27\sqrt{5} = 27\sqrt{5}$$

At right angles to the direction of the maximum directional derivative should be a direction in which the directional derivative is zero.

```
> tz:= tm + Pi/2;
> simplify(subs(t=tz, dd));
```

$$0$$

This should be the direction of the tangent at the point $(2,3)$ to the curve $f(x,y) = C$ passing through the point. Let's plot the curve and its tangent. For the tangent line I'll use the parametric equations with $t = tz$. For the curve I'll use "implicitplot" (although in this case we could easily write $y$ explicitly as a function of $x$). I'll combine the two plots with "display".

```
> with(plots):
> curve:= implicitplot(f(x,y)=f(2,3), x=0..5, y=0..5):
> tang:= plot(subs(t=tz, [xp, yp, s=-1..1])):
> display({curve, tang}, scaling=constrained);
```

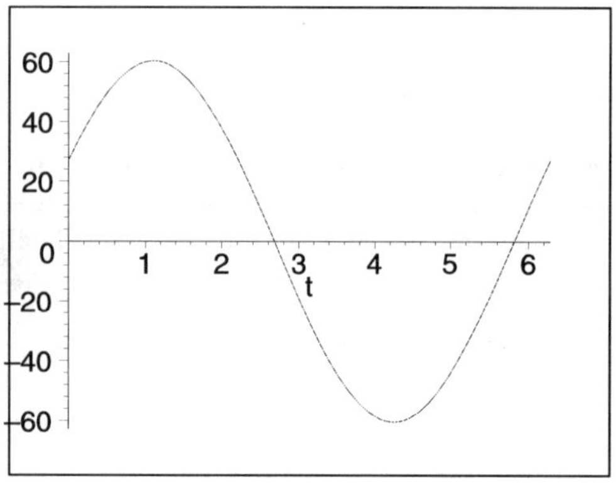


Figure 7.1. Directional derivatives

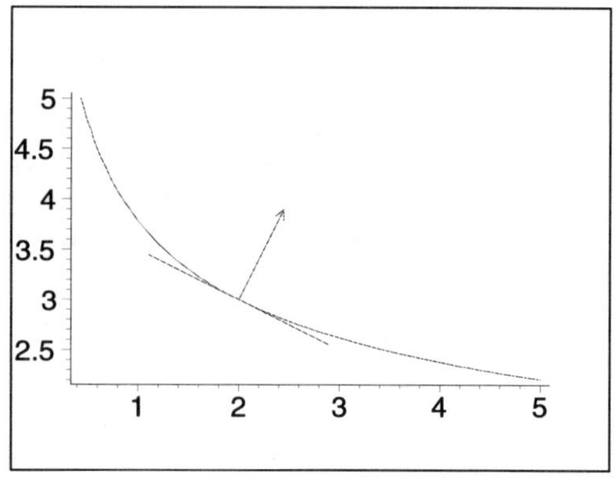


Figure 7.2. Curve, tangent, gradient

There's one more element I'd like to add to the picture: an arrow showing the direction of the gradient. Recall from Lab 14 how we can produce an arrowhead. The "tail" of the arrow will be at the point $\mathbf{P} = (2, 3)$, the "head" at $\mathbf{Q} = \mathbf{P} + \mathbf{V}$, the "barbs" at $\mathbf{R} = \mathbf{Q} - 0.1\mathbf{V} + .05\mathbf{U}$ and $\mathbf{S} = \mathbf{Q} - 0.1\mathbf{V} - 0.5\mathbf{U}$, where $\mathbf{U} = [\cos(tz), \sin(tz)]$ is perpendicular to $\mathbf{V}$.

```
> P:= [2,3]; Q:= evl(P + V);
> U:= [cos(tz),sin(tz)];
> R:= evl(Q - 0.1 * V + .05 * U);
> S:= evl(Q - 0.1 * V - .05 * U);
> arrow:= plot([P,Q,R,Q,S]):
> display({curve, tang, arrow}, scaling=constrained); (Figure 7.2)
```

## Tangent Planes

The tangent plane to the surface $z = f(x, y)$ at $(a,\ b,\ f(a, b))$ has the equation

$$z = f(a, b) + f_1(a, b)(x - a) + f_2(a, b)(y - b)$$

Let's write a function that takes $(a, b)$ and produces a three-dimensional plot of part of this tangent plane, which we will later combine with a plot of the surface itself. This uses the function $f$: if "`f`" has been assigned a value, you should undefine it now.

```
> f:= 'f';
> tanplane:= (a,b) -> plot3d(f(a,b) + D[1](f)(a,b)*(x-a)
>       + D[2](f)(a,b)*(y-b), x=a-.3..a+.3, y=b-.3..b+.3,
>       style=PATCH, colour=red, grid=[7,7]);
```

The red colour should (on a colour display) produce a good contrast between the tangent plane and the original surface. Using "`grid=[7,7]`" with $x$ and $y$ intervals of length $0.6$ means the grid lines will be separated by $0.1$. It's probably best to have approximately the same separation for the grid lines on the surface. If the $x$ and $y$ intervals for that surface are, say, $-1..1$, then we should use "`grid=[21,21]`" for the surface.

**EXAMPLE 4** Plot the surface $z = 1 - x^2 - y^2$ for $-1 \le x \le 1$, $-1 \le y \le 1$, together with its tangent plane at $(1/2, 1/2)$.

**SOLUTION**

```
> f:= (x,y) -> 1 - x^2 - y^2;
> p:= plot3d(f(x,y), x = -1 .. 1, y = -1 .. 1,
>              style = PATCH, grid = [21,21]):
> display({p, tanplane(.5,.5)}); (Figure 7.3)
```

Note that in this case the entire patch of tangent plane is visible. In other cases, some or all of it may be hidden by the surface. In Exercise 8 of Lab 21 we will reconsider the question of when this occurs. You might want to look at the picture from several different angles to understand the geometry of the surface and its tangent plane. ■

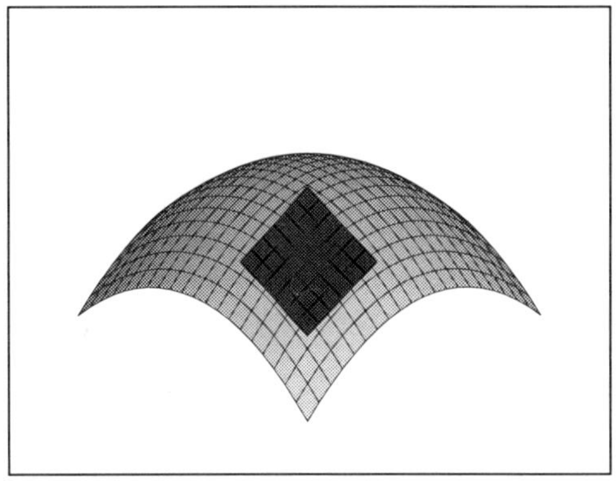

Figure 7.3. Surface and tangent plane

We can also produce an animation of the surface with its tangent plane moving as the point $(a, b)$ moves along the surface. Each frame will be a "display" as above, with "tanplane" called at a different point, and then another "display" with insequence=true will combine them into an animation. In this case, the point of tangency rotates in a circle centred at the origin.

```
> display([ seq(
>    display({p, tanplane(.5*cos(k*Pi/4), .5*sin(k*Pi/4))}),
>       k=-3..3) ], insequence=true);
```

For a surface defined by an equation $g(x, y, z) = constant$, the gradient of $g$ at $(a, b, c)$ on the surface is a normal to the tangent plane (if it is not the zero vector). Therefore the equation of the tangent plane is $(\text{grad}\, g)(a, b, c) \bullet (x - a, y - b, z - c) = 0$.

■ **EXAMPLE 5** Plot part of the surface $x^2y^2 + y^2z^2 + z^2x^2 = 9$ together with its tangent plane at $(1, 1, 2)$.

**SOLUTION** I'll first define the function $g$ and check that $g(1, 1, 2) = 9$.

```
> g:= (x,y,z) -> x^2*y^2 + y^2*z^2 + z^2*x^2;
> g(1,1,2);
```

$$9$$

Now I'll find the gradient of $g$ at the point $(1, 1, 2)$ and the equation of the tangent plane there.

```
> N:= subs(x=1, y=1, z=2, grad(g(x,y,z),[x,y,z]));
> eq:= N &. [x-1,y-1,z-2] = 0;
```

$$eq := 10x - 36 + 10y + 8z = 0$$

For plotting purposes, I'll solve the equation for $z$ as a function of $x$ and $y$. I'll plot the part of the surface for $-1 \le x \le 3, -1 \le y \le 3, 0 \le z \le 4$ and the part of the tangent plane for $1/2 \le x \le 3/2$ and $1/2 \le y \le 3/2$. Note that part of the tangent plane is hidden by the surface: you can rotate the plot to see the hidden part on the other side of the surface.

```
> zt:= solve(eq, z);
> p:= implicitplot3d(g(x,y,z)=9, x=-1..3, y=-1..3, z=0..4,
>                  style= PATCHCONTOUR):
> q:= plot3d(zt, x=1/2 .. 3/2, y=1/2 .. 3/2, colour=red,
>                style=PATCHCONTOUR, grid=[5,5]):
> display({p,q});
```
■

## EXERCISES

1. Use Maple to find $\dfrac{\partial^3 z}{\partial x \partial y^2}$ where $z = x^2y^3$. Compare to $\dfrac{\partial^3 z}{\partial y \partial x \partial y}$ and $\dfrac{\partial^3 z}{\partial y^2 \partial x}$. Try it for some other functions of $x$ and $y$.

**2.** **(a)** Find the gradient of $f(x, y, z) = x^2 + y^2 + 2z^2$ at the point $(1, 2, 3)$.

**(b)** Find the directional derivative of $f(x, y, z)$ in the direction of the gradient at that point, by considering $f$ along a line through the point in that direction, using a parameter $s$ representing distance along the line.

**(c)** Find the second derivative of $f$ with respect to $s$ along the line. This is called the second directional derivative of $f$ in this direction.

**3.** **(a)** Verify that if $g(x)$ and $h(x)$ are differentiable functions, the function $f(x, y) = \exp(g(x) + h(y))$ satisfies the equation $\dfrac{\partial^2}{\partial x\,\partial y} f(x, y) = g'(x)h'(y)f(x, y)$.

**(b)** What similar equation would $f(x, y) = \cos(g(x) + h(y))$ satisfy?

**4.** Consider a function of the form $h(x, y) = f(xy,\, g(x^2, y^2))$ where $f$ and $g$ have continuous partial derivatives of all orders. Find $\dfrac{\partial h}{\partial x}$, $\dfrac{\partial^2 h}{\partial x\,\partial y}$ and (if you want to see a really complicated one) $\dfrac{\partial^3 h}{\partial^2 x \partial y}$.

**5.** **(a)** Find the points on the curve $x^2 y - xy^2 = 1$ where the tangent to the curve is perpendicular to $[x, y]$. *Hint:* first plot the curve to see approximately where these points are. Then use "`fsolve`", specifying intervals for $x$ and $y$.

**(b)** Plot the curve with an arrow in the direction of the gradient of $x^2 y - xy^2$ at each of the points found in (a).

**6.** **(a)** Find all points on the surface $xy^2 + yz^2 + zx^2 = 6$ where the tangent plane is horizontal.

**(b)** Find all points on the surface where the tangent plane is vertical.

**(c)** Plot part of the surface together with a horizontal tangent plane and a vertical tangent plane. *Hint:* you can use "`implicitplot3d`" to produce the vertical tangent plane.

**7.** This is a three-dimensional version of Example 3. Consider the function $f(x, y, z) = (1 + x + y^2 - 2z)^2$.

**(a)** Find the derivative with respect to $s$ of $f(s \sin\phi \cos\theta,\ s \sin\phi \sin\theta,\ s\cos\phi)$ at $s = 0$.

**(b)** Show that this is the directional derivative of $f$ at $(0, 0, 0)$ in a certain direction.

**(c)** For what $\phi$ and $\theta$ is this a maximum? Show that this corresponds to the direction of the gradient of $f$ at $(0, 0, 0)$.

**8.** You can think of the mixed second partial derivative $f_{xy}$ as the rate of change with respect to $y$ of the slope of the graph of $f$ in the $x$ direction. To visualize this, construct an animation. First try $f(x, y) = (x + y)\sin x$ (having $f_{xy}$ depend only on $x$ will make the results easier to see). Each frame of the animation should show $f(x, y)$ and $f_{xy}(x, y)$ for a fixed value of $y$. Observe how the slope of the $f$ curve increases from frame to frame where $f_{xy}$ is positive, and decreases where $f_{xy}$ is negative. Try it with some other functions $f$. You might try one where $f_{xy} = 0$ when $x = 0$ or $y = 0$, or one where $f_{xy}$ is constant.

**9.** The patch of tangent plane plotted by our "`tanplane`" is generally not a square or even a rectangle (its projection on the $xy$ plane is a square, but it is tilted). To produce a square patch, you might try replacing the $x$ and $y$ bounds in "`tanplane`" by something of the form

```
x = a - p(a,b) .. a + p(a,b),
y = b + r(a,b)*(x-a) - q(a,b) .. b + r(a,b)*(x-a) + q(a,b)
```

**(a)** Construct vectors going from one corner of the tangent plane to the two adjacent corners.

**(b)** Determine $r$ so these vectors are orthogonal. With this $r$, determine $p$ and $q$ so the vectors have length .6.

**(c)** Define a new "`tanplane`" command using the new $x$ and $y$ bounds. Plot the surface $z = \sqrt{(x - x^2)(y - y^2)}$ for $0 \le x \le 1$ and $0 \le y \le 1$ together with some of its tangent planes.

**10.** The left side of the three-dimensional Laplace equation in spherical coordinates is

$$\frac{\partial^2 u}{\partial \rho^2} + \frac{2}{\rho}\frac{\partial u}{\partial \rho} + \frac{\cot\phi}{\rho^2}\frac{\partial u}{\partial \phi} + \frac{1}{\rho^2}\frac{\partial^2 u}{\partial \phi^2} + \frac{1}{\rho^2 \sin^2\phi}\frac{\partial^2 u}{\partial \theta^2}$$

Verify this by converting it to rectangular coordinates $x = \rho\sin\phi\cos\theta$, $y = \rho\sin\phi\sin\theta$, $z = \rho\cos\phi$.

# LAB 20 IMPLICIT FUNCTIONS AND APPROXIMATIONS

**Goals:** *To learn to use Maple in calculations involving implicit functions and Taylor series with several variables. To see why the hypotheses of the Implicit Function Theorem include the requirement that the Jacobian determinant be nonzero.*

## Implicit Partial Differentiation

We have already used implicit differentiation when an equation $f(x, y) = C$ defines $y$ as a function of $x$ (see Lab 5). Similar techniques can be used for an equation involving three or more variables.

**■ EXAMPLE 1** The equation $xy^2 + yz^2 + zx^2 = 5$ defines $z$ as a function of $x$ and $y$. Find the partial derivatives $\frac{\partial z}{\partial x}$ and $\frac{\partial^2 z}{\partial x\, \partial y}$.

**SOLUTION**

```
> eq:= x*y^2 + y*z^2 + z*x^2 = 5;
> eq1:= subs(z=z(x,y), eq);
```

$$eq1 := xy^2 + yz(x,y)^2 + z(x,y)x^2 = 5$$

```
> eq2:= diff(eq1, x);
```

$$eq2 := y^2 + 2yz(x,y)\left(\frac{\partial}{\partial x}z(x,y)\right) + \left(\frac{\partial}{\partial x}z(x,y)\right)x^2 + 2z(x,y)x = 0$$

```
> Zx:= solve(eq2, diff(z(x,y),x));
```

$$Zx := -\frac{y^2 + 2z(x,y)x}{2yz(x,y) + x^2}$$

This is $\frac{\partial z}{\partial x}$. To get $\frac{\partial^2 z}{\partial x\, \partial y}$, we differentiate this with respect to $y$.

```
> Zxy:= diff(Zx, y);
```

$$Zxy := -\frac{2y + 2\left(\frac{\partial}{\partial y}z(x,y)\right)x}{2yz(x,y) + x^2} + \frac{(y^2 + 2z(x,y)x)\left(2z(x,y) + 2y\left(\frac{\partial}{\partial y}z(x,y)\right)\right)}{(2yz(x,y) + x^2)^2}$$

Now $\frac{\partial z}{\partial y}$ can be obtained by the same method we used for $\frac{\partial z}{\partial x}$. I will substitute that in this expression. The result is rather complicated, so I'll use "`normal`" to simplify it.

```
> Zy:= solve(diff(eq1,y), diff(z(x,y), y));
> Zxy:= subs(diff(z(x,y),y) = Zy, Zxy);
> normal(Zxy);
```

$$2\frac{-3y^3z(x,y)^2+yx^4+4xz(x,y)^3y+3x^3z(x,y)^2-3y^2z(x,y)x^2-2y^4x}{(2yz(x,y)+x^2)^3}$$

■

**■ EXAMPLE 2** Consider the two equations $xy^2+yz^2+zw^2+wx^2=14$ and $x^2y+y^2z+z^2w+w^2x=22$. Find $\left(\frac{\partial z}{\partial x}\right)_y$ at the point where $x=2$, $y=1$, $z=0$, $w=3$ (note that this does satisfy the equations). The notation $\left(\frac{\partial z}{\partial x}\right)_y$ means that we are to think of $z$ as a function of $x$ and $y$ and take the partial derivative with respect to $x$, keeping $y$ constant.

**SOLUTION** We are to consider $z$ and $w$ as functions of $x$ and $y$. I'll define the equations and differentiate them with respect to $x$.

```
> eq1:= x*y^2 + y*z(x,y)^2 + z(x,y)*w(x,y)^2 + w(x,y)*x^2 = 14;
> eq2:= x^2*y + y^2*z(x,y) + z(x,y)^2*w(x,y) + w(x,y)^2*x = 22;
> eq1x:= diff(eq1,x);
```

$$eq1x := y^2+2yz(x,y)\left(\frac{\partial}{\partial x}z(x,y)\right)+\left(\frac{\partial}{\partial x}z(x,y)\right)w(x,y)^2$$
$$+2z(x,y)w(x,y)\left(\frac{\partial}{\partial x}w(x,y)\right)+\left(\frac{\partial}{\partial x}w(x,y)\right)x^2+2w(x,y)x=0$$

```
> eq2x:= diff(eq2,x);
```

$$eq2x := 2xy+y^2\left(\frac{\partial}{\partial x}z(x,y)\right)+2z(x,y)w(x,y)\left(\frac{\partial}{\partial x}z(x,y)\right)$$
$$+z(x,y)^2\left(\frac{\partial}{\partial x}w(x,y)\right)+2w(x,y)x\left(\frac{\partial}{\partial x}w(x,y)\right)+w(x,y)^2=0$$

Now I could either solve these for $\frac{\partial z}{\partial x}$ and $\frac{\partial w}{\partial x}$ and then substitute the values $x=2$, $y=1$, etc., or substitute first and then solve. Maple can do it either way, but substituting first and then solving will result in less complicated expressions. When using "`subs`", we must remember that the substitutions are made one after the other, so after setting $x$ to 2 and $y$ to 1 we have $z(2,1)$, not $z(x,y)$, to set to 0. The substitutions are also made inside the "`diff`", so to avoid getting results such as "`diff(0,2)`", I'll also substitute names for the derivatives. Note that these names should not have assigned values (unassign them if necessary).

```
> eqs:= subs(diff(z(x,y),x)=dzdx, diff(w(x,y),x)=dwdx, x=2, y=1,
>        z(2,1)=0, w(2,1)=3, {eq1x, eq2x});
```

$$eqs := \{13+9dzdx+4dwdx=0,\ 13+dzdx+12dwdx=0\}$$

```
> solve(eqs, {dzdx, dwdx});
```

$$\{dwdx=-1,\ dzdx=-1\}$$

Although we don't really need $\frac{\partial w}{\partial x}$, we solved the equations for it as well as $\frac{\partial z}{\partial x}$ because we don't want it to appear in the expression for $\frac{\partial z}{\partial x}$.

■

## Jacobian Determinants

Would the method of Example 2 always work? The problem is that two equations in two unknowns don't always have a unique solution.

**EXAMPLE 3** Consider the same equations as in Example 2. Does $\left(\frac{\partial z}{\partial x}\right)_y$ exist at the point where $x = 0$, $y = 3$, $z = 2$, $w = 1$ (which also satisfies the equations)? If not, why not?

**SOLUTION** We try the same method as in Example 2, but using these new values of $x$, $y$, $z$, and $w$.

```
> subs(diff(z(x,y),x)=dzdx, diff(w(x,y),x)=dwdx, x=0, y=3,
>      z(0,3)=2, w(0,3)=1, {eq1x, eq2x});
```

$$\{9 + 13dzdx + 4dwdx = 0,\ 13dzdx + 4dwdx + 1 = 0\}$$

It is clear that this system of equations has no solution.

According to the theory, the condition for the system to have a unique solution is that the Jacobian determinant $\frac{\partial(F,G)}{\partial(z,w)} \neq 0$, where the equations are written $F(x,y,z,w) = 0$ and $G(x,y,z,w) = 0$ (actually the right sides can be constants). The Jacobian matrix can be calculated using the "`jacobian`" command (which is part of the "`linalg`" package), and then "`det`" will calculate the determinant. For $F$ and $G$ we want $w$ and $z$ to be written as variables, not as functions of $x$ and $y$.

```
> with(linalg, jacobian, det):
> F:= subs(z(x,y)=z, w(x,y)=w, op(1,eq1));
> G:= subs(z(x,y)=z, w(x,y)=w, op(1,eq2));
> M:= jacobian([F,G],[z,w]);
```

$$M := \begin{bmatrix} 2yz + w^2 & 2wz + x^2 \\ y^2 + 2wz & z^2 + 2wx \end{bmatrix}$$

```
> detm:=  det(M);
```

$$detm := 2yz^3 + 4wxyz - 3z^2w^2 + 2w^3x - 2wzy^2 - y^2x^2 - 2x^2wz$$

You can check that this is 0 for $x = 0$, $y = 3$, $z = 2$, $w = 1$, but not for $x = 2$, $y = 1$, $z = 0$, $w = 3$.

```
> subs(x=0, y=3, z=2, w=1, detm);
```

$$0$$

The Jacobian determinant not being zero is also needed for the Implicit Function Theorem:

> Consider a system of $n$ equations in $n + m$ variables
>
> $$F_j(x_1, \ldots, x_m, y_1, \ldots, y_n) = 0, \quad j = 1 \ldots n$$
>
> where all the $F_j$ have continuous first partial derivatives in a certain domain $R$. Suppose $x_1 = X_1, \ldots, x_m = X_m$, $y_1 = Y_1, \ldots, y_n = Y_n$ is a solution $P_0$ to these equations and is in the interior of $R$. Suppose the Jacobian determinant $\frac{\partial(F_1, \ldots, F_n)}{\partial(y_1, \ldots, y_n)}$ is not zero at $P_0$. Then there are functions $p_1, \ldots p_n$ of $x_1, \ldots x_m$, defined and differentiable near $X_1, \ldots, X_m$, such that $Y_i = p_i(X_1, \ldots, X_m)$ for $i = 1 \ldots n$, and $y_i = p_i(x_1, \ldots, x_m)$ for $i = 1 \ldots n$ is a solution of the system.

Thus we find that there are functions $z(x,y)$ and $w(x,y)$ defined near $x = 2$, $y = 1$ with $z(2,1) = 0$ and $w(2,1) = 3$ and satisfying equations 1 and 2. We have no such assurance for $x = 0$, $y = 3$ with $z(0,3) = 2$ and $w(0,3) = 1$.

Can we visualize what goes wrong when $x = 0$, $y = 3$, $z = 2$, $w = 1$? It might not be easy, since there are four dimensions involved. I'll eliminate one dimension by looking only at $y = 3$. For any choice of $x$ and $y$, the two equations define two implicit curves in the $zw$ plane. I'll produce an animation, where each frame shows the curves for a certain value of $x$, with $z$ ranging from 1 to 3 and $w$ from 0 to 2. I'll take $x$ values from -0.3 to 0.3.

I first define a function "`eqs`" that takes $k$ and substitutes $x = k/10$ and $y = 3$ into the equations. It produces a set of two equations. There will be seven frames, corresponding to integer values of $k$ from $-3$ to 3. I'll define a function "`frame`" that takes $k$ and produces the corresponding frame, the implicit plot of the equations "`eqs(k)`".

```
> with(plots):
> eqs:= k -> subs(x=.1*k, y=3, {eq1,eq2});
> frame:= k -> implicitplot(eqs(k), z=1..3, w=0..2);
> display([ seq(frame(kk), kk=-3..3)], insequence=true);
```

The first frame (for $x = -.3$) shows two intersections of the curve. As the curves move, the two intersections move closer to each other until in the fourth frame ($x = 0$) they coincide at $z = 2$, $w = 1$ with the two curves tangent. In the later frames, the two curves don't intersect at all (at least in the plot window).

It should now be clear how the conclusion of the Implicit Function Theorem fails for $x = 0$, $y = 3$, $z = 2$, $w = 1$, and why $\left(\dfrac{\partial z}{\partial x}\right)_y$ doesn't exist there. With $y = 3$ and $x$ slightly less than 0, there are not one but two points $(z, w)$ near $(2, 1)$, while for $x$ slightly greater than 0 there are none. ■

## Multivariate Taylor Series

In Lab 16 we saw Taylor series for functions of one variable, which Maple produced with the "`taylor`" command. Maple can also produce Taylor series for functions of several variables, using the "`mtaylor`" command. This must first be read in to Maple with "`readlib`".

```
> readlib(mtaylor):
> f:= exp(x^2*y)*cos(y);
> mt:= mtaylor(f, [x=0,y=Pi], 5);
```

$$mt := -1 + \frac{1}{2}(y-\pi)^2 - x^2\pi - x^2(y-\pi) - \frac{1}{24}(y-\pi)^4 + \frac{1}{2}x^2\pi(y-\pi)^2 - \frac{1}{2}x^4\pi^2$$

The first input to "`mtaylor`" is the expression $f$ for which to find the series. The second is a list or set giving the variables and the values about which to do the expansion. The third is the "truncation order" of the series: if this is $n$, you obtain the Taylor polynomial of degree $n-1$, involving terms in the variables of total degree up to $n-1$. Just as with "`taylor`", the third input is optional: if it is omitted, Maple uses the value of the variable "`Order`", which is initially 6.

Note that in contrast to "`taylor`", "`mtaylor`" returns an ordinary polynomial, not a special "series" data structure. There is no "$O(\ldots)$" term. Our result in this case includes terms of total degree up to 4, those of total degree 4 being $-(1/24)(y-\pi)^4$, $(1/2)x^2(y-\pi)^2$, and $-(1/2)x^4\pi^2$.

Although we will not write down a detailed error formula for a multivariate Taylor series, it is useful to get a rough idea of the size of the error, i.e., the "$O(\ldots)$" term that "`mtaylor`" omits. In the example above, if we change the 5 to a larger number, we would expect to obtain additional terms of total degree 5 or more. In fact, here are the additional terms of total degree 5:

```
> t5:= mtaylor(f, [x=0,y=Pi], 6) - mt;
```

$$t5 := \frac{1}{2}x^2(y-\pi)^3 - x^4\pi(y-\pi)$$

Could we describe this using "$O(\ldots)$" notation? If we have some quantity $p$ with $|x| \le p$ and $|y-\pi| \le p$, we could say

$$\left|\frac{1}{2}x^2(y-\pi)^3 - x^4\pi(y-\pi)\right| \leq \left(\frac{1}{2}+\pi\right)p^5$$

so $t_5 = O(p^5)$. This means that there is some constant $C$ such that whenever $x$ is sufficiently close to 0 and $y$ is sufficiently close to $\pi$, $|t_5| \leq Cp^5$. It is plausible (and true, though not obvious) that this also works for the error in approximating $f$ by the Taylor polynomial "`mtaylor(f, [x=0,y=Pi], 5)`". There are a number of possible choices for $p$. We might take the maximum of $|x|$ and $|y-\pi|$, or their sum. A more geometrically significant choice might be the distance from the point $(x, y)$ to $(0, \pi)$, i.e., $p = \sqrt{x^2 + (y-\pi)^2}$. I'll use this one.

```
> p:= sqrt(x^2 + (y-Pi)^2);
> Er:= f - mt;
```

To see graphically that the error $Er$ is bounded by a constant times $p^5$, I'll make a three-dimensional plot of $z = Er/p^5$ for $x$ in an interval around 0 and $y$ in an interval around $\pi$. Since we're interested in the size of $z$, I'll use an orientation with $\phi = 90°$: this makes the $z$ coordinate correspond to the vertical direction on the screen. The $y$ interval won't be exactly centred at $\pi$, which avoids numerical difficulties that might occur if one of the grid points was almost exactly $(0, \pi)$ (where both $Er$ and $p$ are 0).

```
> plot3d(Er/p^5, x = -.5 .. 0.5, y = 2.6 .. 3.7,
>        orientation=[0,90], axes=BOXED, style=PATCH);
```
(Figure 7.4)

It appears that in this region, $Er/p^5$ is always between about $-4$ and 1.

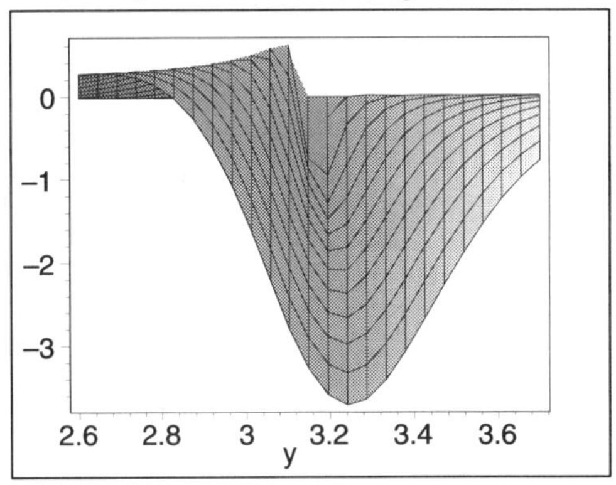


Figure 7.4. Error $= O(p^5)$

In general, suppose we have an expression $f$ involving $m$ variables $x_1, \ldots x_m$, which we expand in powers of $x_1 - a_1, \ldots x_m - a_m$ of total degree up to $n-1$ by "`mtaylor(f, [x1=a1, ..., xm=am], n)`". Then (if $f$ is well-behaved) the error should be $O(p^n)$ where we can take

$$p = \sqrt{(x_1-a_1)^2 + \ldots + (x_m-a_m)^2}$$

**EXAMPLE 4** Find the terms of total degree less than 5 in the Taylor series for

$$\int_0^x \sin(t\cos(y) - y\cos(t))\,dt$$

in powers of $x$ and $y$.

**SOLUTION** First we'll try "`mtaylor`".

```
> mtaylor(Int(sin(t*cos(y)-y*cos(t)), t=0..x), [x=0, y=0], 5);
Error, (in series/int) not implemented yet
```

This should have worked, except that it seems "`mtaylor`" doesn't yet know how to handle integrals. Instead, we can use "`taylor`" to produce a series in powers of $x$ whose coefficients will be functions of $y$, and then use "`mtaylor`" to make this into a series in powers of both $x$ and $y$.

```
> S:= taylor(Int(sin(t*cos(y)-y*cos(t)), t=0..x), x=0, 5);
```

$$S := -\sin(y)x + \frac{1}{2}\cos(y)^2x^2 + \left(\frac{1}{6}y\cos(y) + \frac{1}{6}\sin(y)\cos(y)^2\right)x^3$$
$$+ \left(-\frac{1}{24}\cos(y)^4 + \frac{1}{8}\sin(y)y\cos(y)\right)x^4 + O\left(x^5\right)$$

```
> mtaylor(S, [x=0,y=0], 5);
```

$$-yx + \frac{1}{2}x^2 + \frac{1}{6}y^3x + \frac{1}{3}yx^3 - \frac{1}{2}x^2y^2 - \frac{1}{24}x^4$$

■

## Series for Implicit Functions

In Lab 16 we found a Taylor series (in one variable) for a function defined implicitly by an equation involving two variables. Now we will extend this to multiple Taylor series. First we note a somewhat simpler way of doing the one-variable case, due to the fact that "`taylor`" understands the "`RootOf`".

**EXAMPLE 5** Find the first six terms of the Taylor series for $f(x)$ in powers of $x-1$, if $y = f(x)$ satisfies the equation $x^4 + y^4 = 2xy$ with $y(1) = 1$.

**SOLUTION** We simply ask "`taylor`" for the series of the root of the equation.

```
> eq:= x^4 + y^4 = 2*x*y;
> taylor(RootOf(eq, y), x=1, 6);
```

$$1-(x-1)-7(x-1)^2-49(x-1)^3-449(x-1)^4-4627(x-1)^5+O\left((x-1)^6\right)$$

This is the same answer we obtained (with a good deal more work) in Lab 16. But if you think about it, there's something strange here: we haven't used the condition $y(1) = 1$. How did Maple know that it should take this root of $1 + y^4 = 2y$ to be $y(1)$ rather than another? The answer is that it didn't know! It chose one root, which happened to be the one we wanted. In another case it might not have chosen the "right" one. How can we influence its choice?

Our first method seems to be work reliably only for expressions that are polynomials. This is an additional, optional input to "`RootOf`" that gives the approximate value of the root in question. Thus "`RootOf(eq, y, 1)`" is the root that is near 1, while "`RootOf(eq, y, .544)`" is the root that is near .5436890127 (the other real root of $1 + y^4 = 2y$).

```
> S1:= taylor(RootOf(eq, y, 1), x=1, 6);
```

$$S1 := 1-(x-1)-7(x-1)^2-49(x-1)^3-449(x-1)^4-4627(x-1)^5+O\left((x-1)^6\right)$$

```
> S2:= taylor(RootOf(eq, y, .544), x=1, 6);
```

$$\begin{aligned} S2 := \%1 &+ \left(\frac{17}{11}\%1+\frac{8}{11}\%1^2+\frac{12}{11}\right)(x-1)+\left(\frac{538}{121}+\frac{300}{121}\%1^2+\frac{467}{121}\%1\right)(x-1)^2 \\ &+\left(\frac{40262}{1331}+\frac{21796}{1331}\%1^2+\frac{33771}{1331}\%1\right)(x-1)^3 \\ &+\left(\frac{3410300}{14641}\%1+\frac{2210234}{14641}\%1^2+\frac{4064781}{14641}\right)(x-1)^4 \\ &+\left(\frac{386800652}{161051}\%1+\frac{250565042}{161051}\%1^2+\frac{460849557}{161051}\right)(x-1)^5+O\left((x-1)^6\right) \\ \%1 &= \mathrm{RootOf}\left(\_Z^3+\_Z^2+\_Z-1, .544\right) \end{aligned}$$

This last calculation takes a lot of time and memory. Apparently some of the intermediate steps involve extremely complicated expressions. For five or fewer terms, on the other hand, the calculation is quite rapid.

We can use "`evalf`" to obtain decimal approximations for the coefficients.

```
> evalf(S2);
```

$$\begin{aligned} &.5436890127+2.146135923(x-1.)+7.277537948(x-1.)^2+48.88487606(x-1.)^3 \\ &\quad +448.8944622(x-1.)^4+4627.200591(x-1.)^5+O\left((x-1.)^6\right) \end{aligned}$$

A second method, which should work even for non-polynomials, is to write $y$ as $a_0 + (x-1)v(x)$ where $a_0$ is the constant term $y(1)$, which we can choose, and find $v(x)$ as a Taylor series in $x-1$.

```
> eqv:= subs(y=1+(x-1)*v, eq);
> 1 + (x-1)*taylor(RootOf(eqv, v), x=1, 5);
```

$$1+(x-1)\left(-1-7(x-1)-49(x-1)^2-449(x-1)^3-4627(x-1)^4+O\left((x-1)^5\right)\right)$$

Note that we only need $v$ to order $O\left((x-1)^5\right)$ because multiplying this by $x-1$ makes it $O\left((x-1)^6\right)$. We have the right answer, except that Maple won't multiply the $(x-1)$ into the series. We have to convert the series for $v$ into a polynomial, multiply by $x-1$, and then use "taylor" to make it into a series again.

```
> v1:= convert(taylor(RootOf(eqv,v), x=1, 5), polynom);
```

$$v1 := -1 - 7(x-1) - 49(x-1)^2 - 449(x-1)^3 - 4627(x-1)^4$$

```
> taylor(1 + (x-1) * v1, x=1, 6);
```

$$1 - (x-1) - 7(x-1)^2 - 49(x-1)^3 - 449(x-1)^4 - 4627(x-1)^5$$

This lacks the "$O\left((x-1)^6\right)$" term because the input to the last "taylor" has no terms in higher powers than $(x-1)^5$. We could restore the "$O\left((x-1)^6\right)$" by adding in a fictitious term in $(x-1)^6$ which "taylor" would remove. ■

We can automate the process by writing a function to do it. Given an equation "eq" involving variables "x" and "y" and a solution $x = a$, $y = b$, "imptaylor(eq, y, b, x, a, n)" will produce $n$ terms of the Taylor series for $y$ in powers of $x - a$ with $y(a) = b$. Its definition can be found in the file "imptay.def", which you should now read in if it is available to you. If it is not available, you can type in the definition below. You should then save it in your own "imptay.def" file.

```
> imptaylor:= (eq, y, b, x, a, n) -> taylor(b + (x-a) *
> convert(taylor(RootOf(subs(y=b+(x-a)*_Z, eq)), x=a,n-1),
>           polynom) + (x-a)^n, x=a, n);
```

While "imptaylor" works for rational values of $a$ and $b$, due to limitations of Maple it does not always work for irrational values. Note also that $b$ must be an exact root, not a decimal approximation. In particular, this doesn't work for the root near .544 in the previous example.

Of course, another requirement is that the equation should actually define $y$ as a function of $x$ near the point $x = a$, $y = b$. According to the Implicit Function Theorem, if the equation is written as $F(x, y) = 0$ the requirement is that $\dfrac{\partial F}{\partial y}(a, b) \neq 0$. If this is not true, Maple will return an error message.

```
> imptaylor(x^2 + y^2=1, y, 0, x, 1, 3);
Error, (in imptaylor) does not have a taylor expansion, try
series()
```

Next let's consider an implicit equation involving three variables.

■ **EXAMPLE 6** Find the degree-3 Taylor polynomial for $z$ in powers of $x$ and $y-1$ near $x = 0$, $y = 1$, $z = 1$, where $z^3 + x^3 + y^3 = xy + yz + xz + 1$.

**SOLUTION** The version using the extra input to "RootOf" is very similar to Example 5.

```
> eq:= z^3 + x^3 + y^3 = x*y + y*z + x*z + 1;
> mtaylor(RootOf(eq,z,1), [x=0,y=1], 4);
```

$$2 - y + x + \frac{7}{2}x(y-1) - \frac{7}{2}(y-1)^2 - x^2 + \frac{3}{2}x^3 - \frac{49}{4}(y-1)^3 + \frac{39}{2}x(y-1)^2 - \frac{43}{4}x^2(y-1)$$

How could we proceed without this form of "RootOf"? Let $u(x)$ be the degree-3 Taylor polynomial in $x$ for $z(x, 1)$, obtained by substituting $y = 1$ in the implicit equation. This contains all the terms in the full series that have no $y-1$ factor. The full series is thus of the form $z(x, y) = u(x) + (y-1)w(x, y)$.

First I'll find "u". I need it as a polynomial rather than a series.

```
> u:= convert(imptaylor(subs(y=1, eq), z, 1, x, 0, 4), polynom);
```

$$u := 1 + x - x^2 + \frac{3}{2}x^3$$

Next I substitute $u + (y-1)w$ for $z$ in the equation. I'd like to obtain $w$ as a multivariate power series for the "RootOf" this. However, first I have to subtract the equation with $z = u$ and $y = 1$ (which contains the higher-order terms in $x$ that we want to ignore).

```
> equ:= subs(z = u+(y-1)*w, eq) - subs(z = u, y = 1, eq);
> w:= mtaylor(RootOf(equ, w), [x=0,y=1],3);
```

$$\frac{5}{2} - \frac{7}{2}y + \frac{7}{2}x - \frac{43}{4}x^2 + \frac{39}{2}x(y-1) - \frac{49}{4}(y-1)^2$$

Here, then, is the full series. To make it look more like a series, I'll use "mtaylor" again.

```
> mtaylor(u + (y-1)*w, [x,y=1],4);
```

$$2 + x - y - x^2 + \frac{7}{2}x(y-1) - \frac{7}{2}(y-1)^2 + \frac{3}{2}x^3$$
$$- \frac{43}{4}(y-1)x^2 + \frac{39}{2}x(y-1)^2 - \frac{49}{4}(y-1)^3$$

■

## EXERCISES

1. Find $\dfrac{\partial^2 z}{\partial y^2}$ in Example 1.
2. Suppose $e^{xy} + e^{yz} - e^{xz} = 1$. Which of the first partial derivatives (of one variable with respect to another) exist at $x = 0$, $y = 0$, $z = 1$?
3. Suppose $z$ is defined implicitly as a function of $x$ and $y$ by the equation $F(x, y, z) = 0$. Find $\dfrac{\partial^2 z}{\partial x\, \partial y}$ in terms of $x$, $y$, $z$ and the derivatives of $F$.
4. **(a)** Suppose $g$ is a differentiable function, and let $z(x, y)$ be defined implicitly by the equation $(y - g(z))z = x$. Find $\dfrac{\partial z}{\partial x}$ and $\dfrac{\partial z}{\partial y}$.

   **(b)** Find an equation (a "partial differential equation") involving $z$ and its partial derivatives but not explicitly involving $g$.
5. Produce an animation similar to that of Example 3 but for $x = 2$, $y = 1$, $z = 0$, $w = 3$. Do $z$ and $w$ appear to be functions of $x$ and $y$ near this point?
6. Consider the system of equations

$$s = xy^2 + yz^2 + zx^2$$
$$t = xy - yz + xz$$
$$u = x + y - bz + b$$

   **(a)** For what values of the parameter $b$ does the Implicit Function Theorem guarantee that $x$, $y$, and $z$ can be considered as functions of $s$, $t$, and $u$ near $s = 3$, $t = 1$, $u = 2$, with $x(3,1,2) = y(3,1,2) = z(3,1,2) = 1$?

   **(b)** What is $\left(\dfrac{\partial y}{\partial t}\right)_{s,u}$ at $s = 3$, $t = 1$, $u = 2$?

   **(c)** Substitute into the system: $b = -1$, $s = 3$, $u = 2$, $y = 1 + w$, $z = 1 + pw$, $x = 1 + qw$. Solve the two equations not involving $t$ for $q$ and $p$. Note that there are three possible real values for $q$, with corresponding values for $p$.

   **(d)** Plot the curves in the $y, t$ plane corresponding to each of the $(q, p)$ values found in (c). How is this related to the answer to (a)?
7. **(a)** Find the degree-3 Taylor polynomial for $\ln(1 + x - y^2)$ in powers of $x - 1$ and $y - 1$.

   **(b)** Plot $r^{-5}$ times the error in this approximation, where $r$ is the distance from $(x, y)$ to $(1, 1)$. Does it appear that the error is $O(r^5)$?

**8.** **(a)** Find the degree-6 Taylor polynomials for $y$ in powers of $x$ determined by the equation $x^3y^3 - x + y = y^3$, with each possible value of $y$ when $x = 0$.

**(b)** Produce a 10-frame animation where the $n$'th frame shows the graph of the equation together with the degree-$n$ Taylor polynomials for $y$ in powers of $x$, for $-1 \le x \le 1$ and $-2 \le y \le 2$. On what intervals would you guess that the Taylor series converge?

**9.** Consider the equations $xy^2 + z^3 = 2x^2$ and $yz^2 + x^3 = 2y^2$ near $x = 1, y = 1, z = 1$.

**(a)** Find the degree-4 Taylor polynomial for $z$ as a function of $x$ and $y$ given by the first equation.

**(b)** Substitute the answer to (a) into the second equation, and find the degree-4 Taylor polynomial for $y$ as a function of $x$ determined by both equations.

**(c)** Using the answers to (a) and (b), find the degree-4 Taylor polynomial for $z$ as a function of $x$ determined by both equations.

**(d)** Taking $y$ and $z$ as given by the answers to (b) and (c), for what interval of $x$ values near 1 are both equations satisfied to an accuracy of .01 (i.e., the differences between the two sides of the equations are both less than .01)?

# LAB 21 EXTREME VALUES

**Goals:** *To learn to use Maple in finding the maximum or minimum value of a function of several variables, including problems involving constraints, and in classifying a critical point as a local maximum, local minimum, or saddle point.*

A local (or absolute) minimum or maximum of a function $f$ of several variables can only occur at a point which is either a boundary point of the domain of $f$, a singular point (where grad($f$) doesn't exist), or a critical point (where grad($f$) = 0). We begin by investigating the critical points.

## Critical Points

**EXAMPLE 1** Find and classify the critical points of the function $f(x, y) = x^3 - 2xy + y^2 - x$.

**SOLUTION** Our first task is to find the critical points by solving the equations $\dfrac{\partial f}{\partial x} = 0$ and $\dfrac{\partial f}{\partial y} = 0$.

```
> f:= (x,y) -> x^3 - 2*x*y + y^2 - x;
> eq1:= D[1](f)(x,y)=0;
```

$$eq1 := 3x^2 - 2y - 1 = 0$$

```
> eq2:= D[2](f)(x,y)=0;
```

$$eq2 := -2x + 2y = 0$$

```
> cps:= solve({eq1, eq2}, {x,y});
```

$$cps := \left\{y = \frac{-1}{3},\ x = \frac{-1}{3}\right\},\ \{y = 1,\ x = 1\}$$

We should be careful, because Maple's "solve" doesn't always find all the solutions. There is no problem in this case, in fact we could easily solve the system by hand. The second equation means $y = x$, and substituting that in the first equation will give us a quadratic with two real solutions.

Now we must look at each of our critical points and try to decide whether it is a local maximum, local minimum, or saddle point. Before using more analytic means, we might first try a contour plot near each critical point. For the first one, I am specifying the "contours" option as a list in order to ensure that one of the contour values $(5/27)$ is the value of $f$ at the critical point. The number .0026 was chosen to provide a reasonable number of contours.

```
> with(plots):
> contourplot(f(x,y), x=-.4 ..-.25, y=-.4 ..-.25,
>         axes=BOXED, contours=[seq(5/27 + .0026*kk, kk=-5..5)]);
> contourplot(f(x,y), x=.9 .. 1.1, y=.9 .. 1.1, contours=11,
>         axes=BOXED) ; (Figures 7.5 and 7.6)
```

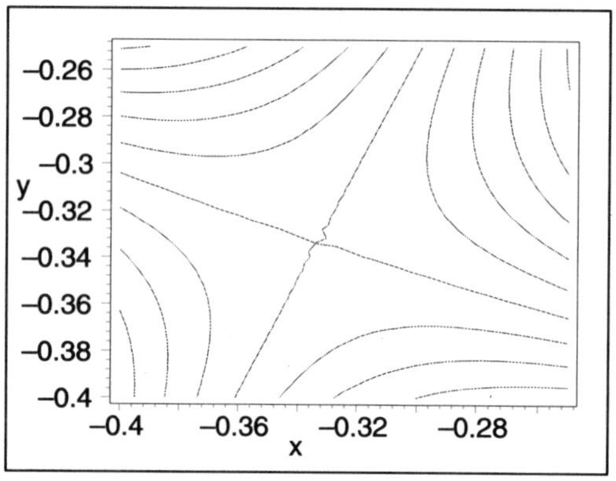


Figure 7.5. Saddle point

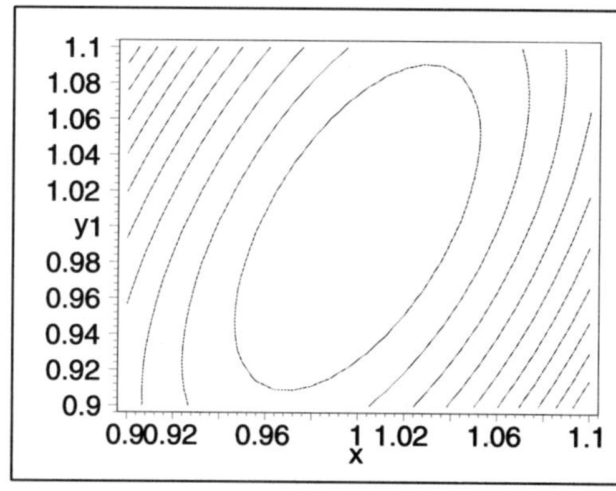


Figure 7.6. Local minimum

These are typical-looking contour plots for a saddle point and a local minimum or maximum respectively. In Figure 7.5, the contour for the value of $f$ at the critical point consists of two curves that cross at the critical point. Maple's picture is a bit confused in the immediate vicinity of the critical point, but they really are two intersecting smooth curves (see Exercise 1). These curves divide the picture up into four regions. In two of the regions $f(x, y)$ is less than $f(-1/3, -1/3)$ and in two it is greater. On a colour monitor you can see this by the colours of the contours: yellower where $f(x, y) > f(-1/3, -1/3)$, redder where $f(x, y) < f(-1/3, -1/3)$. The other contours each have two branches (in opposite regions). Since near $(-1/3, -1/3)$ there are points with lower $f$ values and also points with higher $f$ values, this is not a local minimum or local maximum: it is a saddle point.

In Figure 7.6, the contours near the critical point $(1, 1)$ are concentric closed curves, with the contour for $f(x, y) = f(1, 1)$ itself reduced to a single point. All the other contours shown are for higher values of $f$ (on a colour monitor they go from red for the innermost to yellow for the outermost contour). Thus this critical point is a local minimum.

The usual analytic method of classifying critical points is the second derivative test. The second partial derivatives of $f$ form the Hessian matrix of $f$, where the entry in row $i$ and column $j$ is $D_{i,j}(f)$. Maple can compute it with the "hessian" command, which is in the "linalg" package. The first input to "hessian" is the expression to differentiate, the second is the list of variables.

```
> with(linalg):
> H:=hessian(f(x,y), [x,y]);
```

$$H := \begin{bmatrix} 6x & -2 \\ -2 & 2 \end{bmatrix}$$

In the two-variable case, the Hessian is a $2 \times 2$ matrix. The second derivative test involves the upper left entry $A = D_{1,1}f$ and the determinant $HD = \det H = (D_{1,1}f)(D_{2,2}f) - (D_{1,2}f)^2$ of this matrix. The criteria are as follows:

> If $HD > 0$ and $A > 0$, the critical point is a local minimum.
> If $HD > 0$ and $A < 0$, the critical point is a local maximum.
> If $HD < 0$, the critical point is a saddle point.
> If $HD = 0$, then more information is needed to classify the critical point.

```
> HD:= det(H);
```

$$12x - 4$$

For our first critical point $x = -1/3, y = -1/3$, we have $HD = -8$, confirming that this is a saddle point. For our second critical point $x = 1, y = 1$, we have $HD = 8$ and $A = 6$, confirming that this is a local minimum. ■

For $n$ variables, the second derivative test is more complicated. If the quadratic form $Q(\mathbf{U}) = \mathbf{U}^T H \mathbf{U}$ is positive definite (which means that $Q(\mathbf{U}) > 0$ for all nonzero vectors $\mathbf{U}$ in $\mathbf{R}^\mathbf{n}$) then the critical point is a local minimum. If the form is negative definite (which means that $Q(\mathbf{U}) < 0$ for all nonzero vectors $\mathbf{U}$) then the critical point is a local maximum. If $Q(\mathbf{U}) > 0$ for some vectors and $Q(\mathbf{U}) < 0$ for some vectors then the critical point is a saddle point (there doesn't seem to be a standard adjective to describe this property of a quadratic form: I'll call it "not semidefinite"). The remaining case is where $Q(\mathbf{U}) \geq 0$ for all vectors or $Q(\mathbf{U}) \leq 0$ for all vectors (the form is "positive semidefinite" or "negative semidefinite") but $Q(\mathbf{U}) = 0$ for some nonzero vectors (the form is not positive or negative definite). In this case, more information is needed to classify the critical point.

You can tell whether the quadratic form is positive definite or negative definite by calculating determinants of $H$ and its upper left submatrices (semidefiniteness is slightly more complicated). You may learn how this works in a course on Linear Algebra. But it's easier to use Maple's "`definite`" command, which is in the "`linalg`" package. To test if $H$ is positive definite, negative definite, positive semidefinite, or negative semidefinite, you use "`definite(H, kind)`" where "`kind`" is "`positive_def`", "`negative_def`", "`positive_semidef`", or "`negative_semidef`" respectively. The result will be "`true`" or "`false`".

```
> H2:= subs(cps[2], eval(H));
```

$$H2 := \begin{bmatrix} 6 & -2 \\ -2 & 2 \end{bmatrix}$$

This is the Hessian matrix at the second critical point $x = 1$, $y = 1$. The "`eval`" was needed so that "`subs`" would see the entries of $H$ rather than just the name "`H`". This matrix should be positive definite, indicating a local minimum.

```
> definite(H2, positive_def);
```

true

The Hessian matrix at the first critical point, a saddle point, should not be semidefinite. Note that "`true`" or "`false`" results can be combined with the logical operators "`and`", "`or`" and "`not`".

```
> H1:= subs(cps[1], eval(H));
> definite(H1, positive_semidef)
>   or definite(H1, negative_semidef);
```

false

We can program a single command that will take the Hessian matrix as input and classify the critical point as "loc_min", "loc_max", "saddle", or "uncertain". The definition of "classify" is in the file "classify.def". If it is available to you, read it in to Maple now. If not, you can type it in as shown below, and save it in your own file "classify.def".

```
> classify:= H -> if definite(H, positive_def) then 'loc_min'
>              elif definite(H, negative_def) then 'loc_max'
>              elif definite(H, positive_semidef)
>                or definite(H, negative_semidef) then 'uncertain'
>              else 'saddle' fi;
```

The "definite" command will also work with Hessian matrices containing symbolic parameters where the classification might depend on the values of the parameters. In these cases, instead of "true" or "false" it returns a condition that must be true if the matrix is to be . . . definite.

```
> definite(H, positive_def);
```

$$-x < 0 \text{ and } 1 - 3x < 0$$

**EXAMPLE 2** Find and classify the critical points of

$$f(x, y, z, w) = \frac{xy + yz - zw - wx}{(3 + x^2)(3 + y^2)(3 + z^2)(3 + w^2)}$$

**SOLUTION**

```
> f:= (x,y,z,w) -> (x*y + y*z - z*w - w*x)
>                 /((3+x^2)*(3+y^2)*(3+z^2)*(3+w^2));
> vars:= x,y,z,w;
> eqs:= { seq(D[kk](f)(vars) = 0, kk = 1 .. 4) };
> sols:= solve(eqs,{vars});
```

$$\begin{aligned} sols :=&\{x = -z,\ w = y,\ y = y,\ z = z\},\ \{y = 1,\ x = 1,\ w = -1,\ z = 1\} \\ &\{y = 1,\ w = -1,\ x = -1,\ z = -1\},\ \{y = -1,\ x = 1,\ w = 1,\ z = 1\} \\ &\{y = -1,\ x = -1,\ w = 1,\ z = -1\} \end{aligned}$$

We'll trust Maple that these are all the solutions.

```
> H:= hessian(f(vars), [vars]);
> H1:= subs(sols[1], eval(H));
```

$$H1 := \begin{bmatrix} 0 & \%1 & 0 & -\%1 \\ \%1 & 0 & \%1 & 0 \\ 0 & \%1 & 0 & -\%1 \\ -\%1 & 0 & -\%1 & 0 \end{bmatrix}$$

$$\%1 := \frac{1}{(3 + z^2)^2(3 + y^2)^2}$$

Note that the first solution has $y$ and $z$ arbitrary, and the Hessian there depends on $y$ and $z$. We shouldn't expect "classify" to work in this case, so we'll use "definite" directly.

```
> definite(H1, positive_semidef);
```

$$\frac{1}{(3 + z^2)^4 (3 + y^2)^4} \le 0$$

```
> definite(H1, negative_semidef);
```

$$\frac{1}{(3 + z^2)^4 (3 + y^2)^4} \le 0$$

Since these conditions can never be satisfied, we conclude that the first solution is a saddle point (actually, a two-dimensional surface of saddle points in four-dimensional $xyzw$-space).

We can use "`classify`" for the other four solutions, which have no arbitrary parameters. Using "`map`", we can treat all four at once.

```
> map(s -> classify(subs(s,eval(H))),
>                    [sols[2],sols[3],sols[4],sols[5]]);
```

$$[loc\_max,\ loc\_min,\ loc\_min,\ loc\_max]$$

While we're at it, what are the values of $f$ at these points?

```
> map(s -> subs(s, f(vars)), [sols]);
```

$$\left[0, \frac{1}{64}, \frac{-1}{64}, \frac{-1}{64}, \frac{1}{64}\right]$$

It is possible to show that $f(x, y, z, w) \to 0$ as $(x, y, z, w) \to \infty$ in all directions. Thus the local maxima "`sols[2]`" and "`sols[5]`" (with values $1/64$) and the local minima "`sols[3]`" and "`sols[4]`" (with values $-1/64$) are actually global maxima and minima. ■

## Optimization on a Restricted Domain

Suppose we wish to find the maximum of a function $f(x, y)$ on a domain $D$. The maximum (if it exists) could occur at an interior point of $D$, in which case it is a critical point or singular point of $f$. Or it could occur at a boundary point of $D$. Typically the boundary of $D$ might consist of one or more smooth curves, each defined by an equation $g_j(x, y) = 0$. A maximum on one of these curves might be found using Lagrange multipliers. Or the maximum could occur at the intersection of two boundary curves. All these possibilities must be investigated. For a function of three variables the situation is even more complicated: the boundary consists of surfaces, which intersect on lines, which in turn intersect at points.

■ **EXAMPLE 3** Find the maximum and minimum of $x^2 - xy + y^3 - x$ subject to the constraints $x^2 + xy + y^2 \le 3$ and $x^2 + y^2 \le 4$.

**SOLUTION** Let's first try a plot, to see what the domain and the objective function $x^2 - xy + y^3 - x$ look like, and make an educated guess at the locations of the maximum and minimum. What type of plots shall we use? An "`implicitplot`" would be a good choice to show the boundary of our domain. A "`contourplot`", on the other hand, might be good to show the objective. We can combine these with "`display`". To make the difference between the boundary curves and the contours apparent, I will use different colours: red for the boundaries and black for the contours. If you don't have a colour monitor you can use different line styles or thicknesses: I used "`thickness=3`" for $g_1 = 0$ and $g_2 = 0$ in Figure 7.7. Note that because of the constraint $x^2 + y^2 \le 4$ we don't need to look outside the square $-2 \le x \le 2$, $-2 \le y \le 2$. I'll specify the contour values as a list, so that we can tell what these values are: some experimentation could be used to discover a good range of contour values.

```
> f:= x^2 -x*y + y^3 - x;
> g1:= x^2 + x*y + y^2 - 3;
> g2:= x^2 + y^2 - 4;
> with(plots):
> p1:= implicitplot({g1=0,g2=0}, x=-2..2, y=-2..2, colour=red):
> p2:= contourplot(f, x=-2..2, y=-2..2, colour=black,
>        contours=[seq(2*k,k=-3..5)]):
> display({p1,p2}, axes=BOXED); (Figure 7.7)
```

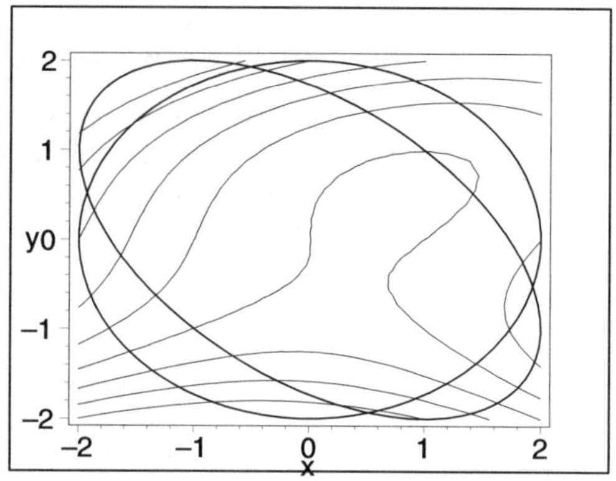


Figure 7.7. Constrained optimization

Our domain is the region inside both red curves. The contour values increase from bottom to top (note, for example, that $f(0, y) = y^3$ is an increasing function of $y$). According to the plot, it appears that the maximum value, approximately 9, occurs on the curve $g_2 = 0$ somewhere between about $(-1, 1.7)$ and the intersection of the two curves near $(-.5, 1.95)$, while the minimum value, slightly less than $-6$, occurs at the intersection of the two curves near $(.5, -1.95)$.

Now we will find the maximum and minimum using calculus. I'll leave the problems of finding critical points of $f$ and critical points on the curve $g_1 = 0$ to Exercises 4 and 5 respectively, and only consider critical points on the curve $g_2 = 0$ and intersections of the curves. Using $t$ as a Lagrange multiplier, we form the Lagrangian $L$, set its derivatives with respect to $x$, $y$, and $t$ equal to 0, and solve the resulting equations.

```
> L:= f + t * g2;
> eqs:= { diff(L, x)=0, diff(L, y)=0, diff(L, t)=0 };
> cps:= solve(eqs, {x,y,t});
```

$$cps := \left\{t = \frac{5857}{584}\%1 + \frac{315}{73} + \frac{369}{584}\%1^5 + \frac{87}{146}\%1^4 - \frac{361}{73}\%1^3 - \frac{278}{73}\%1^2,\ x = \%1,\right.$$

$$\left. y = -\frac{372}{73}\%1 - \frac{196}{73} + \frac{306}{73}\%1^2 - \frac{66}{73}\%1^4 + \frac{223}{73}\%1^3\right\}$$

$$\%1 := \text{RootOf}\left(9\_Z^6 + 12\_Z^5 - 64\_Z^4 - 76\_Z^3 + 113\_Z^2 + 112\_Z + 12\right)$$

Next we find the real roots of the polynomial in the "`RootOf`"[(2)]. Since our solution has $x = \%1$ where $\%1$ is the "`RootOf`", we can get that "`RootOf`" by substituting the solution into "`x`". The polynomial can be extracted from the "`RootOf`" using "`op`".

```
> rootof:= subs(cps, x);
```

$$rootof := \text{RootOf}\left(9\_Z^6 + 12\_Z^5 - 64\_Z^4 - 76\_Z^3 + 113\_Z^2 + 112\_Z + 12\right)$$

```
> rts:= fsolve(op(rootof));
```

$$rts := -.7194278387 - .1237400366, 1.768684665, 1.920638087$$

Each of the four real roots corresponds to a critical point of our Lagrangian on the curve $g_2 = 0$.

```
> s1:= map(r -> subs(rootof=r, cps), [rts]);
```

$$s1 := [\{t = -2.991947720, x = -.7194278387, y = 1.866125287\},$$
$$\{y = -1.996168431, t = 3.025247034, x = -.1237400366\},$$
$$\{t = -.45335706, x = 1.768684665, y = .93367797\},$$
$$\{t = -.88488399, y = -.55780762, x = 1.920638087\}]$$

Some of these points may be outside our domain because $g_1 > 0$ there. Let's look at the $g_1$ values and select those points that are in the domain.

```
> map(s -> subs(s, g1), s1);
```

$$[-.342542480, 1.247005956, 2.651381904, -.071346558]$$

The first and fourth points are in our domain. What are the $f$ values?

(2) Different releases may express the solution in terms of different polynomials, so your results here may be different, although the final answers should be essentially the same.

```
> fvals:= map(s -> subs(s, f), s1);
```

$$fvals := [9.078185451, -8.062063522, .522116891, 2.665997661]$$

We should also check for points on $g_2 = 0$ where the gradient of $g_2$ is zero, which are also candidates for the maximum or minimum.

```
> zps:= solve({ diff(g2, x)=0, diff(g2, y)=0 }, {x,y});
```

$$zps := \{x = 0, y = 0\}$$

This is not on the curve $g_2 = 0$, so we don't need to consider it any further.

Finally, we look at the intersections of the two curves.

```
> ips:= solve({g1=0, g2=0}, {x,y});
```

$$ips := \{y = \text{RootOf}\left(\_Z^4 - 4\_Z^2 + 1\right), \quad x = \text{RootOf}\left(\_Z^4 - 4\_Z^2 + 1\right)\left(-4 + \text{RootOf}\left(\_Z^4 - 4\_Z^2 + 1\right)^2\right)\}$$

Again we find the real roots of the polynomial in the "`RootOf`", which is the value of `y` in "`ips`".

```
> rootof:= subs(ips,y);
```

$$rootof := \text{RootOf}\left(\_Z^4 - 4\_Z^2 + 1\right)$$

```
> r:= fsolve(op(rootof));
```

$$r := -1.931851653, -.5176380902, .5176380902, 1.931851653$$

The four intersection points are obtained by substituting each of these values for the "`RootOf`".

```
> pts:= map(v -> subs(rootof=v, ips), [r]);
```

$$\begin{aligned} pts :=[&\{y = -1.931851653, x = .5176380902\},\\ &\{y = -.5176380902, x = 1.931851653\},\\ &\{y = .5176380902, x = -1.931851653\},\\ &\{y = 1.931851653, x = -.5176380902\}] \end{aligned}$$

What are the corresponding values of $f$?

```
> map(s -> subs(s,f), pts);
```

$$[-6.459457429, 2.661498448, 6.802603170, 8.995355795]$$

You will finish this example in Exercise 5. ■

■ **EXAMPLE 4** Explore the geometry behind Lagrange multipliers, by taking a closer look at what happens at the point (1.920638087, −.55780762) found in Example 3.

**SOLUTION** The point in question is the fourth in our "`s1`".

```
> P:= subs(s1[4], [x,y]);
```

$$P := [1.920638087, -.55780762]$$

We'll plot the contour line $f = f(P)$ together with the boundary curve $g_2 = 0$ on which $P$ lies, and also an arrow indicating the direction of grad$f$ at $P$. If you don't have a colour monitor, use "`linestyle=2`" for "`plot1`".

```
> plot1:= implicitplot(f=fvals[4], x=1.8 .. 2.1,
>            y=-.7 .. -.4, colour=blue):
> plot2:= implicitplot(g2=0, x=1.8 .. 2.1, y=-.7 .. -.4):
```

For the arrow, we need the "`linalg`" package (which was already loaded in the section "Critical Points") and the file "`dot.def`". The construction of an arrow is similar to the one in Figure 7.2 of Lab 19.

```
> read `dot.def`:
> Gf:= subs(s1[4],
> evl(grad(f, [x,y])));
```

$$Gf := [3.399083794, -.9871900643]$$

We take a unit vector $\mathbf{T}$ in the direction of this gradient. The vector $\mathbf{U}$ is also a unit vector, perpendicular to $\mathbf{T}$. For this plot, an arrow of length $0.2$ should be about right.

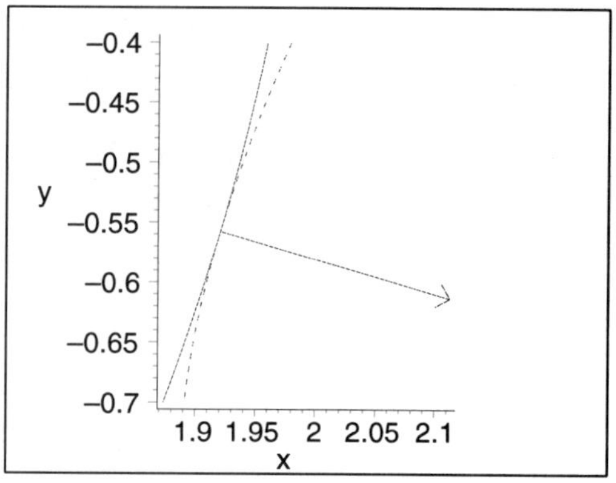


Figure 7.8. Local max on curve

```
> T:= unit(Gf);
> U:= [T[2], -T[1]];
> arrow:= plot([P, evl(P+.2*T), evl(P+.19*T-.01*U),
>                    evl(P+.2*T), evl(P+.19*T+.01*U)]):
> display({plot1, plot2, arrow}, scaling=constrained);
```
(Figure 7.8)

The two curves $f = f(P)$ and $g_2 = 0$ are tangent to each other at $P$, with the region $g_2 \leq 0$ to the left. Therefore the gradient vectors to $f$ and $g_2$ at $P$, which are normals to the curve, are parallel to each other. The Lagrange multiplier method finds points $P$ with this property, where the value of the Lagrange multiplier $t$ in the solution makes $\text{grad} f(P) + t \,\text{grad}\, g_2(P) = 0$. Let's check this. Of course there is some roundoff error, so the answer we get may not be exactly 0.

```
> evl(Gf + subs(s1[2], t * grad(g2, [x,y])));
```

$$[0, .10\,10^{-8}]$$

Since the gradient vector points to the right, while the region $g_2 \leq 0$ is to the left of the contour line $f = f(P)$ (which is dotted in Figure 7.8), it seems that our point $P$ is a local maximum, i.e., $f \leq f(P)$ at all points near $P$ that satisfy the constraints (in fact, for all points in Figure 7.8 satisfying the constraints). ■

## EXERCISES

1. Use the Hessian to classify the critical point at the origin for the functions
   **(a)** $x^2 - 4xy + 3y^2$ **(b)** $3x^2 + 2y^2 + z^2 - xy + xz - yz$
2. For what values of the parameter $c$ does the function $f(x, y, z) = c(x - y - z)^2 + (c - 2)(x + 2y + z)^2 - (x - y)(y + z)$ have a local minimum at the origin? A local maximum?
3. **(a)** To get a better look at the contour $f(x, y) = 5/27$ in Example 1, solve this equation for $y$ and plot the two solutions you obtain.
   **(b)** Make three-dimensional plots of $f(x, y)$ near the two critical points of Example 1, and confirm the classifications of those points.
4. **(a)** Find and classify the critical points of $(x+1)\,(y^2+z^2)+(y+1)\,(x^2+z^2)+(z+1)\,(x^2+y^2)$.
   **(b)** Is there a global minimum or maximum?
5. Consider the critical points of $f = \cos(x) - \cos(y) + (x + y)^2$.
   **(a)** What critical point(s) can you find using "`solve`"? Classify.
   **(b)** Show that there are infinitely many critical points on the line $x + y = 0$. Classify them.
   **(c)** Plot the curves $\partial f/\partial x = 0$ and $\partial f/\partial y = 0$ to approximately locate some other critical points.
   **(d)** Use "`fsolve`", specifying intervals for $x$ and $y$, to find one of the critical points from (c). Classify it.
   **(e)** Why can't $f$ have any local maxima? *Hint:* is the Hessian ever negative semidefinite?

**6. (a)** In Example 3, find the critical points for $f$.

**(b)** Are these points in the domain $D$ which was considered in this example?

**(c)** Classify these critical points.

**7. (a)** In Example 3, find the critical points for $f$ subject to the constraint $x^2 + xy + y^2 = 3$.

**(b)** Which, if any, of these points satisfy $x^2 + y^2 \leq 4$? What are their $f$ values?

**(c)** Putting together the results of Example 3, Exercise 4, and this one, complete Example 3 by finding the maximum and minimum. Make a plot showing the curves $g_1 = 0$ and $g_2 = 0$ with the level curves for the maximum and minimum values of $f$.

**8.** It is possible to use a second derivative test to tell whether a critical point of the Lagrangian is a local maximum or minimum of the objective subject to equality constraints. We will do this for the critical point $(1.768684665, .9336779718)$ of Example 3 with the constraint $g_2 = 0$.

**(a)** Find the second-order Taylor polynomial for $y$ satisfying $g_2 = 0$ in powers of $x - 1.768684665$, with $y = .9336779718$ when $x = 1.768684665$.

**(b)** Substitute this for $y$ in $f$, and find the second-order Taylor polynomial for $f$ on the curve $g_2 = 0$ in powers of $x - 1.768684665$.

**(c)** Is this point a local maximum or a local minimum of $f$ on the curve?

**9.** Find the points in the intersection of the surfaces $x^2 + xyz + y^2 = 3$ and $x^2 + y^2 + 2z^2 = 4$ that are closest to and farthest from the origin. It is usually more convenient to use the square of the distance from the origin, rather than the distance, as the objective function. Use "`solve`" to find the critical points of the Lagrangian. Note that symmetry can reduce the number of points you need to investigate (the constraints and objective are unchanged if you change the signs of any two of $x$, $y$ and $z$).

**10.** In Lab 19 when we plotted surfaces together with their tangent planes, the tangent plane was sometimes completely visible and sometimes partly or completely hidden by the surface. Suppose we are looking from above at the graph of $z = f(x, y)$ with a small patch of the tangent plane at $(x_0, y_0)$.

**(a)** What does the Hessian of $f$ at $(x_0, y_0)$ have to do with the visibility of the tangent plane?

**(b)** Test your answer to (a) by plotting the surface and tangent plane at various points for $f(x, y) = \cos(x)\cos(y)$.

# LAB 22 NEWTON'S METHOD

**Goals:** *To see how Newton's method can be used to solve systems of equations in several variables, and investigate the convergence of the method.*

## Solving Systems of Equations

In Lab 7 we used Newton's method to solve an equation involving one variable. Suppose now we wish to solve a system of $m$ equations in $m$ variables:

$$f_1(x_1, \ldots, x_m) = 0, \ \ldots \ f_m(x_1, \ldots, x_m) = 0$$

The basic idea is the same: start with an initial guess, and consider linear approximations to the functions $f_1, \ldots f_m$ near the initial guess. Find a point where the linear approximations are all 0, and use that as the next guess. Repeat this process until it appears to converge.

In the one-variable case, if the initial guess is $x = a$ the linear approximation is $f(x) \approx f(a) + f'(a)(x - a)$ so the next guess is $a - f(a)/f'(a)$. Now the linear approximations are

$$f_j(x_1, \ldots, x_m) \approx f_j(a_1, \ldots, a_m) + \frac{\partial f_j}{\partial x_1}(a_1, \ldots, a_m)(x_1 - a_1) + \ldots + \frac{\partial f_j}{\partial x_m}(a_1, \ldots, a_m)(x_m - a_m)$$

This can be written more conveniently in vector notation as

$$\mathbf{F}(\mathbf{X}) \approx \mathbf{F}(\mathbf{A}) + J(\mathbf{A})(\mathbf{X} - \mathbf{A})$$

where $J$ is the Jacobian matrix with entries $J_{ij} = \dfrac{\partial F_i}{\partial x_j}$. We can therefore write the new guess symbolically as $\mathbf{A} - J(\mathbf{A})^{-1}\mathbf{F}(\mathbf{A})$. However, in practice I'll just use "`solve`" on the system of linear equations coming out of the linear approximations.

■ **EXAMPLE 1** Solve the system of equations $x^2y + xy^2 = 2$, $x + \sin(y) = 2$.

**SOLUTION** I'll use "`mtaylor`" to find the linear approximations: "`ap1`" for the first equation (written as "`f1=0`") and "`ap2`" for the second. The starting point will be $(1, 1)$ (which is a solution to the first equation).

```
> f1:= x^2 *y + x*y^2 - 2;
> f2:= x + sin(y)  - 2;
> readlib(mtaylor):
> ap1:= mtaylor(f1, [x=1,y=1], 2);
```

$$ap1 := 3y - 6 + 3x$$

```
> ap2:= mtaylor(f2, [x=1, y=1], 2);
```

$$ap2 := -2 + \sin(1) + x + \cos(1)(y - 1)$$

Now we find our next guess by solving the system of equations $3y - 6 + 3x = 0$, $-2 + \sin(1) + x + \cos(1)(y - 1) = 0$.

```
> S:= solve({ap1=0, ap2=0}, {x,y});
```

$$S := \left\{ y = -\frac{\sin(1) - \cos(1)}{-1 + \cos(1)},\ x = \frac{\sin(1) + \cos(1) - 2}{-1 + \cos(1)} \right\}$$

We want the result as a decimal number rather than an exact expression, so we use "`evalf`". Later in the sequence we won't need this "`evalf`" because we will be starting from decimals.

```
> p[1]:= evalf(S);
```

$$p_1 := \{y = .6551450721, x = 1.344854927\}$$

Note that the result can be used as the input to "`mtaylor`" for the next iteration. We can use a "`for`" loop to run through several more iterations.

```
> for nn from 2 to 10 do
>      p[nn]:= solve({mtaylor(f1, p[nn-1],2),
>                    mtaylor(f2, p[nn-1],2)}, {x,y}) od;
```

$$p_2 := \{y = .7300611308,\ x = 1.331320310\}$$
$$p_3 := \{y = .7258272652,\ x = 1.336239609\}$$
$$p_4 := \{y = .7258430227,\ x = 1.336233791\}$$
$$p_5 := \{x = 1.336233790,\ y = .7258430239\}$$

Note the switch from $\{y = \ldots,\ x = \ldots\}$ to $\{x = \ldots,\ y = \ldots\}$: Maple can write a set in any order, and the order is not predictable. I haven't bothered to include the last five iterations, which are almost identical to $p_5$ except for some changes in the last one or two decimal places (which can be attributed to roundoff error). As in the one-variable version of Newton's method, the number of correct decimal places approximately doubles at each iteration. Let's check that the result satisfies $f_1 = 0$ and $f_2 = 0$.

```
> eval(subs(p[10],[f1, f2]));
```

$$[0,\ .8\,10^{-9}]$$

This method is correct, but seems not very efficient since it requires finding new Taylor expansions at each iteration. It might be better to have a formula calculated just once. Moreover, instead of working with a set $\{x = \ldots,\ y = \ldots\}$ I'll construct a function that takes a list $L = [x,\ y]$ and returns the next iteration as another list.

```
> S:= solve({mtaylor(f1, {x=L[1], y=L[2]}, 2),
>               mtaylor(f2, {x=L[1], y=L[2]}, 2)}, {x,y});
> newt:= unapply(subs(S, [x,y]), L);
```

I didn't reproduce the outputs, which were quite long. This method wouldn't be practical for more than two variables, as the formulas would get very complicated.

Here are a few iterations of "`newt`" with a different starting point. Don't forget the decimal points in "`P`", which ensure that the results will be numerical rather than very complicated expressions.

```
> P:= [1.,-2.];
> for nn from 1 to 7 do P:= newt(P) od;
```

$$P := [2.909297427, -1.999999999]$$
$$P := [2.431596243, -3.147914980]$$
$$P := [2.388892443, -2.752692350]$$
$$P := [2.411953052, -2.717264872]$$
$$P := [2.411882346, -2.717073858]$$
$$P := [2.411882347, -2.717073846]$$
$$P := [2.411882347, -2.717073848]$$

This time the iterations converged to a different solution.

To see if there could be other solutions, we can plot the curves $f_1 = 0$ and $f_2 = 0$.

```
> plots[implicitplot](
> {f1=0, f2=0}, x=0..4, y=-4..4);
```

(Figure 7.9)

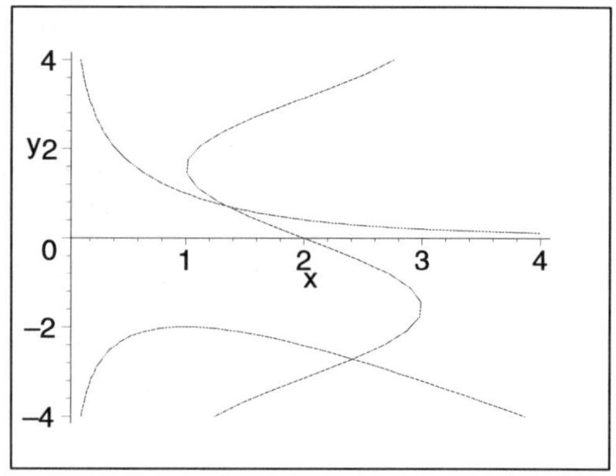


Figure 7.9. $f_1 = 0$ and $f_2 = 0$

The two curves intersect in two places on the graph, which are the two solutions Newton's method found. Could there be any others, outside the rectangle we plotted? If $f_2 = 0$ it is clear that $1 \le x \le 3$. On the other hand, $f_1 = 0$ is quadratic in $y$, so any vertical line can intersect this curve in at most two points. Now on any vertical line between $x = 1$ and $x = 3$ there are two intersections with $f_1 = 0$ in our graph. Therefore the answer is no: there are only two solutions to this system of equations. ■

## Convergence of Newton's Method

According to the theory, when the initial point is close enough to one of the two solutions of Example 1, the successive iterates should converge to that solution. How close is "close enough"?

One way to investigate this question is to plot circles of various radii centred at the solution together with their images under "newt". Suppose we find that for every radius $r$ with $0 < r \le R$, the image of the circle is always inside that circle[(3)]. Then the iterates of any point with distance from the centre less than $R$ will converge to the centre. We can't try **all** $r$ in the interval $0 < r \le R$, but we can try several of them. The result will not be a rigorous proof, but the answer will have a very good chance of being correct.

We can use the parametric equations of a circle: $x = x_0 + r\,\cos(t)$, $y = y_0 + r\,\sin(t)$ for $0 \le t \le 2\pi$. Note that "newt" produces a two-element list. To make a parametric plot (as in Lab 14) we need the parameter interval to be a third element in the list. This can be obtained as follows: if L is a list, "op(L)" is the sequence of members of the list, so "[op(L),a]" is the list L with "a" appended to it.

```
> x0:= subs(p[10],x): y0:= subs(p[10],y):
> r:= 0.1;
> plot({ [x0 + r*cos(t), y0 + r*sin(t), t=0 .. 2*Pi],
>      [op(newt([x0 + r*cos(t), y0 + r*sin(t)])), t=0 .. 2*Pi]});
```

(Figure 7.10)

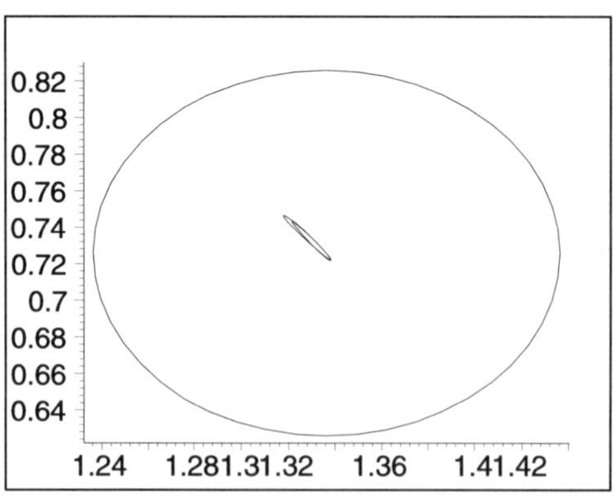


Figure 7.10. Circle and its image

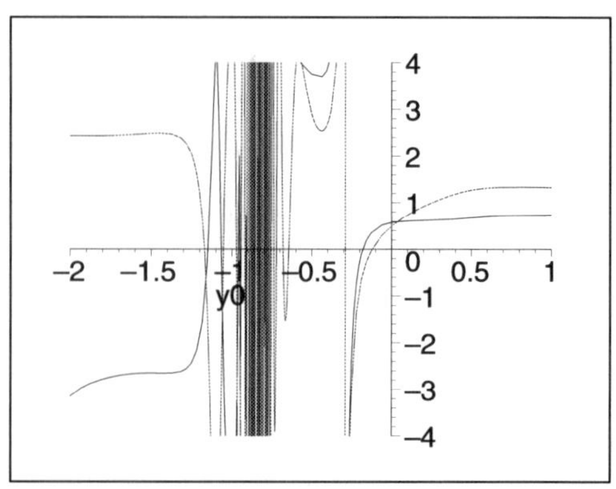


Figure 7.11. Two iterations with $x_0 = 1$

(3) Technically, I should specify that the image of the circle of radius $r$ is inside the concentric circle of radius $(1 - \epsilon)r$ for some constant $\epsilon > 0$.

Trying various values of $r$, it appears that the image of the circle is inside the circle when $r < 0.3$. Thus we conclude that Newton's method converges to the solution that is approximately $p_{10}$ whenever the initial guess is within distance 0.3 of that point. Of course, this is only a lower bound, and it might converge to this point for substantially larger $r$ values.

Now let's consider what happens when the starting point $(x_0, y_0)$ is on the line $x_0 = 1$. As we have seen, when $y_0 = 1$ the iterates converge to one solution of the set of equations, when $y_0 = -2$ to a different one. We can plot the $x$ and $y$ coordinates of various iterates of "`newt`."

```
> x0:= 'x0': y0:= 'y0':
> plot({'(newt@@2)([1.0,y0])[1]', '(newt@@2)([1.0,y0])[2]'},
>            y0 = -2.0 .. 1.0, -4..4); (Figure 7.11)
```

The quotes are inserted to prevent "premature evaluation": we do not want Maple to evaluate "`(newt@@2)([1.0,y0])`" with "`y0`" as a symbolic variable, because this would result in a very complicated expression. We only want it to evaluate it with numbers substituted for "`y0`". This is not really critical for "`(newt@@2)`", but would be very important for higher iterates. Also important are the decimal points, which ensure that the numbers substituted for "`y0`" are decimals rather than integers: exact expressions involving quantities such as "`sin(1)`" could also produce very complicated expressions when iterated a large number of times. The interval `-4..4` is specified for the vertical axis of the graph, because otherwise some very large values might result (note that the iteration formulas involve a denominator that may be 0 for some $y_0$ values). Note also that there will be two curves on the same set of axes, one for the $x$ coordinate and one for the $y$ coordinate.

We could even produce an animation of some of the iterates (this could take a rather long time, depending on the speed of your computer):

```
> plots[display]([seq(
>       plot({'(newt@@n)([1.,y0])[1]', '(newt@@n)([1.,y0])[2]'},
>            y0 = -2.0 .. 1.0, -4..4),
>         n=1..6)], insequence=true);
```

The first frame (n=1) is not very remarkable, except for a singularity near $y = -.8$ where the denominator of both components is 0. Let's look at that denominator (which is the Jacobian determinant of $(f_1, f_2)$).

```
> d:= denom(newt([1,y0])[1]);
```

$$d := -1 - 2y + 2\cos(y0)y0 + \cos(y0)y0^2$$

And where is it zero in our interval $-2 \leq y_0 \leq 1$?

```
> fsolve(d, y0, y0=-2..1);
```

$$-.8281374233$$

At n=2 parts of the curves have already started to flatten out at the two solutions. For $-2 \leq y_0 \leq -1.3$ (approximately) we have one curve (representing the $x$ coordinate) near 2.4 and the other (the $y$ coordinate) near $-2.7$. This means that in this interval of $y_0$, the iterates appear to be settling down to our second solution. Similarly, for $.5 \leq y \leq 1$ they appear to be approaching the first solution. Things appear very confused for $-1.2 \leq y \leq -.6$, and there is a singularity near $y = -.3$. If we look at higher iterates, some parts of the "confused" interval seem to settle down to one or the other of the two solutions. At n=15 the first interval where the iterates are near our second solution has grown a bit to $-2 \leq y_0 \leq -1.2$. There seem to be intervals of the first solution near $-1$, $-.5$, $-.2$ and just to the right of 0, and of the second solution near $-1.05$ and $-.25$. But there are some regions that still seem "confused". Let's try a $y_0$ value in one of these regions.

```
> P:= [1., -.1];
> for nn from 1 to 20 do P:= newt(P) od;
```

I'll just show the last few iterates:

$$P := [.008811499168, 14.75700038]$$
$$P := [.01159058230, 12.73524962]$$
$$P := [.008958074164, 14.58452225]$$
$$P := [.01247745797, 12.07418501]$$

It's certainly not anywhere near a solution. What is happening? Let's give the process some more time to settle down, then save and plot the next 500 points.

```
> for nn from 1 to 500 do
>     P:= newt(P) od:
> for nn from 1 to 500 do
>     P:= newt(P): Q[nn]:= P od:
> plot([seq(Q[nn], nn=1..500)], style=POINT, symbol=POINT);
```

(Figure 7.12)

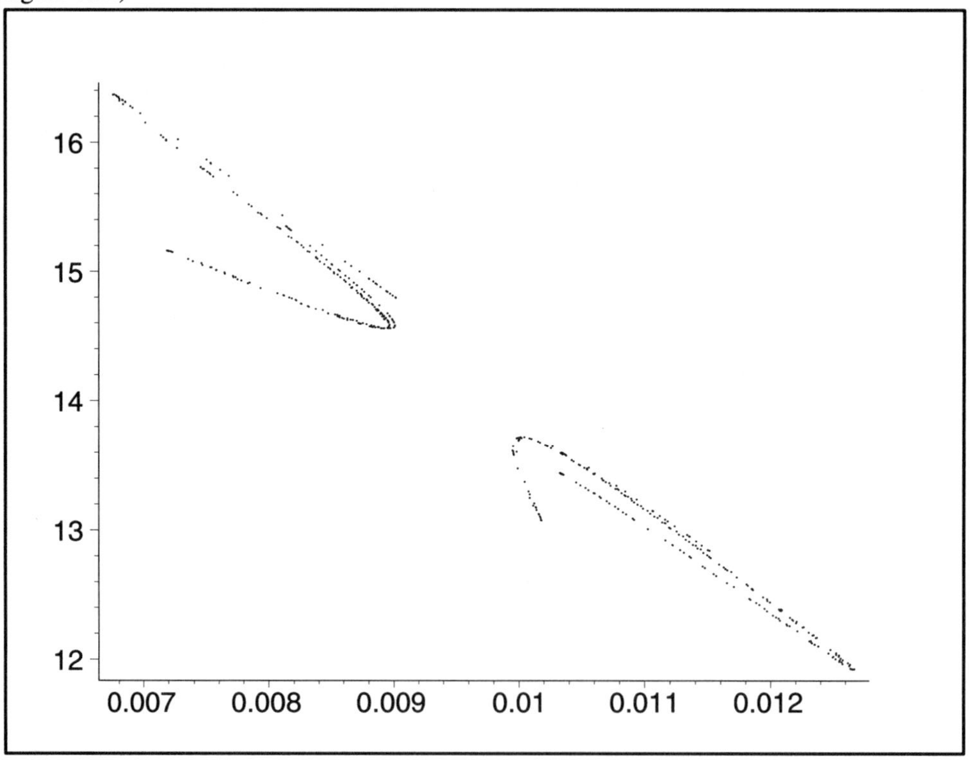


Figure 7.12. A strange attractor

The points appear to lie on two curves that are folded over on themselves. On closer examination, it turns out that there seem to be an infinite number of "folds": this is an example of a so-called "strange attractor". It is a small hint of an important and active area of mathematics (nonlinear dynamical systems) about which it is impossible to say very much here. But some connections with Lab 8 can be noted.

Our "`newt`" can be considered as a nonlinear mapping in the $xy$ plane, which we are iterating. Fixed points of this mapping correspond to solutions of our system of equations. Just as in the one-dimensional case, there can also be periodic points. We can use the Jacobian matrix of a mapping and some linear algebra to investigate the stability of such points[(4)]. In the case of "`newt`", its Jacobian

(4) A fixed point is an attractor if all eigenvalues of the Jacobian matrix at that point have absolute value less than 1.

matrix[5] at a fixed point turns out to be $0$, so the fixed points are attractors. However, there may be periodic points which might be stable or unstable. And there are more exotic possibilities as well, such as the strange attractor we just found.

## EXERCISES

**1.** Consider the system of equations $x^3 + y^3 = 10$, $xy^2 - x^2y = 2$. Use "`implicitplot`" to find how many solutions there are, and their approximate locations. Then use Newton's method to find the solutions.

**2.** Consider the system of equations $x^3+y^3+z^3 = 12$, $xy^2-yz^2+x^2z = -1$, $x^2y+y^2z+xz^2 = 10$. Find a solution using Newton's method, starting with initial guess $x = y = 1$, $z = 2$.

**3.** Consider the problem of finding a point in the plane where the sum of the distances to the four points $(0,0)$, $(1,0)$, $(0,1)$, and $(2,1)$ is a minimum.

**(a)** What two equations should the solution $(x, y)$ satisfy?

**(b)** Use Newton's method to solve these equations, choosing a reasonable starting point. Note: the equations are so complicated that the approach we used to construct "`newt`", solving the system of linear approximations to $f_1$ and $f_2$ at a general point $L$, may not work. Instead you can use the first method from Example 1, which invokes "`solve`" only after substituting in the numerical values for $L$.

**(c)** Can you use geometric insight to see why the solution is what it is?

**4.** One solution of the system of equations $x^3 + y^3 = 9$, $xy^2 - x^2y = 2$ is $x = 1$, $y = 2$. Having an exact solution makes it somewhat easier to study convergence to this solution. Let "`newt`" be the function that produces the next iteration of Newton's method.

**(a)** Find the degree-2 Taylor polynomials of the components of $\mathtt{newt}([x, y])$ in powers of $x - 1$ and $y - 2$. What is the significance of the lack of first-degree terms?

**(b)** Let $x = 1 + r\cos\theta$ and $y = 2 + r\sin\theta$. Plot the implicit curve $|\mathtt{newt}([x,y]) - [1,2]| = r$ in the $r\theta$ plane, for $0 < r < 0.1$. Does this allow you to conclude anything about convergence of Newton's method in this example? What if you used $0.9r$ on the right side instead of $r$?

**(c)** Using graphical methods, find a constant $B$ such that $|\mathtt{newt}([x,y]) - [1,2]| < Br^2$ when $r \leq 0.071$. What does this, together with the result of (b), tell you about the number of iterations needed to obtain a solution with an error less than $10^{-20}$ in both variables, starting at a point with $r \leq 0.071$? Ignore roundoff error.

**(d)** How many iterations are actually necessary to obtain a solution with error less than $10^{-20}$ in both variables, starting at $x = 1.05$, $y = 1.95$? Use `Digits=30`.

**5. (a)** Find $a$, $b$, and $c$ so that the curves $x^3 + y^3 = a$ and $5xy^2 + bx^2y = c$ are tangent to each other at $x = 1$, $y = 2$. *Hint:* consider a Jacobian determinant to find $b$.

**(b)** Take $a$, $b$, and $c$ as in (a), and let "`newt`" be the Newton's method iteration function corresponding to the system of equations defining the two curves. Find the degree-1 Taylor polynomials of the components of $\mathtt{newt}([x, y])$ in powers of $x - 1$ and $y - 2$.

**(c)** Let $x = 1 + r\cos\theta$ and $y = 2 + r\sin\theta$. For what positive constants $K$ is it true that $|\mathtt{newt}([x,y]) - [1,2]| < Kr$ for all $\theta$ and all sufficiently small $r$? What does this tell you about convergence of Newton's method for this system?

**(d)** How many iterations are necessary to obtain a solution with error less than $10^{-20}$ in both variables, starting at $x = 1.05$, $y = 1.95$? Use `Digits=30`.

**(e)** This example is rather special: the Jacobian determinant is 0 on the line $y = 2x$, and the two functions $x^3 + y^3 - a$ and $5xy^2 + bx^2y - c$ are constant multiples of each other there. Verify

(5) Not the same as the Jacobian matrix of the system of equations!

these facts. What happens to $\texttt{newt}([x, y])$ as $y \to 2x$?

**(f)** Change the first equation to $x^3 + y^3 + (x + 4y)^2 = 90$. Show that $x = 1, y = 2$ is still a solution, and the Jacobian determinant is still 0 on the line $y = 2x$, but the two functions are no longer constant multiples of each other there. What happens now to $\texttt{newt}([x, y])$ as $y \to 2x$?

**6. (a)** Instead of plotting circles centred at the solution and their images under "newt", as we did in Example 1, an alternative would be to use "plot3d" to graph $g(r, \theta)/r$, where if $(x_0 + r\cos\theta,\ y_0 + r\sin\theta) = P$ then $g(r, \theta)$ is the distance from newt(P) to the solution $(x_0,\ y_0)$. What orientation should be used in order to see the maximum distance for a given $r$?

**(b)** What might be the advantages and disadvantages of the two methods?

**(c)** In Example 1, find (approximately) the largest $R$ such that $(\texttt{newt@@2})(P)$ is closer to the solution than $P$ whenever $0 < r < R$, where $r$ is the distance from $P$ to the solution.

**(d)** Use "plot3d" with $g(r, \theta)/r^2$ in Example 1, to find a constant $B$ such that $g(r, \theta) < Br^2$ when $r$ is small. Repeat for Exercise 5 to show that no such $B$ exists in that case.

*CHAPTER 8*

# *Multiple Integration*

## LAB 23 DOUBLE AND TRIPLE INTEGRALS

**Goals:** *To use Maple in setting up and evaluating multiple integrals, with particular attention to the limits of integration for regions specified geometrically.*

Maple evaluates double and triple integrals as iterated integrals. Thus to compute the double integral $\int_2^4 \int_1^2 x^2 y \, dy \, dx$ you could use

```
> int(int(x^2*y, y = 1 .. 2), x = 2 .. 4);
                                    28
```

Maple does this in exactly the same way as we do when we compute the iterated integral by hand. First the inner "`int`" integrates $x^2 y$ with respect to $y$ on the interval 1 to 2, treating $x$ as a symbolic constant, and obtaining an expression depending on $x$: $3x^2/2$. The outer "`int`" then integrates this with respect to $x$ on the interval 2 to 4.

One point to watch out for: normal mathematical notation, writing the integral sign before the integrand, shows the outer integral sign with its endpoints before the inner integral sign and its endpoints; in Maple we write the interval of integration after the integrand, so the inner one comes first. If you think in terms of "inner" and "outer" instead of "first" and "second", you will avoid confusion.

Apart from the actual evaluation of the integrals, the major source of difficulty is usually in setting up the intervals of integration. Typically an integration over a geometric region, in the plane or in three-dimensional space, must be written as an iterated integral. Sometimes we begin with an iterated integral and must change the order of integration. Maple does not have specific commands for doing these tasks automatically, but its graphical capabilities can be very useful tools to help us do them.

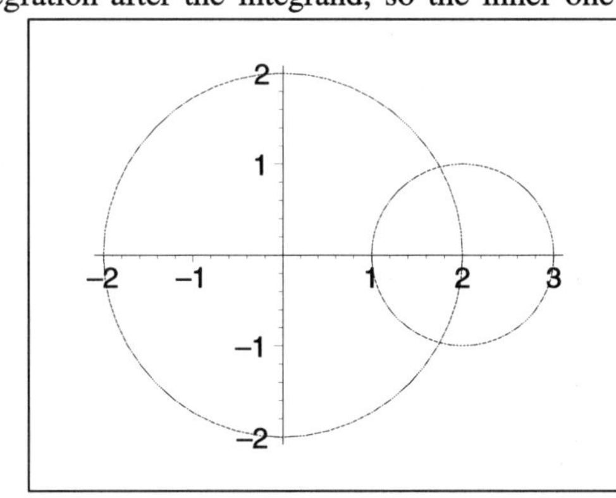

Figure 8.1. Intersecting circles

■ **EXAMPLE 1** Calculate $\iint_R xy^2 \, dA$, where $R$ is the region common to the circle of radius 1 centred at $(2, 0)$ and the circle of radius 2 centred at $(0, 0)$.

**SOLUTION** We begin by plotting the curves forming the boundary of the region $R$. We could use "`implicitplot`", but a parametric plot might be quicker and look better.

```
> plot({[2+cos(t), sin(t), t = 0 .. 2*Pi],
>       [2*cos(t), 2*sin(t), t = 0 .. 2*Pi]},
>      scaling = constrained); (Figure 8.1)
```

It is clear from the graph that the intersection of a horizontal line with $R$ (if it does intersect) is a line segment with left endpoint on the smaller circle ($(x-2)^2+y^2=1$) and right endpoint on the larger circle ($x^2+y^2=4$). In each of these curves, we can solve for $x$ (it's easy enough that I won't use Maple to do it, but we have to be careful of the signs). The horizontal line intersects $R$ when $y$ is between the two $y$ values at the intersection of the two circles (approximately $\pm 0.95$ according to the graph). Note that the region is symmetric about the $x$ axis. Thus the integral is

$$\int_{-b}^{b}\int_{2-\sqrt{1-y^2}}^{\sqrt{4-y^2}} xy^2\,dx\,dy$$

where $(a, \pm b)$ are the intersections of the two circles. We can use Maple to find that intersection:

```
> S:= solve(2-sqrt(1-y^2)=sqrt(4-y^2), y);
```

$$S := \frac{1}{4}\sqrt{15},\ -\frac{1}{4}\sqrt{15}$$

The positive solution is our $b$.

```
> b:= S[1];
```

Now we can set up the integral, and let Maple evaluate it.

```
> J:= int(int(x*y^2, x=2-sqrt(1-y^2) .. sqrt(4-y^2)), y=-b..b);
```

$$J := -\frac{13}{256}\sqrt{15} + \frac{1}{2}\arcsin\left(\frac{1}{4}\sqrt{15}\right)$$

```
> evalf(J);
```

$$.4623831000$$

Just for practice, let's try putting the $y$ integral on the inside. The intersection of a vertical line with our region is a line segment with both endpoints on the smaller circle if $x < a$ or on the larger circle if $x > a$. So our integral becomes

$$\int_{1}^{a}\int_{-\sqrt{1-(x-2)^2}}^{\sqrt{1-(x-2)^2}} xy^2\,dy\,dx + \int_{a}^{2}\int_{-\sqrt{4-x^2}}^{\sqrt{4-x^2}} xy^2\,dy\,dx$$

```
> a:= sqrt(4-b^2);
```

$$a := \frac{7}{4}$$

```
> int(int(x*y^2, y = -sqrt(1-(x-2)^2) .. sqrt(1-(x-2)^2)),
>      x = 1 .. a)
> + int(int(x*y^2, y = -sqrt(4-x^2) .. sqrt(4-x^2)),
>       x = a .. 2);
```

$$-\frac{13}{256}\sqrt{15} - \frac{1}{2}\arcsin\left(\frac{1}{4}\right) + \frac{1}{4}\pi$$

Of course, this turns out to be the same as the first answer. ■

■ **EXAMPLE 2** Find $\iiint_R z\,dV$ where $R$ is the region inside the sphere $x^2+y^2+z^2=1$ and above the paraboloid $z=(x-1)^2+y^2$.

**SOLUTION** Let's look at the cross-sections of the region $R$ for $x =$ constant. The cross-section of the sphere is a circle of radius $\sqrt{1-x^2}$ centred at the origin (if $-1 \le x \le 1$) and the cross-section of the paraboloid is a parabola. The best way to plot the circle is as a parametric curve $y = r\cos\theta$, $z = r\sin\theta$.

I'll use "animate" to show cross-sections at different values of $x$. Now "animate" can plot a set of curves, but if one is parametric the others must also be parametric; we can use $y$ as the parameter for the parabola. An alternative way would be to plot the circle and the parabola separately, and combine them using "display", but that would be more complicated. Note that the axes in these plots are $y$ and $z$: there is a "labels" option for the plotting routines to specify axis labels. You may want to use "scaling=constrained" so that the circles look circular, although this will result in a very tall, narrow animation.

```
> curves:= {
>         [sqrt(1-x^2)*cos(t), sqrt(1-x^2)*sin(t), t = 0 .. 2*Pi],
>         [y, (x-1)^2 + y^2, y = -1 .. 1]};
> plots[animate](curves, x=-1 .. 1, frames=9, labels=[y,z]);
> plot(subs(x=0.5, curves), labels=[y,z], scaling=constrained);
```

(Figure 8.2)

At $x = -1$ the circle is reduced to a single point and the parabola extends upwards from $z = 4$. As $x$ increases, the circle grows and the parabola moves down. The two curves first meet at $x = 0$, when the circle is at its largest. Thereafter, the circle shrinks as the parabola continues to move down, but the lowest point of the parabola stays within the circle until at $x = 1$ the circle has shrunk to a single point.

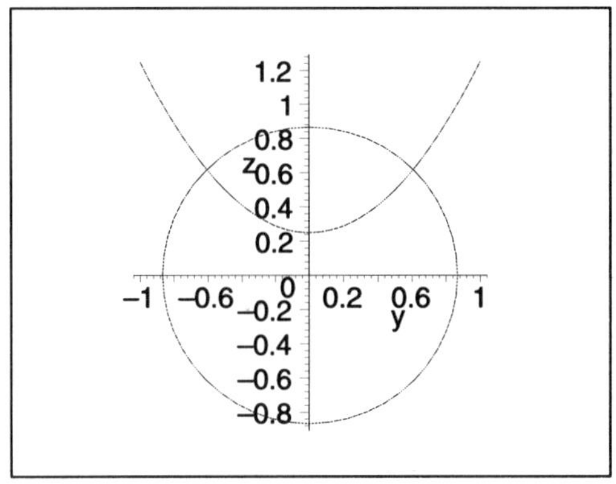

Figure 8.2. Parabola and circle

In the outer integral, $x$ will go from 0 to 1. It seems simplest to take the middle integral over $y$, as otherwise we would have to split the inner integral into two separate pieces. We need to know the two $y$ values at which the parabola and the circle intersect.

```
> S:= solve(x^2 + y^2 + ((x-1)^2 + y^2)^2 = 1, y);
```

$$S := \frac{1}{2}\sqrt{-6-4x^2+8x+2\sqrt{9-8x}},\ -\frac{1}{2}\sqrt{-6-4x^2+8x+2\sqrt{9-8x}},$$
$$\frac{1}{2}\sqrt{-6-4x^2+8x-2\sqrt{9-8x}},\ -\frac{1}{2}\sqrt{-6-4x^2+8x-2\sqrt{9-8x}},$$

Two of these four solutions are the ones we want, while presumably the other two are not real. The ones with $-\sqrt{9-8x}$ could involve the square root of a negative number. We might confirm this by substituting a value for $x$.

```
> subs(x=0.5, [S]);
```
(1)

$$[.6066580495, -.6066580495, 1.366760400\, I, -1.366760400\, I]$$

We extract the positive solution and call it $y_0$. Thus in the middle integral $y$ goes from $-y_0$ to $y_0$.

```
> y0:= S[1];
```

$$y0 := \frac{1}{2}\sqrt{-6-4x^2+8x+2\sqrt{9-8x}}$$

In the inner integral $z$ will go from a value on the paraboloid ($z = (x-1)^2 + y^2$) to its positive value on the sphere ($z = \sqrt{1-x^2-y^2}$). Our integral is

$$\int_0^1 \int_{-y_0}^{y_0} \int_{(x-1)^2+y^2}^{\sqrt{(1-x^2-y^2)}} z\, dz\, dy\, dx$$

```
> V:= int(int(int(z, z=(x-1)^2 + y^2 .. sqrt(1-x^2-y^2)),
>          y=-y0 .. y0), x=0..1);
```

(1) Note that I must make S into a list with "[ ]" in order to do the substitution.

$$V := \int_0^1 -\frac{1}{160}\%1^{5/2} + \frac{1}{12}\left(-\frac{1}{2} - (x-1)^2\right)\%1^{3/2} + \frac{1}{2}\sqrt{\%1} - \frac{1}{2}x^2\sqrt{\%1} - \frac{1}{2}(x-1)^4\sqrt{\%1}\,dx$$
$$\%1 := -6 + 2\sqrt{9-8x} - 4x^2 + 8x$$

Perhaps not surprisingly, Maple can't evaluate the outer integral in "closed form". Well, we can still do it numerically.

```
> evalf(V);
```

$$.1722182285$$

■

## Numerical Approximation

As we have seen in Lab 13, Maple's "`evalf`" uses numerical methods to approximate an integral that can't be evaluated in "closed form". This applies to iterated integrals as well. Just as with single integrals, you can use "`Int`" rather than "`int`" to ensure that Maple doesn't try to evaluate the integral symbolically. In Example 1, we could have used:

```
> evalf(Int(Int(x*y^2, x=2-sqrt(1-y^2)..sqrt(4-y^2)), y=-b..b));
```

$$.4623831003$$

Numerical evaluation of double integrals can be quite slow. Numerical evaluation of triple integrals can be **very** slow(2). Here are some typical times on a rather fast computer.

`> evalf(Int(x^2+x^3+x^4, x=0 .. 1));` took less than 1/100 second.

`> evalf(Int(Int(x^2+y^3+y^4, y=0 .. 1), x=0 .. 1));` took almost a second.

`> evalf(Int(Int(Int(x^2+y^3+z^4, z=0 .. 1), y=0 .. 1),x=0 .. 1));` took over 23 seconds.

Why are double and triple integrals so much slower? In order to evaluate an integral $\int_a^b f(x)\,dx$, "`evalf`" must evaluate the integrand $f(x)$ at a number of different $x$ values. The number of evaluations is at least 19, but can be larger if Maple determines that more are needed to get an accurate result. For a double integral such as $\int_a^b \int_c^d f(x,y)\,dy\,dx$, "`evalf`" will evaluate $\int_c^d f(x,y)\,dy$ for at least 19 different values of $x$. Each of these evaluations will require evaluating $f(x,y)$ for at least 19 values of $y$. Thus to evaluate a double integral requires a total of at least $19 \times 19 = 361$ evaluations of the integrand. Similarly, a triple integral requires at least $19^3 = 6859$ evaluations of the integrand.

It may be worthwhile to give some thought to reducing the amount of computation time required for an iterated integral, especially in the case of a triple integral. If possible, you should arrange matters so that the inner integral can be found in closed form (unless that closed form is extremely complicated). Sometimes this can be done by arranging the order of integrations properly.

■ **EXAMPLE 3** Calculate $\int_1^2 \int_0^1 \ln(\cos(x)+y)\,dx\,dy$.

**SOLUTION** It is clear that the inner integral should be simpler if it is with respect to $y$ rather than $x$. You could probably do $\int_1^2 \ln(\cos(x)+y)\,dy$ by hand, but $\int_0^1 \ln(\cos(x)+y)\,dx$ looks much more difficult. On the other hand, the result of integrating $\int_1^2 \ln(\cos(x)+y)\,dy$ would also be very difficult or impossible to integrate in "closed form" with respect to $x$. Thus it would be best to use

(2) You may need to interrupt Maple's calculations when it takes too long. See the section "Quitting and Interrupting" in Lab 1.

```
> evalf(Int(int(ln(cos(x)+y), y=1..2), x=0..1));
```

$$.8411812218$$

This took about 4 seconds in Release 4 on a Pentium II computer. On the other hand,

```
> evalf(int(int(ln(cos(x)+y), y=1..2), x=0..1));
```

took about 6 seconds to arrive at the same answer: apparently Maple spent the extra 2 seconds trying unsuccessfully to do the outer integral symbolically(3).

Doing the double integral numerically with

```
> evalf(Int(Int(ln(cos(x)+y), y=1..2), x=0..1));
```

also took about 6 seconds.

As it happens, Maple can evaluate $\int_1^2 \ln(\cos(x) + y)\,dy$ in closed form.

```
> int(ln(cos(x)+y), x = 0 .. 1);
```

This closed form is a horribly complicated expression involving complex numbers and the "`dilog`" function(4). The expression is so complicated that it is much slower to evaluate it than to do the integral numerically. But there's an extra twist to this story: the result, after about 83 seconds, is a wrong answer:(5)

```
> evalf(Int(int(ln(cos(x)+y), x=0 .. 1), y=1..2));
```

$$10.91117751 + 39.47841760\,I$$

So besides questions of speed, there is another important consideration: be careful with symbolic integrals with complicated results (especially those involving square roots or logarithms), because they may lead to wrong answers.

## EXERCISES

**1.** Evaluate the following integrals using Maple.

**(a)** $\displaystyle\int_1^2 \int_3^4 \cos(xy)\,dy\,dx$ **(b)** $\displaystyle\int_1^2 \int_3^4 \int_5^6 x^2 y \cos(xyz)\,dz\,dy\,dx$

**2.** Evaluate the integral $\displaystyle\iint_R y\,dA$, where $R$ is the region containing the point $(0, 1)$ and bounded by a loop of $y^2 + 2y\cos x + \sin^2 x = 0$.

**3.** Find the centroid of the region of space satisfying $x \ge 0$, $y \le 1$, and $0 \le z \le xy - x^3$. *Hint:* the centroid of a solid region $R$ has coordinates $\bar{x} = \displaystyle\iiint_R x\,dV \Big/ \iiint_R dV$, $\bar{y} = \displaystyle\iiint_R y\,dV \Big/ \iiint_R dV$, $\bar{z} = \displaystyle\iiint_R z\,dV \Big/ \iiint_R dV$. To save typing, you can define a function that performs the triple integral of an arbitrary expression over $R$, and then obtain all three coordinates in one command.

---

(3) A difference of only two seconds in this example is not worth bothering about, but there are cases where the difference in time is significant. It was more impressive in Release 3 on my old '386 computer, where these times were 31 seconds and 168 seconds.

(4) This function is itself defined by an integral: $\operatorname{dilog}(x) = \displaystyle\int_1^x \frac{\ln t}{1-t}$, although that's not what Maple uses to calculate it.

(5) The reason for this bug is that Maple's "closed-form" answer for the antiderivative $\displaystyle\int \ln(\cos(x) + y)\,dx$ has a discontinuity at $x = 0$ when $y > 1$, which Maple fails to deal with correctly when evaluating the definite integral for $x$ from 0 to 1.

4. Find the volume of the region inside both the cylinder $(x-1)^2+y^2=1$ and the sphere $x^2+y^2+z^2=1$.

5. Write each of the following as iterated integrals with the outer integral over $z$ and the inner integral over $y$.

   (a) $\displaystyle\int_0^1\int_0^{y^2} f(y,z)\,dz\,dy+\int_1^2\int_{1-(y-2)^2}^{1} f(y,z)\,dz\,dy.$

   (b) $\displaystyle\int_0^1\int_{2x^2-x}^{3x-2x^2}\int_0^{\cos(\pi(x-y))} f(x,y,z)\,dz\,dy\,dx.$

6. Consider two circular cylinders of radius 1. The axis of one cylinder is the $x$ axis. The other cylinder's axis is in the $xy$ plane, passing through the origin at a 60 degree angle to the $x$ axis. Note that the surface of the second cylinder satisfies the equation $(\sqrt{3}x/2-y/2)^2+z^2=1$.

   (a) Write an expression for the integral of a function $f(x,y,z)$ over the region of space inside both of the cylinders, using iterated integrals with $x$ in the outer integral. *Hint:* be sure to consider all cases. An animation of the cross-sections $x = constant$ may help.

   (b) Write another expression for the same integral, where the variable in the outer integral is $z$.

7. In Lab 14 we found equations of the form $y=f(x)$ for the different branches of the limaçon of Pascal. Integrate $xy^2$ over the region enclosed by the large loop of the limaçon, obtaining a numerical answer (this will be much simpler using polar coordinates in the next lab).

8. What command would have Maple evaluate numerically $\displaystyle\iint_R x\sin(x^2+y^2)\,dA$ as quickly as possible, where $R$ is the region $0\le x\le y\le 1$?

# LAB 24 CHANGE OF VARIABLES

**Goals:** *To use Maple in setting up and evaluating multiple integrals involving non-Cartesian coordinates, in particular polar, cylindrical, and spherical coordinates.*

It is often more convenient to use some other coordinate system instead of the Cartesian $x$, $y$ or $x$, $y$, $z$. The most common of these alternative coordinate systems are polar coordinates in the plane and cylindrical and spherical coordinates in space.

When making a change of variables in a double integral, we express the old variables, say $x$ and $y$, in terms of the new variables $u$ and $v$: $x=x(u,v)$, $y=y(u,v)$. The element of area $dx\,dy$ is replaced by $J(u,v)\,du\,dv$ where $J(u,v)$ is the absolute value of the Jacobian determinant

$$J(u,v)=\left|\frac{\partial(x,y)}{\partial(u,v)}\right|=\left|\frac{\partial x}{\partial u}\frac{\partial y}{\partial v}-\frac{\partial x}{\partial v}\frac{\partial y}{\partial u}\right|$$

Similarly, for a change of variables in a triple integral, the element of volume $dx\,dy\,dz$ is replaced by $J(u,v,w)\,du\,dv\,dw$, where $J(u,v,w)$ is the absolute value of the Jacobian determinant for the transformation expressing the old variables $x$, $y$, $z$ in terms of the new variables $u$, $v$, $w$. The difficulty, if any, is often connected with finding the appropriate limits of integration for the new variables. This often involves inverting the transformation, i.e., solving for the new variables in terms of the old variables.

■ **EXAMPLE 1** Express $\int_0^1 \int_0^z \int_0^y f(x, y, z)\, dx\, dy\, dz$ in terms of $u$, $v$, $w$ with $x = uv + w$, $y = u - w$, $z = u + w$.

**SOLUTION** In Maple, the "`jacobian`" and "`det`" functions are part of the "`linalg`" package.

```
> with(linalg):
> det(jacobian([u*v+w, u-w, u+w], [u,v,w]));
```

$$-2u$$

Thus the integrand will be $2|u|f(uv + w,\ u - w,\ u + w)$. To help find the limits of integration, we first solve for $u$, $v$, $w$ in terms of $x$, $y$, and $z$ (actually, since $y$ and $z$ just depend on $u$ and $w$, it would suffice to solve for $u$ and $w$ in terms of $y$ and $z$).

```
> solve({x=u*v+w, y=u-w, z=u+w}, {u,v,w});
```

$$\left\{ v = \frac{2x - z + y}{y + z},\ u = \frac{1}{2}y + \frac{1}{2}z,\ w = \frac{1}{2}z - \frac{1}{2}y \right\}$$

Since $0 \le x \le y \le z \le 1$, it is clear that $u$ can go from 0 to 1 and $w$ from 0 to 1/2; $v$ is not so clear. We'll put the $u$ integral on the outside. Now the cross-section of our region of integration for $u = constant$ is defined by the following inequalities in the $vw$ plane:

$x = uv + w \ge 0$, i.e., $v \ge -w/u$

$x \le y$, i.e., $uv + w \le u - w$, or $v \le 1 - 2w/u$

$y \le z$, i.e., $u - w \le u + w$ or $w \ge 0$

$z = u + w \le 1$, i.e., $w \le 1 - u$.

Let's look at these cross-sections. Since we know an interval for $w$ but not for $v$, we will plot the lines $v = -w/u$ and $v = 1 - 2w/u$ for $0 \le w \le 1 - u$ (taking $w$ on the horizontal axis, and $v$ on the vertical). We'll also plot the vertical line $w = 1 - u$ between its intersection with $v = -w/u$ and its intersection with $v = 1 - 2w/u$.

```
> with(plots, display):
> usec:= u -> display({plot({-w/u, 1-2*w/u}, w=0..1-u),
>                plot([[1-u, -(1-u)/u], [1-u, 1-2*(1-u)/u]])},
>                labels=[w,v]);
> usec(0.7);
```
(Figure 8.3)

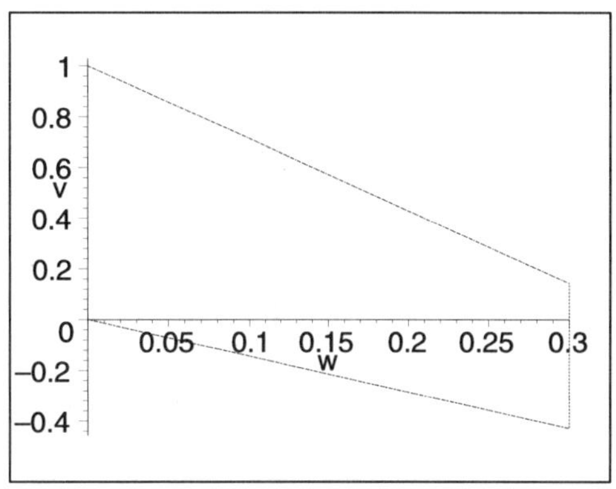


Figure 8.3. Cross-section $u = 0.7$

We might try an animation of the cross-sections for different values of $u$. I'll leave out the singular cases $u = 0$ and $u = 1$.

```
> display([ seq(usec(k/10),
> k=1..9) ], insequence=true);
```

Our region is below the slanted line through $w = 0$, $v = 1$, above the line through $w = 0$, $v = 0$, left of the vertical line $w = 1 - u$, and right of the $v$ axis. For $0 < u \le 1/2$, $w$ goes from 0 to $u$ where the two slanted lines intersect ($1 - 2w/u = -w/u$). For $1/2 < u < 1$, on the other hand, $w$ goes from 0 to $1 - u$ on the vertical line. In each case, $v$ goes from $-w/u$ to $1 - 2w/u$. Thus the integral is

$$\int_0^{1/2} \int_0^u \int_{-w/u}^{1-2w/u} 2uf(uv + w,\ u - w,\ u + w)\, dv\, dw\, du$$

$$+ \int_{1/2}^1 \int_0^{1-u} \int_{-w/u}^{1-2w/u} 2uf(uv + w,\ u - w,\ u + w)\, dv\, dw\, du$$ ■

## Polar Coordinates

In polar coordinates $r$, $\theta$, the element of area is $dA = r\,dr\,d\theta$. Thus when integrating an expression $f$ over a region expressed in polar coordinates, we might use

```
> int(int(r*f, r=...), theta=...);
```

**EXAMPLE 2** Integrate $xy^2$ over the region between the two loops of the limaçon of Pascal (see Example 3 of Lab 14 and Exercise 7 of Lab 23).

**SOLUTION** Look at the picture of the limaçon in Figure 4.8. The limaçon is described by the polar equation $r = 1 + 2\cos\theta$. The large loop is the portion of the curve with $-2\pi/3 \le \theta \le 2\pi/3$, where $r > 0$. The small loop is the portion with $2\pi/3 \le \theta \le 4\pi/3$, where $r < 0$. However, for purposes of integration we will want to write this in terms of a positive $r$. If we write $r_1 = -r_2$ and $\theta_1 = \theta_2 + \pi$, then $[r_1, \theta_1]$ and $[r_2, \theta_2]$ represent the same point, with Cartesian coordinates $(r_1\cos\theta_1,\ r_1\sin\theta_1) = (r_2\cos\theta_2,\ r_2\sin\theta_2)$. The equation $r_1 = 1 + 2\cos\theta_1$ corresponds to $r_2 = -1 + 2\cos\theta_2$. Thus the small loop of the limaçon is the polar graph of $r = -1 + 2\cos\theta$ for $-\pi/3 \le \theta \le \pi/3$.

Since we are given $r$ as a function of $\theta$ for the two curves, it is most natural to write the integral over $\theta$ on the outside and the integral over $r$ on the inside. We could split up the $\theta$ integration using three intervals: $-2\pi/3 \le \theta \le -\pi/3$ and $\pi/3 \le \theta \le 2\pi/3$, where $r$ goes from 0 to $1 + 2\cos\theta$, and $-\pi/3 \le \theta \le \pi/3$, where $r$ goes from $-1 + 2\cos\theta$ to $1 + 2\cos\theta$. However, it is simpler to just calculate the integral over the region enclosed by the large loop ($J_1$) and subtract the integral over the region enclosed by the small loop ($J_2$). It is convenient to define "xp" and "yp" as the polar equivalents of $x$ and $y$, and express the integrand in terms of these.

```
> xp:= r*cos(theta); yp:= r*sin(theta);
> J1:= int(int(r*xp*yp^2, r = 0 .. 1+2*cos(theta)),
>          theta = -2*Pi/3 .. 2*Pi/3);
```

$$J1 := \frac{405}{224}\sqrt{3} + 2\pi$$

```
> J2:= int(int(r*xp*yp^2, r = 0 .. -1+2*cos(theta)),
>          theta = -Pi/3 .. Pi/3);
```

$$J2 := -\frac{405}{224}\sqrt{3} + \pi$$

```
> J1-J2;
```

$$\frac{405}{112}\sqrt{3} + \pi$$

■

## Cylindrical Coordinates

In cylindrical coordinates, we use the polar coordinates $r$ and $\theta$ in the $xy$ plane in addition to the third Cartesian coordinate $z$. The element of volume is $dV = r\,dr\,d\theta\,dz$.

**EXAMPLE 3** Find the volume inside both the sphere $x^2 + y^2 + z^2 = 9$ and the cylinder $(x-2)^2 + y^2 = 4$.

**SOLUTION** The sphere can be written in cylindrical coordinates as $r^2 + z^2 = 9$. What about the cylinder? We could probably do it by hand, but let's use Maple. We substitute $x = r\cos\theta$ and $y = r\sin\theta$ in the equation, then solve for $r$.

```
> solve(subs(x=r*cos(theta), y=r*sin(theta), (x-2)^2+y^2=4), r);
```

$$0, 4\frac{\cos(\theta)}{\cos(\theta)^2 + \sin(\theta)^2}$$

We can disregard the 0, since $r = 0$ is included in the second solution (for $\theta = \pi/2$). We don't need Maple to tell us that the denominator of the second solution is 1 (of course, we could use "`simplify`" to do this). So the equation in cylindrical coordinates is $r = 4\cos\theta$, where $-\pi/2 \le \theta \le \pi/2$ to make $r \ge 0$.

In visualizing a three-dimensional region, three-dimensional graphs are often helpful. There is a "cylinderplot" command (in the "plots" package) that can be used to plot surfaces in cylindrical coordinates. For a surface $r = f(\theta, z)$ where $a < \theta < b$ and $c < z < d$, the syntax is

```
> cylinderplot(f(theta,z), theta = a..b, z = c..d);
```

Note that the $\theta$ interval must always come first and the $z$ interval second: Maple doesn't ascribe any special significance to the names "theta" and "z", it just takes the first "variable=range" to refer to the angular coordinate and the second to refer to the vertical coordinate. The bounds $a$, $b$, $c$, and $d$ must be constants (this is in contrast to "plot3d", where the bounds for the second variable can depend on the first variable).

The "cylinderplot" command can handle several surfaces at once, all with the same $\theta$ and $z$ intervals, if you replace "f(theta,z)" by a set of functions. In our case, however, we want a different $\theta$ interval for the sphere than for the cylinder. Therefore I'll produce two different plots and combine them with "display".

```
> sph:= cylinderplot(sqrt(9-z^2),
> theta = 0 .. 2*Pi, z = -3 .. 3):
> cyl:= cylinderplot(4*cos(theta),
> theta=-Pi/2 .. Pi/2, z=-3 .. 3):
> display({sph, cyl},
>   scaling=constrained);
```

It may be helpful to look at this plot from several different viewpoints. Particularly useful is a viewpoint on the positive $z$ axis: orientation $[-90, 0]$ looks down on the $xy$ plane, with $x$ and $y$ axes in their usual orientations (Figure 8.4). The cylinder is seen "edge-on", as a circle of radius 2.

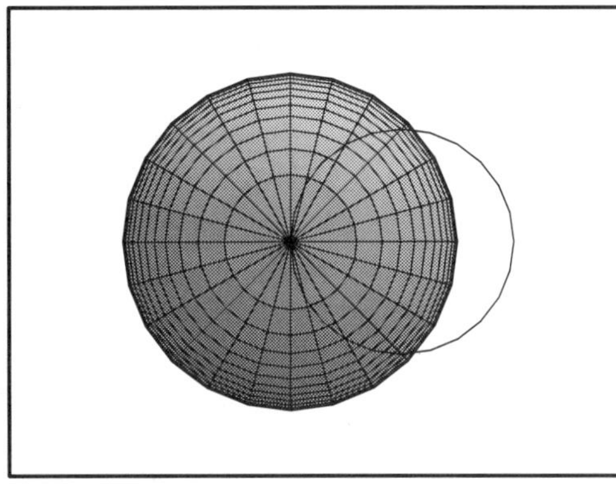

Figure 8.4. Sphere and cylinder

Which variable shall we take on the outside in our integral? Any of the three is a possibility. However, we might not want to use $\theta$, since we would have to break the integral up into pieces (some cross-sections at constant $\theta$ intersect the surface of the cylinder, others do not). Let's use $r$, which runs from 0 to 3. This is a particularly good choice since the endpoints of the $\theta$ and $z$ integrals will depend only on $r$ (because the equation of the sphere doesn't involve $\theta$, and the equation of the cylinder doesn't involve $z$).

```
> V:= int(int(int(r, z = -sqrt(9-r^2) .. sqrt(9-r^2)),
>               theta = -arccos(r/4) .. arccos(r/4)),
>          r = 0 .. 3);
```

$$V := \int_0^3 4r\sqrt{9-r^2}\,\arccos\left(\frac{1}{4}r\right)\,dr$$

```
> evalf(V);
```

$$39.74123918$$

■

## Spherical Coordinates

Spherical coordinates are $[\rho, \phi, \theta]$, where $\rho$ is the distance from a point to the origin, $\phi$ (the "colatitude") is the angle between the line from the point to the origin and the positive $z$ axis, and $\theta$ (the "longitude") is the same as in cylindrical coordinates. These are related to Cartesian coordinates by the equations

$$x = \rho\sin\phi\cos\theta, \quad y = \rho\sin\phi\sin\theta, \quad z = \rho\cos\phi$$

The element of volume in spherical coordinates is $\rho^2\sin\phi\,d\rho\,d\phi\,d\theta$. Let's confirm this. The definitions of "xs", "ys", and "zs" we make here will also be useful later on.

```
> xs:= rho*sin(phi)*cos(theta); ys:= rho*sin(phi)*sin(theta);
> zs:= rho*cos(phi);
> J:= det(jacobian([xs, ys, zs], [rho, phi, theta]));
```

$$J := \sin(\phi)^3 \cos(\theta)^2 \rho^2 + \sin(\phi)^3 \sin(\theta)^2 \rho^2 + \rho^2 \cos(\phi)^2 \cos(\theta)^2 \sin(\phi) + \rho^2 \sin(\phi) \sin(\theta)^2 \cos(\phi)^2$$

```
> simplify(J);
```

$$\sin(\phi)\rho^2$$

■ **EXAMPLE 4** Use spherical coordinates to find the centroid of the region where $\sqrt{x^2+y^2} \le z \le 1$ and $x \ge 0$.

**SOLUTION** We begin by defining the boundary of the region in spherical coordinates. The conical surface $\sqrt{x^2+y^2} = z$ is $\phi = \pi/4$ (it may be easier to see this for yourself rather than getting Maple to come up with it). Thus we want $0 \le \phi \le \pi/4$. The plane $z = 1$ is $\rho \cos\phi = 1$, so $0 \le \rho \le 1/\cos\phi$. The plane $x = 0$ is $\theta = \pi/2$ and $\theta = -\pi/2$, so $-\pi/2 \le \theta \le \pi/2$.

Since we have several different quantities to integrate over our region, it may save some typing to first define a function "J" that takes an arbitrary expression and integrates it over the region. Thus in ordinary mathematical notation,

$$J(f) = \int_0^{\pi/4} \int_{-\pi/2}^{\pi/2} \int_0^{1/\cos(\phi)} f\rho^2 \sin(\theta)\, d\rho\, d\theta\, d\phi$$

```
> J:= f -> int(int(int(f * rho^2 * sin(phi),
> rho=0 .. 1/cos(phi)), theta=-Pi/2 .. Pi/2), phi=0 .. Pi/4);
```

To find the volume, we integrate 1 over the region.

```
> V:= J(1);
```

$$V := \frac{1}{6}\pi$$

Now to find the $x$, $y$, and $z$ coordinates of the centroid we integrate $x$, $y$, and $z$ (as expressed in spherical coordinates by "xs", "ys", and "zs") over the region and divide by $V$. We could actually use symmetry to conclude that the $y$ coordinate will be 0.

```
> CM:= [J(xs)/V, J(ys)/V, J(zs)/V];
```

$$CM := \left[\frac{1}{\pi}, 0, \frac{3}{4}\right]$$ ■

## EXERCISES

1. Integrate $xy^2$ over the parallelogram with vertices $(0,0)$, $(1,2)$, $(3,3)$, and $(2,1)$. Use a transformation to new variables $u$ and $v$ for which the sides of the parallelogram are of the form $u = constant$ and $v = constant$.
2. Find the centroid of the loop of the lemniscate $r^2 = \cos(2\theta)$ in the right half plane.
3. Use polar coordinates to integrate $\sqrt{x^2+y^2}$ over the triangle bounded by $x = 0$, $x = y$, and $y = 1$.
4. Use cylindrical coordinates to find the volume inside both a sphere of radius 1 and a sphere of radius 2, where the centre of the small sphere is on the surface of the large sphere. *Hint:* place one sphere's centre at the origin. Where should the other be in order to have cylindrical symmetry?
5. Use cylindrical coordinates to find the centroid of the region inside both the torus $(r-2)^2+z^2 = 1$ and the cylinder $(x-2)^2+y^2 = 1$. *Hint:* put the $r$ integral on the outside.

**6.** The gravitational potential at a point $P$ due to a massive body is obtained by integrating $G\sigma(x, y, z)/d(x, y, z)$ over the body, where $G$ is the gravitational constant, $\sigma(x, y, z)$ the density at the point $(x, y, z)$, and $d(x, y, z)$ the distance from $(x, y, z)$ to $P$. Consider a solid cylinder of height 2, radius 1, and constant density 1. Find the gravitational potential at a point on the axis of the cylinder at a distance $d$ from the cylinder.

**7.** In this exercise you will find the volume of the region inside the torus $(r - 2)^2 + z^2 = 1$ and above the plane $z = 1 + x$.

**(a)** Plot the torus and part of the plane in three dimensions. You can use "`cylinderplot`" for the torus, "`plot3d`" for the plane, and combine them with "`display`". Note that you can plot both the inner and outer surfaces of the torus in one "`cylinderplot`", using a set of two expressions instead of one expression as the first input to the command. Look at the graph from several different viewpoints, in particular "`orientation=[-90,0]`" (looking at it from the top, with $x$ going from left to right) and "`orientation=[-90,45]`" (looking directly along the plane $z = 1 + x$).

**(b)** Plot some of the sections $r =$ constant for our region, where $1 \leq r \leq 3$, using $\theta$ on the horizontal axis and $z$ on the vertical axis. An animation may be useful. Note that the section of our region lies between two horizontal lines (coming from the torus) and above a curve (coming from the plane).

**(c)** At what values of $\theta$ does the curve intersect the upper line?

**(d)** For what interval of $r$ does the curve intersect the lower line, and at what $\theta$ values does this occur?

**(e)** Find the volume of the region below the top surface of the torus and above the plane. Find the volume of the region below the bottom surface of the torus and above the plane. Subtract to find the volume of the region inside the torus and above the plane. Note: the $r$ integral must be done numerically.

**8.** There is also a "`sphereplot`" command in the "`plots`" package. This is analogous to the "`cylinderplot`" command, but using spherical coordinates (where $\rho$ is given as a function of $\phi$ and $\theta$). The syntax is "`sphereplot(f(phi,theta), theta=a..b, phi=c..d)`". Note that $\theta$ comes before $\phi$. Use this to examine the surface $\rho = 1 + \cos\theta$. Why does the surface contain a sharp edge? Find the volume of the region enclosed by this surface.

## CHAPTER 9

# Surfaces, Curves, and Fields

## LAB 25 SPACE CURVES

**Goals:** *To use Maple in visualizing three-dimensional parametric curves. To analyze the velocity and acceleration vectors for motion in three-dimensional space. To calculate and visualize the Frenet frame and osculating circle for a curve.*

We have discussed parametric curves in the plane in Lab 14. Now we wish to consider a curve in three-dimensional space. We can think of it as the trajectory of a moving point, where the parameter $t$ represents time:

$$x = x(t), \quad y = y(t), \quad z = z(t)$$

Often it is more convenient to put the $x$, $y$, and $z$ components together into a vector $\mathbf{R} = \mathbf{R}(t) = [x(t), y(t), z(t)]$.

We will be using the "`plots`" package, as well as the file "`dot.def`" which we used in Labs 14 and 17. If this file is not available to you, you can type in the definitions from Appendix I.

```
> with(plots): read `dot.def`;
```

### Plotting Space Curves

■ **EXAMPLE 1** Plot the curve $x(t) = (2 + \sin(5t))\cos(t)$, $y(t) = (2 + \sin(5t))\sin(t)$, $z(t) = \cos(5t)$ for $0 \le t \le 2\pi$. This curve winds around the torus $(r - 2)^2 + z^2 = 1$ (in cylindrical coordinates).

**SOLUTION**

```
> R:= t -> [(2+sin(5*t))*cos(t), (2+sin(5*t))*sin(t), cos(5*t)];
```

Some of our results will be somewhat simpler if we put this in a different form with "`combine(...,trig)`".

```
> R:= unapply(combine(R(t),trig),t);
```

$$R := t \to \left[2\cos(t) + \frac{1}{2}\sin(6t) + \frac{1}{2}\sin(4t),\ 2\sin(t) + \frac{1}{2}\cos(4t) - \frac{1}{2}\cos(6t),\ \cos(5t)\right]$$

The "`spacecurve`" command in the "`plots`" package plots a parametric curve in three dimensions. Since I'll want to use the plot again, I'll first save it to a variable, then show it on the screen.

```
> curve:= spacecurve(R(t), t = 0 .. 2*Pi, numpoints = 200,
>              scaling = constrained, shading = Z):
> curve;
```

Note the difference between this and the "`plot`" command for a parametric curve in two dimensions: in "`plot`", the parameter interval "`t=0..2*Pi`" must be included in the list with the expressions for the $x$ and $y$ coordinates, while here it is a separate input to "`spacecurve`"(1).

(1) In "`spacecurve`" you can also include the parameter interval in the list with the $x$, $y$, and $z$ coordinates.

As with any three-dimensional plot, you can look at this from various viewpoints to get a better picture of what is going on. This is particularly useful for space curves, since the picture of the curve itself has few visual clues for depth. Using the "`shading=Z`" option may help a bit, on a colour monitor: the higher parts of the curve (larger $z$ values) are redder, while the lower parts are bluer. You might try "`ZGREYSCALE`", which makes the higher parts light and the lower parts dark. However, for plots with white backgrounds(2), the lighter parts of the curve may not show up well.

Since our curve is on a torus, it may be helpful to plot it together with the torus. We can produce the torus using "`cylinderplot`" (see Lab 24), and combine this with the space curve using "`display`". However, when the curve lies right on the surface Maple may have difficulty deciding whether it should be hidden by the surface or not. To make sure that the curve is visible when it should be, it is better to have it raised slightly above the surface. Therefore instead of plotting the torus $(r-2)^2+z^2=1$, I will use $(r-2)^2+z^2=0.95$.

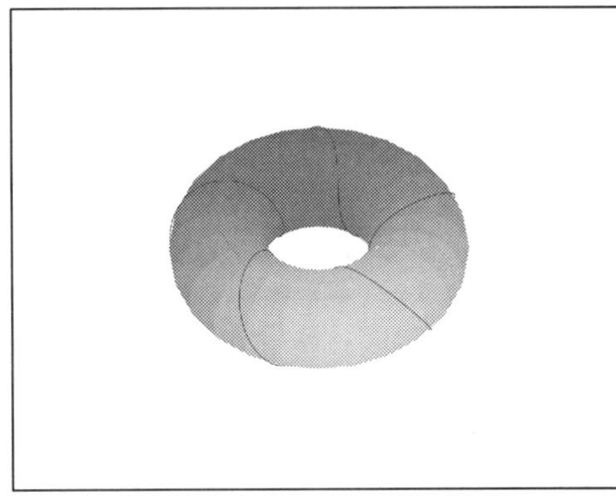

Figure 9.1. Curve and torus

```
> torus:= cylinderplot({sqrt(.95-z^2)+2, 2-sqrt(.95-z^2)},
>          t = 0 .. 2*Pi, z = -sqrt(.95) .. sqrt(.95),
>          style = PATCHNOGRID):
> display({torus,curve}, scaling=constrained); (Figure 9.1)
```

Another command that helps to visualize a curve in three dimensions is "`tubeplot`"(3). Instead of plotting the curve itself, this plots a tube of a certain radius centred on the curve. The radius can be specified as one of the options to "`tubeplot`" (the default is 1). Another option is "`tubepoints`": the cross-section of the tube is a regular polygon of "`tubepoints`" sides (the default is 10). Here is our curve $\mathbf{R}(t)$ made into a tube.

```
> tubeplot(R(t), t=0..2*Pi, numpoints=100, scaling=constrained,
>               radius=0.2);
```

■

## Velocity and Acceleration

If $t$ represents time, the derivative of the position $\mathbf{R}(t)$ with respect to $t$ is the velocity. We can't use "`D`" for this, but "`diff`" works. Of course, "`diff`" works only with the expression "`R(t)`", not with a function "`R`" directly, so I'll use "`unapply`" to make the expression "`diff(R(t),t)`" into a function. I'll use the $\mathbf{R}$ from Example 1.

```
> V:= unapply(diff(R(t), t), t);
```

$$V := t \to [-2\sin(t)+3\cos(6t)+2\cos(4t),\ 2\cos(t)-2\sin(4t)+3\sin(6t),\ -5\sin(5t)]$$

Similarly, we differentiate $\mathbf{V}$ to obtain the acceleration:

```
> A:= unapply(diff(V(t), t), t);
```

$$A := t \to [-2\cos(t)-18\sin(6t)-8\sin(4t),\ -2\sin(t)-8\cos(4t)+18\cos(6t),\ -25\cos(5t)]$$

(2) It is possible to change the background colour of "window" plots, but not inline plots. For the Windows interface, you must edit the file "`maplev`$n$`.ini`" where $n$ is the release number (4 or 5), which should be in your Windows directory. You can use Notepad or any other text editor. Be sure to make a backup copy of the file first. In the "`[Options]`" section there will be a line that starts "`PlotBGColor=`". If what comes next is "`default`" or "`255 255 255`", the plot backgrounds will be white. If it is "`0 0 0`", they will be black. You can specify any three numbers from 0 to 255 for red, green, and blue components. This only takes effect when you start a new Maple session; you can't change the colour in the middle of a session.

(3) Also in the "`plots`" package.

We can produce a visual representation of the velocity and acceleration vectors at various points on the curve, using arrows. A convenient way to produce these arrows is with "`arrow3d`", defined in the file "`arrow3d.def`" which should be available to you. The command "`arrow3d(R,V)`" draws an arrow representing vector `V` starting at `R`. Other options such as "`colour`" can also be added. I'll use the colour to distinguish the velocity and acceleration vectors. If you don't have a colour monitor, you could use "`linestyle`" instead. The following function plots the curve (which we previously saved) together with two arrows starting at $\mathbf{R}(t)$: a blue one for the velocity $\mathbf{V}(t)$ and a red arrow for the acceleration $\mathbf{A}(t)$. The scaling factors $1/4$ and $1/16$ are chosen so that these arrows have reasonable lengths. The `thickness` (of the arrow stem) is set to 1 instead of the default 3, because it looks better with arrows of this size.

```
> plotav:= t -> display(
>         {curve,  arrow3d(R(t), V(t)/4, colour=blue, thickness=1),
>          arrow(R(t), A(t)/16, colour=red, thickness=1)},
>          scaling=constrained);
> plotav(1);
```

We could also try an animation where $\mathbf{R}(t)$ traverses part of the curve.

```
> display([seq(plotav(k*Pi/20), k=0..10)], insequence=true);
```

## The Skier

**EXAMPLE 2** Part of the surface of a ski hill satisfies the equation

$$z = -(x+y)\left(.13 + .1\,\sin\left(\frac{x+y}{8}\right)\right) - .15\,x\sin\left(\frac{x}{8} - \frac{3y}{40}\right)$$

(where $x$, $y$ and $z$ are in metres). A downhill skier starts from rest at $x = 0, y = 0, z = 0$, and follows a path on the surface with $y = .7\,x - 4\,\sin(x/7)$ for $0 < x < 40$. Study the forces felt by the skier. In particular, does she leave the surface of the snow?

**SOLUTION** We begin by defining the hill and the skier's path. The most convenient parameter to use for the path is $x$ (we would prefer the time $t$, but it's not readily available). I'll define the surface using a function of $x$ and $y$. I'll also use functions of $x$ for quantities along the path. Many of these functions are going to be quite complicated, so we won't bother displaying the results except numerically or graphically.

```
> f:= (x,y) -> -(x+y) * (.13 + .1 * sin((x+y)/8))
>                 - .15 * x * sin(x/8 - 3*y/40);
> yp:= .7 * x - 4 * sin(x/7); R:= unapply([x, yp, f(x, yp)], x);
```

Let's plot the hill and the path. In order to make sure the path is not hidden by the surface, I'll lower the surface slightly.

```
> surface:= plot3d(f(x,y)-.2, x = 0 .. 40, y = -5 .. 40,
>                style = PATCHNOGRID, ambientlight=[.2,.2,.5],
>                light=[60,60,.75,.75,.75], shading = ZHUE):
> curve:= spacecurve(R(x), x = 0 .. 40, colour = red):
> display({surface, curve}, orientation = [-15,45],
>              scaling=constrained, axes=BOXED); (Figure 9.2)
```

You should examine this plot from various angles. It might also be useful to see a contour plot with the curve added. Since we already have the curve in three dimensions, I'll use the three-dimensional version "`contourplot3d`" rather than the two-dimensional "`contourplot`". Note that options such as "`orientation`" must be specified in the "`display`", not just in the "`contourplot3d`".

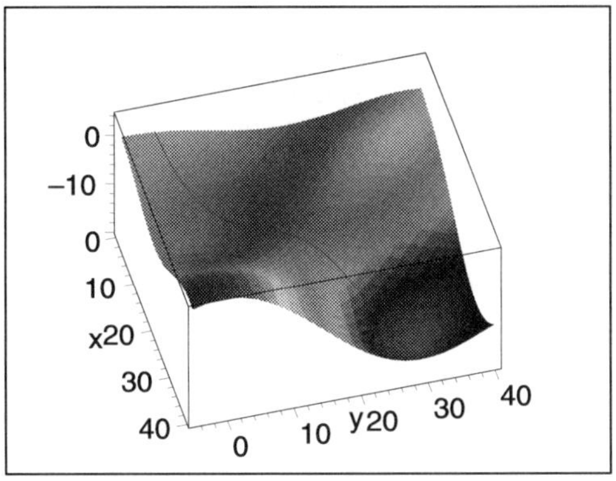


Figure 9.2. Ski hill and skier's path

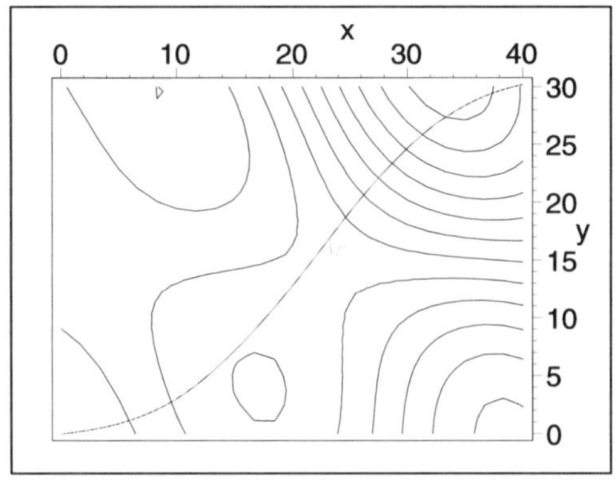


Figure 9.3. Contours and skier's path

```
> cplot:= contourplot3d(f(x,y), x=0..40, y=0..30):
> display({cplot, curve}, orientation = [-90,0], axes = BOXED,
>  scaling = constrained, shading = ZHUE,
>  style = CONTOUR); (Figure 9.3)
```

At first the skier's path runs downwards at a fairly gentle slope (note that on the contour plot, slopes are gentle when the contours are far apart and steep when they are close together). From about $x = 11$ to $x = 17$ the path actually slopes slightly upwards. It runs across a saddle-shaped "pass" between two peaks. Then after about $x = 25$ the slope becomes steep, and the skier descends into a deep "pit".

I will assume that the only force on the skier that has a component in the direction of the velocity vector is gravity (under ideal conditions, when the skier is trying to go as fast as possible, this may be a fairly good approximation to reality), and that the skier does not raise or lower her centre of gravity (e.g., by bending or straightening the knees and hips). Under these assumptions, the skier's total energy (potential + kinetic) $mv^2/2 + mgz$ is constant, where $m$ is the skier's mass, $v$ is the speed, and $g = 9.8\,\mathrm{m/s^2}$. Since $v = 0$ and $z = 0$ at the start, the constant is 0. We can use this to determine $v$.

```
> g:= 9.8; v:= x -> sqrt(-2*g*R(x)[3]);
```

To find the velocity vector $\mathbf{V}$, we can first differentiate $\mathbf{R}$ with respect to $x$(4). The unit vector in this direction is the tangent vector $\mathbf{T}$ to the path. We then multiply by $v$ to obtain $\mathbf{V}$.

```
> T:= unapply(evl(unit(diff(R(x),x))),x);
|> V:= x -> evl(v(x)*T(x));
```

Now for the acceleration. According to the Chain Rule, $\mathbf{A} = \dfrac{d\mathbf{V}}{dt} = \dfrac{dx}{dt}\dfrac{d\mathbf{V}}{dx} = V_1\dfrac{d\mathbf{V}}{dx}$.

```
> A:= unapply( evl(V(x)[1] * diff(V(x),x)),x);
```

Subtracting the gravitational force $[0,\,0,\,-mg]$ from $m\mathbf{A}$ leaves us with the force $\mathbf{F}_s$ exerted by the snow on the skis. Actually, since we don't know the skier's mass, I won't bother including it in our expressions. All the forces we obtain should really be multiplied by $m$. We'll compare them to the skier's weight $mg$.

```
> Fs:= x -> evl(A(x) + [0,0,g]);
```

The component of this vector in the direction of the tangent vector $\mathbf{T}$ should be zero (by our assumption). Most of the force between the skis and the snow surface should be perpendicular to the surface, though there can be some sideways force (which is necessary to keep the skier on the right path). A normal to the surface $z = f(x,y)$ is $[-\partial f/\partial x,\ -\partial f/\partial y,\ 1]$ (to avoid confusing this with the principal normal to the curve, I'll use $\mathbf{N}_s$ to denote the unit vector in this direction). Let $F_n$ be the component of the force in this direction.

(4) Since "`D`" doesn't work for a list-valued function, we must evaluate `diff(R(x),x)` and make a function using "`unapply`".

```
> Ns:= (x,y) -> evl(unit([-D[1](f)(x,y), -D[2](f)(x,y), 1]));
> Fn:= x -> Fs(x) &. Ns(x, R(x)[2]);
```

The "sideways" force is in a direction perpendicular to both the tangent vector to the curve and the normal to the surface. The cross product can be used to obtain a unit vector **U** in this direction. According to the "right-hand rule", **U** points to the skier's right. I'll call the component of the force in this direction $F_r$.

```
> U:= x -> evl(T(x) &x Ns(x, R(x)[2]));
> Fr:= x -> Fs(x) &. U(x);
```

Now let's plot $F_n$, $F_r$, and $|\mathbf{F}_s|$ as functions of $x$. These functions would be extremely complicated (even for Maple) if written out as expressions in the symbolic variable $x$, so we don't want to have Maple evaluate "`Fn(x)`", etc. Instead, we use a form of `plot` that works with functions rather than expressions.

```
> plot({Fn, Fr, len @ Fs}, 0 .. 40);
```
(Figure 9.4)

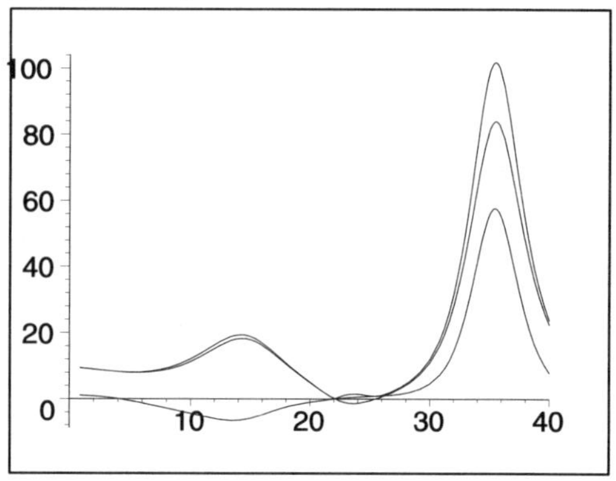

Figure 9.4. Forces on the skier

There are three curves here, which (on a colour monitor) are plotted in different colours. The top curve is $|\mathbf{F}_s|$, which of course is always greater than or equal to the other two (which are the components of $\mathbf{F}_s$ in certain directions). For most of the graph, it is almost indistinguishable from $F_n$, as $F_r$ is quite small in comparison to $F_n$. This is as it should be, for the skier would not be able to exert a sideways force that was too large a fraction of the force normal to the surface. In general, whenever forces are a result of friction between two surfaces pressed together, the sideways force is no greater than some fraction (the "coefficient of friction") of the normal force. Thus $|F_r| \le cF_n$ for some $c$. An attempt to violate this would result in the skier slipping off the desired path.

The most obvious feature in this graph is the big peak in all three curves around $x = 35$, where $|\mathbf{F}_s|$ rises to over 100. That means that at this point the skier's legs would have to support more than 10 times her normal weight ($10g = 98$). This would seem to be beyond human capabilities, so the actual result might be a crash, possibly resulting in serious injury to the unfortunate skier. However, before this point something else of interest occurs: near $x = 22$ both $F_n$ and $F_r$ pass through 0. It is no accident that both functions are 0 at nearly the same point (it required work on my part to ensure this when I made up the example): if, as mentioned above, $|F_r| \le cF_n$, then $F_r \to 0$ when $F_n \to 0$[5]. Now the skis can push on the snow but can't pull it, so $F_n$ can never actually be negative. Instead, what happens when $F_n$ hits 0 is that the skis leave the snow surface and the skier flies through the air. After this point our formulas are no longer valid (to see whether that can save the skier from a crash, see Exercise 8).

Now let's consider in detail the situation at $x = 10$. Where is the skier, and what is happening?

```
> R0:= evalf(R(10)); v0:= evalf(v(10)); T0:= evalf(T(10));
V0:= evalf(V(10)); U0:= evalf(U(10));
```

$$R0 := [10., 3.040387694, -4.276707539]$$

$$v0 := 9.155515702$$

$$T0 := [.7712439854, .4774018353, -.4210346804]$$

$$V0 := [7.061136418, 4.370859999, -3.854789627]$$

$$U0 := [.4927744201, -.8664870368, -.0798347394]$$

(5) Actually, if you plot $F_r$ and $F_n$ in a small interval around $x = 22$ you will see that the $x$ values where they are 0 do not quite coincide. The difference is only about 6 cm, which for our purposes is negligible.

The skier has descended 4.28 m from her starting point, and is travelling about 9.16 m/s (or almost 33 km/h). Let's express the direction of motion (given by $\mathbf{T}_0$) in terms of the angles $\phi$ and $\theta$ of spherical coordinates (in degrees). We will use these shortly: I'll draw a picture of the skier and the forces on her as seen from this direction, so these will be used to set the "`orientation`" option for that plot.

```
> phi0:= evalf(arccos(T0[3])*180/Pi);
> theta0:= evalf(arctan(T0[2], T0[1]) * 180/Pi);
```

$$\phi 0 := 114.8999284$$

$$\theta 0 := 31.75760468$$

Thus at this moment the skier's direction is about 24 degrees below the horizontal and 32 degrees counterclockwise from the positive $x$ axis.

```
> A0:= evalf(A(10));
```

$$A0 := [2.342335425, 5.478990179, .7031539721]$$

The component of the skier's acceleration in the direction of the tangent vector should be the rate of change of her speed, which (by the chain rule) can also be calculated as $\frac{dv}{dt} = \frac{dv}{dx}\frac{dx}{dt}$. This should also be the component of the gravitational acceleration $[0, 0, -g]$ in the direction of the tangent vector, since under our assumption the non-gravitational forces have component 0 in this direction. Let's calculate it all three ways. Note that again we must use "`diff`" rather than "`D`" do find $dv/dx$.

```
> dvdx:= unapply(diff(v(x),x),x);
> [A0 &. T0,  evalf(dvdx(10) * V0[1]), [0,0,-g] &. T0];
```

[4.126139867, 4.126139870, 4.126139868] (close enough!)

Now we find the force $\mathbf{F}_s$ of the snow on the skis, its magnitude, and its components $F_n$ normal to the snow surface and $F_r$ along the surface.

```
> Fs0:= evalf(Fs(10)); len(Fs0); Fn0:= evalf(Fn(10));
> Fr0:= evalf(Fr(10));
```

$$Fs0 := [2.342335425, 5.478990179, 10.50315397]$$

$$12.07568267$$

$$Fn0 := 11.23306395$$

$$Fr0 := -4.431747544$$

Let's plot the forces acting on the skier. Again I'll use the "`arrow3d`" function, with the arrows for vectors $\mathbf{A}$, $\mathbf{F}_s$, $[0, 0, -g]$, and $F_n\mathbf{N}_s$ based at the origin, and $F_r\mathbf{U}$ based at the tip of the last arrow (showing that $F_n\mathbf{N}_s + F_r\mathbf{U} = \mathbf{F}_s$). In order to have the arrowheads appear at a reasonable size, I'll multiply all the vectors by $1/4$.

As an extra aid to visualization, I'll also include a crude sketch of the skier (in the crouching "tuck" position to minimize air resistance). Note that the skier will be leaning so that her body is aligned with the vector $\mathbf{F}_s$. This allows her to be balanced, with equal forces on each ski. For this sketch, we calculate the unit vector $\mathbf{B}$ in the direction of $\mathbf{F}_s$ and a vector $\mathbf{C}$ perpendicular to both $\mathbf{F}_s$ and $\mathbf{T}$ (which thus should be the direction between the skier's shoulders). The skier is drawn as a collection of ellipsoids: the code for this is in the file "`skier.def`" which should be available to you.

The vectors are labelled using "`textplot3d`"(6), which is similar to "`textplot`" but uses a three-dimensional coordinate system. Note the back quotes, which are used to make a "string" containing a space. In `` ` Fs ` `` and `` ` A ` `` the blank spaces are used because we don't want Maple to interpret these labels as the variables "`Fs`" and "`A`"(7).

(6) In the "`plots`" package.

(7) In Release 5 we could make these strings using double quotes, and then the spaces would not be needed.

```
> read `skier.def`; B:= evl(unit(Fs0)); C:= evl(VP &x B);
>  Ns0:= evalf(Ns(10, R(10)[2]));
> labels:= textplot3d({[op(evl(.15*Fn0*Ns0 + .4*U0)), `Fn Ns`],
>      [op(evl(.28*Fn0*Ns0 - .45*U0)), `Fr U`],
>      [op(evl(.15 * Fs0 - .4*U0)), ` Fs `],
>      [op(evl(.2*A0 - .4*U0)), ` A `],
>      [op(evl([0,0,-.15*g] - .4*U0)), `-g k`]}, colour=black);
> curve:= spacecurve(R(x),x=8..12,colour=red);
> surf:= plot3d(f(x, y), x = 8 .. 12, y = R0[2]-2 .. R0[2]+2,
>                      style = wireframe);
> display({ arrow3d(R0, A0/4, colour=red, thickness=1),
>            arrow3d(R0, Fs0/4, colour=blue, thickness=1),
>            arrow3d(R0, [0,0,-g/4], colour=blue, thickness=1),
>            arrow3d(R0, Fn0*Ns0/4, colour=green, thickness=1),
>            arrow3d(Fn0*Ns0/4, Fr0*U0/4, colour=green,
>                 thickness=1),  curve, surf,
>            skier(R0,T0,C,B,U0), labels},
>        orientation=[theta0, phi0], scaling=constrained);
```

(Figure 9.5)

We are looking at the skier from a position directly in her path. Although the snow surface slopes slightly downward to the skier's right, she is exerting enough lateral force to cause her path to curve to the left. There is also a net upward component to her acceleration, so the path must be curving upward.

It is useful to look at this three-dimensional graph from various angles, especially using real-time rotation. You can also try plotting a similar picture at various other points of the skier's path, or even making an animation of part of the path. ■

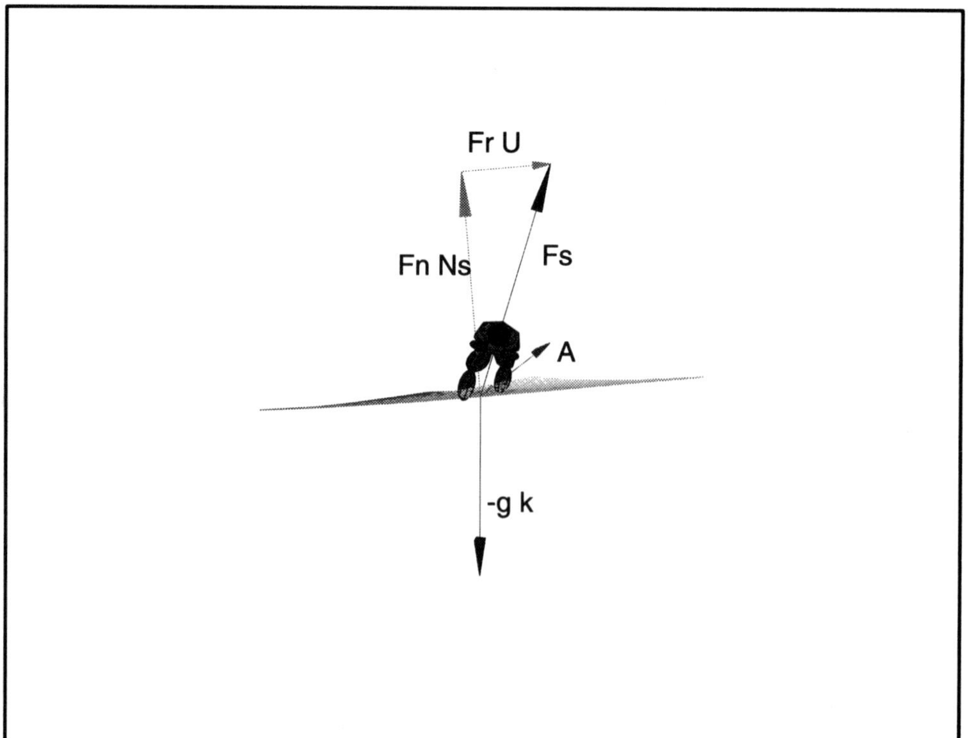


Figure 9.5. Forces and the skier

## Curvature, Torsion, and the Frenet Frame

In this section we will use Maple to calculate and examine the curvature and torsion of a curve, and the three vectors (tangent, principal normal, and binormal) forming the **Frenet frame**. We have already used the tangent vector **T**, which is a unit vector in the direction of the velocity **V**. The others are usually defined in terms of the arc length parameter $s$:

| | |
|---|---|
| Curvature $\kappa = \left\|\dfrac{d\mathbf{T}}{ds}\right\|$. | Principal normal $\mathbf{N} = \dfrac{1}{\kappa}\dfrac{d\mathbf{T}}{ds}$. |
| Binormal $\mathbf{B} = \mathbf{T} \times \mathbf{N}$. | Torsion $\tau$ such that $\dfrac{d\mathbf{B}}{ds} = -\tau\mathbf{N}$. |

Usually our curves are given in terms of parameters other than arc length. We could still use these formulas: for some other parameter $t$, $\dfrac{d}{ds} = \dfrac{1}{v}\dfrac{d}{dt}$ by the Chain Rule (where $v = \dfrac{ds}{dt} = |\mathbf{V}|$ is the speed). However, there is an alternative set of formulas which usually lead to less complicated calculations:

$$\mathbf{V} \times \mathbf{A} = v^3\kappa\mathbf{B} \qquad \mathbf{N} = \mathbf{B} \times \mathbf{T} \qquad \tau = \frac{(\mathbf{V} \times \mathbf{A}) \bullet (d\mathbf{A}/dt)}{|\mathbf{V} \times \mathbf{A}|^2}$$

**EXAMPLE 3** Find the curvature, torsion, and Frenet frame for the curve $x = \cos(t), y = \sin(t)$, $z = \sin(2t)/2$, and verify the Frenet-Serret formulas in this case.

**SOLUTION** The calculation is quite straightforward using the formulas above. We start by defining the curve **R** (as an expression) and successively differentiating with respect to $t$ to find the velocity **V**, acceleration **A**, and derivative of the acceleration.

```
> R:= [cos(t), sin(t), sin(2*t)/2];
> V:= diff(R, t); A:= diff(V, t); dAdt:= diff(A, t);
```

$$V := [-\sin(t),\ \cos(t),\ \cos(2t)]$$
$$A := [-\cos(t),\ -\sin(t),\ -2\sin(2t)]$$
$$dAdt := [\sin(t),\ -\cos(t),\ -4\cos(2t)]$$

Next the speed. Note that I'm assuming that the file "`dot.def`" has already been read. The output of "`len(V)`" in this case can be simplified. But in this case it's best to use "`combine(...,trig)`" to write products or powers of sines and cosines in terms of sines and cosines of multiples of $t$.

```
> v:= combine(len(V), trig);
```

$$v := \frac{1}{2}\sqrt{6 + 2\cos(4t)}$$

We could use "`unit`" to get **T**, but since we already have $v$ we may as well use it.

```
> T:= evl(V/v);
```

$$T := \left[-2\frac{\sin(t)}{\sqrt{6 + 2\cos(4t)}},\ 2\frac{\cos(t)}{\sqrt{6 + 2\cos(4t)}},\ 2\frac{\cos(2t)}{\sqrt{6 + 2\cos(4t)}}\right]$$

Since $\mathbf{V} \times \mathbf{A}$ occurs several times in the formulas, let's calculate it next.

```
> VxA:= combine(V &x A,trig);
```

$$VxA := \left[-\frac{1}{2}\sin(3t) - \frac{3}{2}\sin(t),\ -\frac{3}{2}\cos(t) + \frac{1}{2}\cos(3t),\ 1\right]$$

Now according to our formula $\mathbf{V} \times \mathbf{A} = v^3\kappa\mathbf{B}$, this vector has length $v^3\kappa$ and the unit vector in its direction is $\mathbf{B}$. We use this to calculate $\mathbf{B}$ and $\kappa$.

```
> B:= combine(unit(VxA),trig);
```

$$B := \left[\frac{-\sin(3t) - 3\sin(t)}{\sqrt{14 - 6\cos(4t)}}, \frac{-3\cos(t) + \cos(3t)}{\sqrt{14 - 6\cos(4t)}}, 2\frac{1}{\sqrt{14 - 6\cos(4t)}}\right]$$

```
> kappa:= combine(len(VxA)/v^3,trig);
```

$$\kappa := 4\frac{\sqrt{14 - 6\cos(4t)}}{(6 + 2\cos(4t))^{3/2}}$$

Finally, we calculate $\mathbf{N}$ and $\tau$ using our formulas.

```
> N:= combine(B &x T, trig);
```

$$N := \left[\frac{-6\cos(t) - 3\cos(3t) + \cos(5t)}{\sqrt{14 - 6\cos(4t)}\sqrt{6 + 2\cos(4t)}}, \frac{-6\sin(t) + \sin(5t) + 3\sin(3t)}{\sqrt{14 - 6\cos(4t)}\sqrt{6 + 2\cos(4t)}}, -8\frac{\sin(2t)}{\sqrt{14 - 6\cos(4t)}\sqrt{6 + 2\cos(4t)}}\right]$$

```
> tau:= combine( VxA &. dAdt / len(VxA)^2, trig);
```

$$\tau := 6\frac{\cos(2t)}{-7 + 3\cos(4t)}$$

Next, we verify the Frenet-Serret formulas, which express the derivatives of $\mathbf{T}$, $\mathbf{N}$, and $\mathbf{B}$ in terms of these vectors and the curvature and torsion, in this example.

```
> eq1:= diff(T,t)/v = kappa * N;
```

I won't show the result, which looks rather complicated. Let's try to simplify the difference between the two sides. The commands "`rhs`" and "`lhs`" return the right and left sides, respectively, of an equation.

```
> simplify(evl(lhs(eq1)-rhs(eq1)), trig);
```

$$[0, 0, 0]$$

We can do the other formulas similarly, with the same result.

```
> simplify(evl(diff(N,t)/v - (- kappa * T + tau * B)),trig);
> simplify(evl(diff(B,t)/v - (-tau * N)),trig);
```

■

**■ EXAMPLE 4** Produce visual representations of the curve of Example 3 with its Frenet frames at various points, and of the motion of the osculating circle around the curve.

**SOLUTION** We'll use the "`arrow3d`" command from the file "`arrow3d.def`". We'll also use the "`plots`" package, so load those now if necessary.

In order to evaluate the Frenet frame, curvature, and torsion at particular values of $t$ it is convenient to make these into functions using "`unapply`".

```
> R:= unapply(R,t); T:= unapply(T,t);
> N:= unapply(N,t); B:= unapply(B,t);
> kappa:= unapply(kappa,t); tau:= unapply(tau,t);
```

Next we make a plot of our curve and save it in a variable, for use in all of our pictures.

```
> curve:= spacecurve(R(t), t=0..2*Pi, colour=black):
```

We will draw arrows for the three vectors $\mathbf{T}(t)$, $\mathbf{N}(t)$, and $\mathbf{B}(t)$, all starting at $\mathbf{R}(t)$ on the curve. It is helpful to use colours to distinguish them: red for $\mathbf{T}$, green for $\mathbf{N}$, and blue for $\mathbf{B}$. For convenience, we can define a command "`frenet`" that displays the curve and the arrows corresponding to any given $t$.

```
> frenet:= t -> display({curve,
>                        arrow3d(R(t), T(t), colour=red),
>                        arrow3d(R(t), N(t), colour=green),
>                        arrow3d(R(t), B(t), colour=blue)});
```

Using "frenet", we can produce an animation showing the Frenet frame moving all around the curve[(8)].

```
> display([seq(frenet(k*Pi/8), k=0..15)], insequence=true,
>          scaling=constrained);
```

Examine this from several viewpoints. Note how the red vector **T** is always turning in the direction of the green vector **N**. It may be harder to spot, but the blue vector **B** should be turning in the direction of either **N** or $-\mathbf{N}$ (depending on the sign of $\tau$). This should be easiest to see when $|\tau|$ is near its maximum, looking from the direction of **T**. By looking at a graph of $\tau(t)$, we see that one maximum is at $t = 0$. Since $\mathbf{T}(0) = [0, \sqrt{2}/2, \sqrt{2}/2]$, you might look at the first few frames of this animation from an orientation of `[90,45]`.

An **osculating circle** at a point $\mathbf{R}(t)$ on the curve is a circle that best approximates the curve near $\mathbf{R}(t)$. It has the same tangent, principal normal, and curvature as the curve. The radius of this circle is $1/\kappa(t)$, and its centre lies at that distance from $\mathbf{R}(t)$ in the direction of $\mathbf{N}(t)$, i.e., at $\mathbf{R}(t) + \mathbf{N}(t)/\kappa(t)$. The circle lies in the **osculating plane** determined by the vectors $\mathbf{T}(t)$ and $\mathbf{N}(t)$, so it has a parametric representation

$$\mathbf{R}(t) + \frac{(1 + \cos(s))\mathbf{N}(t) + \sin(s)\mathbf{T}(t)}{\kappa(t)}, \quad 0 \le s \le 2\pi$$

Here is a function that, for any $t$, produces the expression for the osculating circle at $\mathbf{R}(t)$.

```
> ocirc:= t ->
>   evl(R(t) + ((1+cos(s)) * N(t)
>   + sin(s) * T(t))/kappa(t));
```

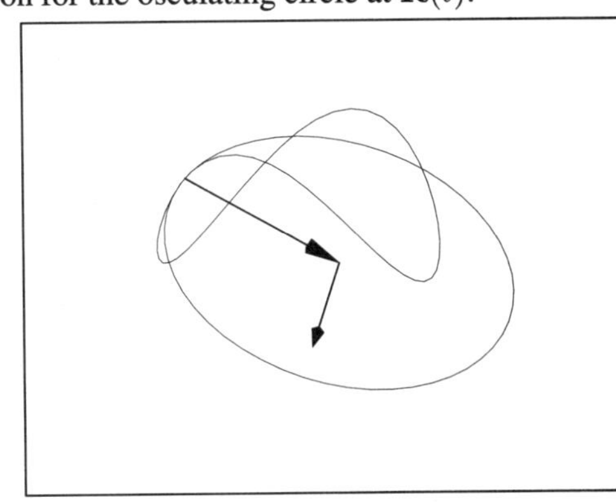

Figure 9.6. Osculating Circle.

We can use this to produce an animation of the osculating circle moving around the curve. I'll draw the circle in red, to distinguish it from the original curve. In addition, I'll draw an arrow from $\mathbf{R}(t)$ to the centre of the osculating circle (in green), which will help us see where the osculation occurs, and an arrow from the centre of the circle representing the vector $\tau(t)\mathbf{B}(t)$ (in blue), which shows how the osculating plane is rotating. Here is the command to produce a frame of the animation for a given $t$. Note the "evalf", which may result in improved speed because the input to "spacecurve" will be simpler.

```
> frame:= t -> display({curve,
>     spacecurve(evalf(ocirc(t)), s = 0 .. 2*Pi, colour = red),
>     arrow3d(R(t), evl(N(t)/kappa(t)), colour=green),
>     arrow3d(evl(R(t) + N(t)/kappa(t)), evl(tau(t)*B(t)),
>         colour = blue)});
> frame(.3);
```

(Figure 9.6, with "constrained" scaling and orientation [70, 80])

It is better to use a smaller interval between frames than we used for the previous animation, as the osculating circle changes size rather rapidly. I'll just go 1/3 of the way around the curve (with a fast computer and lots of memory, you might try going all the way around).

```
> display([seq(frame(k*2*Pi/45), k=0..14)], insequence=true,
>          scaling=constrained);
```

■

(8) If your machine is slow or has serious memory limitations, you may want to reduce the number of frames in this animation, e.g., just use "k=0..7" to show the frame moving around half the curve.

## EXERCISES

**1.** **(a)** Plot several turns of the helix

$$x = t + \sin(t) + \sqrt{2/3}\cos(t), \quad y = t - \sin(t) + \sqrt{2/3}\cos(t), \quad z = 2t - \sqrt{2/3}\cos(t)$$

**(b)** Find a viewing direction from which the curve appears as a circle. *Hint:* the projection perpendicular to a certain vector will get rid of the "$t$" terms.

**(c)** Find the curvature, torsion, and Frenet frame for this curve.

**2.** **(a)** For what value of the parameter $p$ does the curve

$$x = 2t^2 - 3t, \quad y = 7t - 4t^3, \quad z = t^2 - p(t + 1)$$

intersect itself? At what point does that occur? *Hint:* use "`solve`".

**(b)** Plot the self-intersecting curve.

**(c)** Find the angle between the tangents to the curve at the point where it intersects itself.

**3.** **(a)** Write the intersection of the surfaces $z = x^2 + y^2$ and $yz = x^2 - y^2$ for $y > 0$ as a parametric curve. *Hint:* use $\theta$ of cylindrical coordinates as a parameter.

**(b)** Plot the curve together with pieces of the surfaces, in order to see its shape.

**4.** An airplane performs an aerobatic manoeuvre in which its velocity vector is

$$\mathbf{V} = (24 + 6\cos(t/3))[1 + 2\cos(t/3),\ 1 + 2\cos(t/3 - 4\pi/3),\ 1 + 2\cos(t/3 - 2\pi/3)]\ \mathrm{m/s}$$

for time $0 \le t \le 6\pi$ seconds.

**(a)** Find the position vector as a function of $t$, if the airplane starts out at $[0, 0, 100]$. Plot it, and view the plot from several directions. Describe the motion in words.

**(b)** Find the non-gravitational forces that must act on the airplane during the motion. How many "gees" does the pilot experience?

**(c)** Produce an animation of the airplane doing the manoeuvre with the forces on the airplane shown as arrows. Let the airplane be oriented so that its "vertical" axis is in the direction of the principal normal to its path (the direction of $d\mathbf{T}/dt$). You will probably need to make the airplane at least 50 metres long in order to see it and the whole path well on your monitor (although it might not be a good idea to try this manoeuvre in a real plane the size of a large airliner).

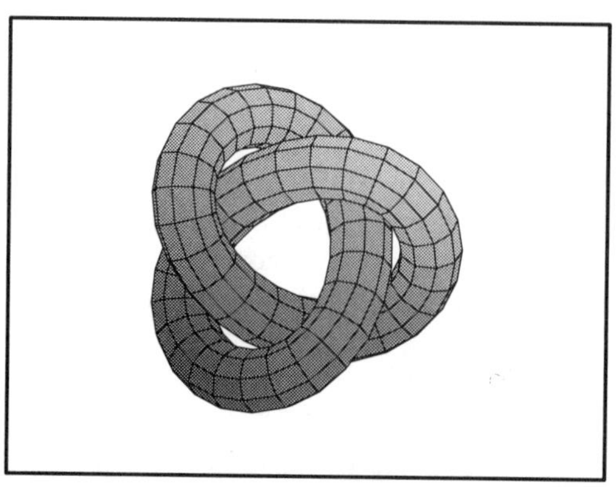

Figure 9.7. Trefoil knot

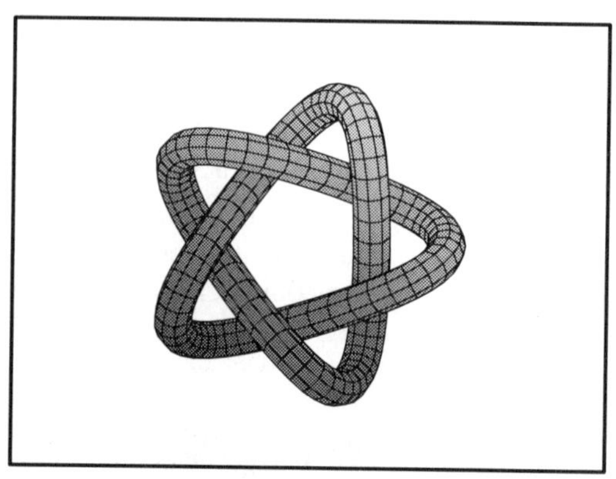

Figure 9.8. Cinquefoil knot

**5.** **(a)** The "trefoil knot" in Figure 9.7 was made using "`tubeplot`", with a curve of the form $[a\cos(t) + b\cos(2t), -a\sin(t) + b\sin(2t), c\cos(3t)]$ for $0 \le t \le 2\pi$, where $a$, $b$, and $c$ are constants. Find $a$, $b$, $c$ and the tube radius to produce a picture as similar to this as possible. Try to ensure that your tubes don't intersect. *Hint:* consider the square of the distance from $\mathbf{R}(t)$ to $\mathbf{R}(s)$ for $0 \le t \le \pi/3$ and $\pi \le s \le 4\pi/3$. If the minimum value of this quantity is at least $(2r)^2$, then a tube radius $r$ will work. (Why?)

**(b)** Show that the trefoil-knot curve has the property that $\mathbf{R}(t + 2\pi/3)$ is $\mathbf{R}(t)$ rotated about the $z$ axis by the angle $-2\pi/3$. *Hint:* to rotate any vector $\mathbf{V}$ by an angle $\alpha$ about the $z$ axis, perform the matrix multiplication $A\mathbf{V}$ where

$$A = \begin{pmatrix} \cos\alpha & -\sin\alpha & 0 \\ \sin\alpha & \cos\alpha & 0 \\ 0 & 0 & 1 \end{pmatrix}$$

You may want to look back at Lab 17, where we dealt with matrices and their multiplication.

**(c)** Construct a "cinquefoil knot" (Figure 9.8). *Hint:* this time you want $\mathbf{R}(t + 2\pi/5)$ to be $\mathbf{R}(t)$ rotated by a certain angle.

The next four exercises refer to Example 2.

**6.** **(a)** Plot the slope of the skier's path as a function of $x$. Where is the path horizontal? Where are the local maxima and local minima of the slope?

**(b)** Plot the curvature of the skier's path as a function of $x$. Where are the local maxima and minima of the curvature?

**7.** **(a)** A more prudent skier would slow down. The skier, travelling on the same path, might survive if she arrives at $x = 35$ with a lower speed. What speed at this point would produce a more manageable $|\mathbf{F}_s| = 20m$?

**(b)** Find a point $P$ on the path such that if the skier started with speed 0 at $P$ (rather than at the origin) and otherwise satisfies our assumptions, she would arrive at $x = 35$ with the speed found in (a).

**8.** **(a)** Find (to 2 decimal place accuracy) the point at which the skier leaves the surface and begins to fly through the air. *Note:* "`fsolve`" has difficulty with this, probably because $F_n$ is so complicated an expression. Graphical methods work well here.

**(b)** Neglecting air resistance (and the considerable aerodynamic effect that the skis can have), find the skier's path through the air. How long is the flight, and where does she land?

**(c)** Assume that when the skier lands, the component of her velocity normal to the surface disappears but the components tangent to the surface are unchanged. Assume that the projection of the remainder of her trajectory on the $xy$ plane lies in a straight line (at least until $x = 40$). What is that trajectory? What is the maximum of $|\mathbf{F}_s|$ on it? Is she still likely to crash?

**9.** **(a)** Consider a skier like ours, moving along a path $y = yp(x)$ on a surface $z = f(x, y)$ (all functions assumed to be smooth). Suppose that at a certain time $t_0$ the skier is at a certain point $[x_0, yp(x_0), f(x_0, yp(x_0))]$ and $\mathbf{F}_s = 0$ (so that the skier could begin a flight through the air at this point). Show that $yp''(x_0) = 0$, and find an equation that must be satisfied by $yp'(x_0)$, the speed $v(t_0)$, and the first and second partial derivatives of $f$ at $(x_0, yp(x_0))$.

*Hint:* use $\dfrac{d^2x}{dt^2} = 0$, $\dfrac{d^2y}{dt^2} = 0$, $\dfrac{d^2z}{dt^2} = -g$, and $\left(\dfrac{dx}{dt}\right)^2 + \left(\dfrac{dy}{dt}\right)^2 + \left(\dfrac{dz}{dt}\right)^2 = v^2$.

**(b)** Design a ski hill and a skier's path so that the skier, starting from rest at the origin, becomes airborne at the point $[10, 0, -5]$. *Hint:* first choose derivatives of $f$ at $x = 10$, $y = 0$ so that the equation from (a) (with the proper value for $v$) has real solutions $y'(10)$. Once you know $y'(10)$, choose some $y(x)$ so $y(0) = 0$, $y(10) = 0$, $y'(10)$ has the value you wish and $y''(10) = 0$. Finally, find $f$ that has the derivatives you want at $(10, 0)$ with $f(10, 0) = -5$ and $f(0, 0) = 0$.

# LAB 26 VECTOR FIELDS

**Goals:** *To visualize and investigate vector fields, their field lines and potentials, using Maple.*

A (three-dimensional) **vector field** associates a three-component vector $\mathbf{F}(x, y, z)$ with each point $[x, y, z]$ of a region of space. You can also have a two-dimensional vector field, defined on a region of the plane.

Just as with parametric curves, it is often best to use lists rather than Maple's vectors to represent a vector field. Thus we could define vector fields in the following style:

```
> F:= (x,y,z) -> [y*z, x + y, z];
```

In most cases, rather than using a function of the variables $x$, $y$, and $z$, it is simpler to let our field be a list of expressions depending on those variables.

```
> G:= [y*z, x + y, z];
```

## Plotting Vector Fields

Maple's "`plots`" package contains two commands that can produce a visual representation of a vector field: "`fieldplot`" (for two-dimensional vector fields) and "`fieldplot3d`" (for three-dimensional fields). These will plot arrows representing the vectors at the points of a certain grid. Of course the vector field produces a vector at every point of our region, but only a finite number (hopefully a representative sample) can be plotted.

```
> with(plots):
> fieldplot([x+y,x*y], x=-4..4, y=-3..3);
```
(Figure 9.9)

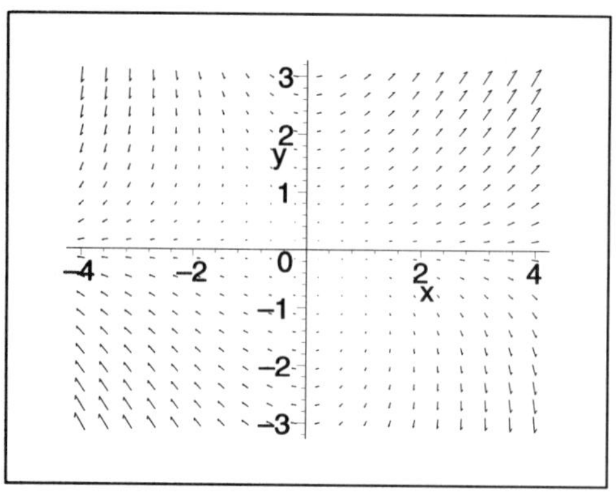


Figure 9.9. 2-D vector field

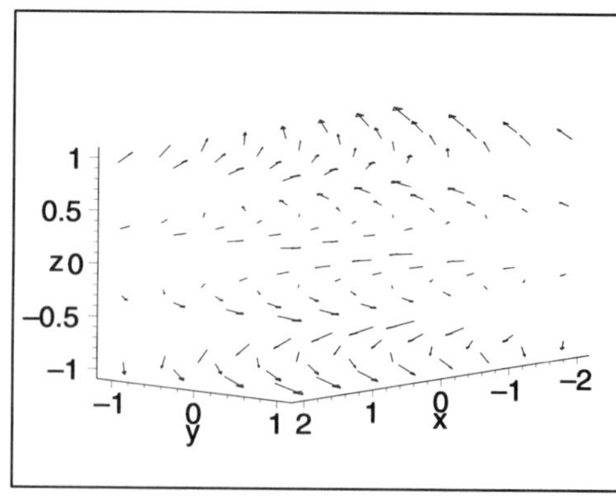


Figure 9.10. 3-D vector field

The first input to "`fieldplot`" is a list representing the vector field, the second and third are the ranges for the $x$ and $y$ coordinates. The arrows are automatically scaled in length so that the longest one is a reasonable length in relation to the grid spacing. The usual "`plot`" options such as "`colour`", "`scaling`", and "`axes`" can be used. A very useful option is "`grid`", which specifies the number of rows and columns of arrows to be plotted. Thus "`grid=[6,8]`" will produce a $6 \times 8$ rectangular grid of arrows. The default is $[20, 20]$. Another option, "`arrows`", controls the appearance of the arrows: the possibilities are "`arrows=LINE`" (the arrows are just straight lines with no arrowheads), "`THIN`" (the default, with only one barb for an arrowhead), "`SLIM`" (with a triangular arrowhead), and "`THICK`" (fat, outlined arrows, which can be filled with colour if the "`colour`" option is used).

The "fieldplot3d" command is similar to "fieldplot", but produces a three-dimensional plot of a three-dimensional vector field. The default "grid" is [8,8,8].

```
> fieldplot3d(F(x,y,z), x = -1..1, y = -1..1, z = -1..1);
```

With the default settings, this produces a tangled mess. You are looking at a three-dimensional grid of arrows in space, but the closer arrows are mixed up with the farther ones, and with no clues as to depth this is very hard to sort out. On a colour monitor, it may help to use "shading=ZHUE". Then the layers of arrows at different $z$ values will have different colours, from red at the top to violet at the bottom, and your eye may be able to pick them out individually. Even more helpful is to use an orientation near one of the coordinate planes ($\phi$ or $\theta$ near 0 or 90), so that the different layers are separated.

```
> fieldplot3d(F(x,y,z), x=-2..2, y=-1..1, z=-1..1, grid=[8,4,4],
>             axes=FRAMED, orientation=[40,82], shading=ZHUE,
>             scaling=constrained); (Figure 9.10)
```

## Field Lines

A **field line** for a vector field is the path of a particle that moves so that its velocity at any point is the value of the vector field at that point. Depending on the type of field, these may also be called **integral curves, streamlines, lines of force,** or **flux lines**. To find field lines requires solving differential equations. This will be covered in more detail in Lab 30. What I'll use here is the command "DEplot" from the "DEtools" package, which plots field lines in the two-dimensional case.

```
> with(DEtools, DEplot):
```

**EXAMPLE 1** Find and plot the field lines of the two-dimensional vector field $\mathbf{F}(x,y) = [x+y, x-y]$, together with the field itself, for $-4 \le x \le 4$ and $-3 \le y \le 3$.

**SOLUTION** The command is

```
> DEplot([D(x)(t)=x+y, D(y)(t)=x-y], [x(t),y(t)], t=-1 .. 1,
>        {seq([x(0)=0, y(0)=k], k=-3 .. 3)});
```

The first input to "DEplot" is the system of differential equations corresponding to the vector field $\mathbf{F}(x,y)$: $\dfrac{dx}{dt} = F_1(x,y)$, $\dfrac{dy}{dt} = F_2(x,y)$. Second is the list of variables. Third is an interval for the independent variable $t$, and fourth is a set of initial values. Each initial value is a list of the form $[x(0) = x_0, y(0) = y_0]$. The result is a plot of the parametric curves $x = x(t)$, $y = y(t)$ satisfying the system of differential equations with the given initial conditions, and for the given interval of $t$[9]. In addition, "DEplot" draws arrows (in red on a colour monitor) showing the direction of the field at each point on a certain grid. However, these arrows are a bit different from those of "fieldplot": they are all the same length[10]. Thus they don't show the magnitude of the field (only the direction is relevant in determining the field lines). You can suppress the arrows by using the option "arrows=NONE". Of course, if you want the arrows of "fieldplot" you can combine a "fieldplot" with a "DEplot" using "display".

The command, written as above, displays a rectangle big enough for all of the curves from $t = -1$ to $t = 1$, rather than the rectangle $-4 \le x \le 4$, $-3 \le y \le 3$ that we're actually interested in. We can specify this rectangle by including "x=-4..4" and "y=-3..3" as options to "DEplot".

This time I'll choose the initial conditions on the left boundary of our rectangle, with integer values of $y$. Since it's easy to see that the $x$ component of the vector field is negative in this part of the graph, there's no need to use positive values of $t$ here.

---

(9) The curves are thick, and shown in yellow. You can change this with the "linecolour" and "thickness" options (e.g., "linecolour=red, thickness=1").

(10) But they may not *look* like the same length unless you use "scaling=constrained". Also, a vector that is exactly 0 is not shown.

```
> DEplot([D(x)(t)=x+y, D(y)(t)=x-y], [x(t),y(t)], t=-1 .. 0,
>         {seq([x(0) = -4, y(0) = k], k=-3..3)},
>          x = -4 .. 4, y = -3 .. 3);
```

Some of the curves seem to stop in the middle of the plot. Apparently, $-1 \le t \le 0$ is too small an interval. We might try $-2 \le t \le 0$. However, it appears that no interval of $t$ will extend the curves into the top right of the picture. To take care of this region we could use another set of initial points, along the right edge of the picture.

```
> DEplot([D(x)(t)=x+y, D(y)(t)=x-y], [x(t),y(t)], t=-2 .. 0,
>         {seq([x(0)=-4, y(0)=k], k=-3..3),
>          seq([x(0)= 4, y(0)=k], k=-3..3)},
>         x = -4 .. 4, y = -3 .. 3, arrows=NONE,
>         linecolour=red, thickness=1);
```

This is a satisfactory picture, except that it doesn't show what happens near the origin $[0, 0]$. The origin is an **equilibrium point** (also called a **critical point** or **stagnation point**) of the vector field, i.e., a point where the vector field has the value $[0, 0]$. This corresponds to our particle staying motionless at that point, or in other words a constant solution of the system of differential equations. These points, and what happens near them, are particularly important in analyzing a vector field. We need some more initial values near the origin (not exactly at the origin, of course, because then we'd never get anywhere!). Rather than recalculate what we already have, I'll save the last result and combine it with the new curves.

```
> P1:= ": (Release 4) or > P1:= %: (Release 5)
> DEplot([D(x)(t) = x+y, D(y)(t) = x-y], [x(t), y(t)],
>       t = -2 .. 2, {[x(0)=0.01, y(0)=0]}, x = -4 .. 4,
>       y = -3 .. 3, arrows=NONE);
```

This one doesn't get very far from the origin for $-2 \le t \le 2$. That should perhaps be expected, since the velocity vector is near 0 at points near 0. Some experimentation shows that the interval $-5 \le t \le 5$ is enough to have the field line extend beyond the plotting window (in both forward and backward directions). I'll also take the initial value $[x(0) = -0.01, y(0) = 0]$ which produces a field line similar to this one on the other side of the origin (note that our vector field has the symmetry $\mathbf{F}(-x, -y) = -\mathbf{F}(x, y)$, so that the reflection of a field line across the origin is again a field line).

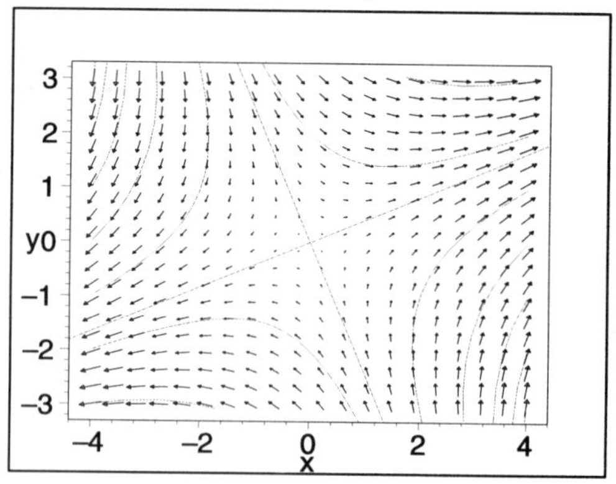


Figure 9.11. Field and field lines

```
> P2:= DEplot([D(x)(t) = x+y, D(y)(t) = x-y], [x(t), y(t)],
>       t=-5 .. 5, {[x(0)= 0.01, y(0)=0], [x(0)= -0.01, y(0)=0]},
>       x = -4 .. 4, y = -3 .. 3, arrows = NONE, linecolour = red,
>       thickness = 1): P2;
```

Finally, we combine the two "`DEplot`" plots and a "`fieldplot`" using "`display`".

```
> P3:= fieldplot([x+y,x-y], x=-4..4, y=-3..3, arrows=SLIM):
> display({P1, P2, P3}, axes=BOXED, scaling=constrained);
```

(Figure 9.11)

■

There is one more option to "`DEplot`", which we didn't need here but will sometimes be necessary: "`stepsize`". In "`DEplot`" the solutions of our differential equations are approximated using numerical methods. The $t$ interval is divided into subintervals of equal length (the step size) at points $t_0, \ldots, t_n$, the method calculates approximate values of the dependent variables for each $t_j$, and the points thus obtained are joined by straight lines. The default step size is $1/20$ of the $t$ interval. If the step size is too large, the "curve" will not look smooth(11). The disadvantage of a small step size, on the other hand, is that it requires more time and memory.

## Potentials for Fields

Some vector fields are gradients of scalar fields. We have already seen the "`grad`" command in Lab 19. This is in the "`linalg`" package.

```
> with(linalg):
> F:= grad((x+1)^2-y*z, [x,y,z]);
```

$$F := [2x + 2 \quad -z \quad -y]$$

If the vector field **F** is the gradient of a scalar field $\phi$ in a certain domain $D$, we say that **F** is a **conservative** vector field in $D$, and that $\phi$ is a **potential** (or **scalar potential**) for **F** in $D$.

We would like to check whether or not a given vector field is conservative, and to construct a potential if it is. The "`linalg`" package has a command for this, called "`potential`".

```
> potential(F, [x,y,z], 'V');
```

true

The first input to "`potential`" is the vector field. The second is the list of coordinate names. The third is the name of a variable(12) The result is either "`true`" (if there is a potential) or "`false`" (if not). If the result is "`true`", a potential for the field is assigned to the variable. If it is "`false`", the value of the variable is unchanged. Let's see if the potential that Maple found was the same as what we started with.

```
> V;
```

$$x^2 - yz + 2x$$

This differs by a constant from the scalar field $(x + 1)^2 - yz$ that we started with. That is to be expected, since you can always add a constant to a potential and obtain another potential for the same vector field.

Unfortunately, in some cases there are problems with "`potential`". Consider the following two-dimensional example:

```
> F:= [ sin(x)^2 * y^2, (x - cos(x)*sin(x))*y ];
> potential(F, [x,y], 'V');
```

false

Yet here is a potential for $F$:

```
> V1:= y^2 * (x - cos(x)*sin(x))/2;
> G:= grad(V1, [x,y]);
```

$$G := \left[\frac{1}{2}y^2(1 + \sin(x)^2 - cos(x)^2), (x - \cos(x)\sin(x))y\right]$$

The gradient doesn't look quite the same as **F**, but a bit of simplification shows that they are the same. Since `G` is a Maple vector, we'll use our "`evl`" procedure to evaluate the difference of the two as a list, and then "`simplify`"(13). The result should be **0**.

---

(11) Also the solution will not be as accurate. But the lack of smoothness is usually what is noticeable in a plot.

(12) Enclosing the variable name in quotes prevents it from being evaluated, since we want the input to "`potential`" to be the name of the variable and not any value it might currently have.

(13) Another way would be `map(simplify, evalm(G-F))`. Applying "`simplify`" directly to the difference wouldn't work.

```
> read(`dot.def`):
> simplify(evl(G - F));
```

$$[0, 0]$$

What's happening here is that the trigonometric identity $\cos^2 x + \sin^2 x = 1$ is needed to see that $\mathbf{F} = \mathbf{G}$. The same trigonometric identity is needed to see that $\mathbf{F}$ is conservative: the necessary condition for a two-dimensional field is $\frac{\partial}{\partial y}F_1 = \frac{\partial}{\partial x}F_2$.

```
> diff(F[1], y) = diff(F[2], x);
```

$$2\sin(x)^2 y = (1 + \sin(x)^2 - \cos(x)^2)y$$

Maple's "`simplify`" can handle this identity, but unfortunately "`potential`" doesn't use that: it only uses "`normal`", which provides a more basic level of simplification, not including trigonometric identities.

```
> normal(diff(F[1], y) - diff(F[2], x));
```

$$\sin(x)^2 y - y + y\cos(x)^2$$

For this reason, I would have to be very suspicious of "`false`" results from "`potential`", unless only rational functions are involved.

Another reason for caution is that, like most Maple functions, "`potential`" doesn't worry about domains.

```
> F:= [-y/(x^2+y^2), x/(x^2+y^2)];
> potential(F, [x,y], 'V');
```

$$\text{true}$$

```
> V;
```

$$-\arctan\left(\frac{x}{y}\right)$$

This is the standard example of a two-dimensional vector field that satisfies the necessary condition $\frac{\partial}{\partial y}F_1 = \frac{\partial}{\partial x}F_2$ to be conservative, but is not conservative on its domain of definition (the plane with the origin removed). On the other hand, it is conservative on other domains, such as the plane with the ray $\{[x, 0] : x \leq 0\}$ removed. The potential that Maple found is undefined on the $x$ axis (using the two-variable form of arctan, we can define potentials such as $\arctan(y, x)$ which are defined in the plane with a ray removed).

Since "`potential`" won't always give us a potential when there is one, and there's no guarantee that a potential it does give us will be defined on the domain we want, let's try constructing a potential without its help.

**EXAMPLE 2** Find a potential for the vector field

$$\mathbf{F}(x, y, z) = \left[x\ln\left(y^2 + z^2\right), \frac{\left(x^2 - z^2\right)y}{y^2 + z^2}, z\left(\frac{x^2 + y^2}{y^2 + z^2} - \ln\left(y^2 + z^2\right)\right)\right]$$

**SOLUTION**

```
> F:= [ x*ln(y^2+z^2), (x^2-z^2)*y/(y^2+z^2),
>       z*((x^2+y^2)/(y^2+z^2)-ln(y^2+z^2))];
```

Two of the necessary conditions, $\frac{\partial F_1}{\partial y} = \frac{\partial F_2}{\partial x}$ and $\frac{\partial F_1}{\partial z} = \frac{\partial F_3}{\partial x}$, are immediate, but the third needs some simplification.

```
> diff(F[2],z) = diff(F[3],y);
```

$$-2\frac{zy}{y^2+z^2}-2\frac{\left(-z^2+x^2\right)yz}{\left(y^2+z^2\right)^2}=-2\frac{z\left(x^2+y^2\right)y}{\left(y^2+z^2\right)^2}$$

```
> normal(diff(F[2],z) - diff(F[3],y));
```

$$0$$

Now to find a potential. We first try finding an antiderivative of $F_1$ with respect to $x$.

```
> V1:= int(F[1], x);
```

$$V1 := \frac{1}{2}x^2\ln\left(y^2+z^2\right)$$

This might not be our potential $V$, but we should have $\dfrac{\partial}{\partial x}(V - V_1) = 0$, so $V - V_1$ should be a function of $y$ and $z$ but not $x$. If we now differentiate $V_1$ with respect to $y$, we should have $\dfrac{\partial}{\partial y}(V - V_1) = F_2 - \dfrac{\partial V_1}{\partial y}$.

```
> s:= F[2] - diff(V1, y);
```

$$s := \frac{\left(-z^2+x^2\right)y}{y^2+z^2}-\frac{x^2y}{y^2+z^2}$$

At first sight this has an $x$ in it, but that can be removed by simplifying.

```
> s:= normal(s);
```

$$s := -\frac{yz^2}{y^2+z^2}$$

Our next try for a potential adds $V_1$ to the antiderivative of this with respect to $y$. That should differ from the true potential by at most a function of $z$ alone.

```
> V2:= V1 + int(s, y);
```

$$V2 := \frac{1}{2}x^2\ln\left(y^2+z^2\right)-\frac{1}{2}z^2\ln\left(y^2+z^2\right)$$

You might check that the partial derivatives of $V_2$ with respect to $x$ and $y$ are $F_1$ and $F_2$ respectively. Now we compare $F_3$ to $\partial V_2/\partial z$. As you might expect, we'll have to do some simplification to see that the difference is a function of $z$.

```
> s:= normal(F[3] - diff(V2, z));
```

$$s := z$$

To obtain our potential we add to $V_2$ an antiderivative of this.

```
> V:= V2 + int(s, z);
```

$$V := \frac{1}{2}x^2\ln\left(y^2+z^2\right)-\frac{1}{2}z^2\ln\left(y^2+z^2\right)+\frac{1}{2}z^2$$

Finally, we verify that this is a potential for $\mathbf{F}$.

```
> normal(evl(grad(V, [x,y,z]) - F));
```

$$[0, 0, 0]$$

■

## EXERCISES

1. Plot the vector fields $\mathbf{F}(x, y) = [x, y]$ and $\mathbf{F}(x, y) = [y, -x]$. Predict from these plots what the field lines for these vector fields should look like. Then use "`DEplot`" to verify your predictions.

2. **(a)** Find a vector field in the plane whose field lines are the curves $y = c(x^2 + 1)$. *Hint:* write the equation as $c = \ldots$ and differentiate implicitly with respect to $t$. The vector field should give values $F_1(x, y)$ and $F_2(x, y)$ to $\dfrac{dx}{dt}$ and $\dfrac{dy}{dt}$ respectively that satisfy the resulting equation.

   **(b)** Plot the field and some of its field lines.

3. For what values of $n$ is the vector field $\mathbf{F}(x, y) = [x,\ y]/r^n$ conservative, where $r = \sqrt{x^2 + y^2}$? Find a potential in the case(s) where it is conservative.

4. **(a)** An electrical dipole has potential $\phi = -z/\rho^3$ where $\rho = \sqrt{x^2 + y^2 + z^3}$. Find the corresponding vector field $\mathbf{F}$.

   **(b)** Plot $\mathbf{F}$ and the equipotential surfaces $\phi = 1$ and $\phi = -1$.

   **(c)** Plot the cross-section $y = 0$ of $\mathbf{F}$ and of the equipotential surfaces $\phi = 1$ and $\phi = -1$. Note that at any point $\mathbf{p}$ in the $xz$ plane, the field $\mathbf{F}(\mathbf{p})$ is in this plane.

5. **(a)** Plot the vector field $\mathbf{F}(x, y) = [x^3 - 3xy^2, 3x^2y - y^3]$ in a window centred at the origin.

   **(b)** In most of the plotting window, the arrows in (a) are very short compared to those near the corners. Why? To see the directions of the arrows better, plot a vector field of unit vectors in the same direction as $\mathbf{F}(x, y)$. Describe the vector field in words. In particular, where are the vectors pointing toward the origin, away from the origin, or at right angles to the direction of the origin? What symmetries does the picture have?

   **(c)** Show that the lemniscates $r^2 = a\sin(2\theta)$ are field lines for the vector field $\mathbf{F}$. *Hint:* convert the lemniscate equation to rectangular coordinates, make $x$ and $y$ into functions of $t$, and differentiate. Substitute for $x'(t)$, $y'(t)$, and $a$.

   **(d)** Plot some of the lemniscates (using "`plot`" with polar coordinates) together with the unit vector field from (b). Note how the properties of the vector field you saw in (b) are related to the shapes of the lemniscates.

   **(e)** Make a plot similar to that of (d), but using "`DEplot`".

6. **(a)** The vector field $\mathbf{F}(x, y)$ of Exercise 3 is not conservative. However, $(x^2 + y^2)^p\mathbf{F}(x, y)$ is conservative for a certain constant $p$. What is $p$?

   **(b)** Using this $p$, find a potential for the field $(x^2 + y^2)^p\mathbf{F}(x, y)$.

   **(c)** Plot some equipotential curves, together with the unit vector field from Exercise 3(b). Describe these curves. *Hint:* express them using polar coordinates. If you've done Exercise 3, the shape of the curves should look familiar!

   **(d)** When two lemniscates centred at the origin, one rotated by 45 degrees with respect to the other, intersect (away from the origin), they do so at right angles. Why does this follow from the results of (c) and 3(c)? Plot some lemniscates of different sizes that illustrate this property.

7. **(a)** The field lines that go near the origin in Figure 9.11 look like straight lines through the origin. Find all field lines for $\mathbf{F}$ that are (parts of) straight lines through the origin. *Hint:* if the straight line is $y = ax$, what is $F_2 - aF_1$?

   **(b)** For what values of the constants $a$, $b$, $c$, and $d$ is the parabola $y = x^2$ a field line of $\mathbf{F} = [ax + by,\ cx + dy]$?

8. Consider the vector field

$$\mathbf{F}(x, y, z) = [6x + 4y - 5z,\ -5x - 3y + 5z,\ 4x + 4y - 3z]$$

**(a)** Find all unit vectors $[x, y, z]$ such that $\mathbf{F}(x, y, z)$ is parallel to $[x, y, z]$. Note: if you've taken linear algebra, you may recognize this as asking for the eigenvectors of a matrix. But no knowledge of eigenvectors is necessary here: simply ask Maple to solve a system of four equations in four unknowns.

**(b)** $\mathbf{F}(x, y, z)$ is also parallel to $[x, y, z]$ if $[x, y, z]$ is any scalar multiple of one of the vectors found in (a). Why?

**(c)** Make a three-dimensional plot of the vector field, viewed from the direction of one of the vectors found in (a). The plot should be less confused-looking than it is for other viewing directions, because vectors along the same line of sight should look the same. Explain.

**9. (a)** Find a potential for the conservative vector field $\left(1 + \left(x^2 + y + 1\right)^2\right)^{-1} [2x,\ 1]$ without using "`potential`".

**(b)** Describe the equipotential curves, and plot some of them.

**(c)** Plot some field lines.

**10.** Consider the vector field

$$\mathbf{F}(x, y, z) = \frac{2}{4z^2 + (x^2 + y^2 + z^2 - 1)^2} \left[-2xz,\ -2yz,\ x^2 + y^2 - z^2 - 1\right]$$

**(a)** Express $\mathbf{F}(x, y, z)$ in terms of cylindrical coordinates, using $\widehat{\mathbf{r}} = \cos\theta\,\mathbf{i} + \sin\theta\,\mathbf{j}$, $\widehat{\theta} = -\sin\theta\,\mathbf{i} + \cos\theta\,\mathbf{j}$, and $\mathbf{k}$. Note that the $\widehat{\theta}$ component is 0, and the other components don't depend on $\theta$. What does that tell you about the field?

**(b)** One consequence of the result of (a) is that in order to visualize the field we need only look at a cross-section containing the $z$ axis. Plot the two-dimensional field you get by taking $x = 0$ and ignoring the $x$ component of $\mathbf{F}$, for $y$ from $-2$ to 2 and $z$ from $-1$ to 1. For best results, use a "`grid`" setting that will avoid points too close to $y = 1$, $z = 0$. Why is this advisable?

**(c)** Find a potential $V$ for $\mathbf{F}$. Is it defined everywhere that $\mathbf{F}$ is defined? Where it is not defined, does it have a limit? Note: you should tell Maple (using "`assume`") that $x$, $y$, and $z$ are real to avoid complicated expressions involving "`csgn`", which is a version of the signum function.

**(d)** Consider a sphere with centre on the $z$ axis and with the circle $[\cos\theta, \sin\theta, 0]$ on its surface. Show that $V$ is constant on this sphere. Give a geometrical interpretation of $V$.

**(e)** Let $q(x, y, z) = \dfrac{x^2 + y^2 + z^2 + 1}{\sqrt{x^2 + y^2}}$. Show that $\mathbf{F}(x, y, z)$ is perpendicular to the gradient of $q(x, y, z)$. Why does this imply that $q(x, y, z)$ is constant on the field lines of $\mathbf{F}$? Show that the field lines (except for the $z$ axis) are circles $q =$ constant, $\theta =$ constant. Note that $\mathbf{T} \bullet \mathbf{F} > 0$ everywhere on one of these circles, where $\mathbf{T}$ is the tangent vector to the circle. Could this be true for a vector field that had a potential on a domain containing the circle?

# LAB 27 PARAMETRIC SURFACES

**Goals:** *To study parametric representations of surfaces, and to plot them and their tangent planes and normal vectors using Maple.*

Up to now we have represented surfaces mainly by equations $z = f(x, y)$, or sometimes by implicit equations $f(x, y, z) = 0$. Now we turn our attention to parametric surfaces, which are defined by vector-valued functions of two variables: $\mathbf{R}(u, v) = [x(u, v), y(u, v), z(u, v)]$, i.e., $x = x(u, v)$, $y = y(u, v)$, $z = z(u, v)$. This is analogous to parametric curves, but with two parameters $u$, $v$ instead of one parameter $t$.

■ **EXAMPLE 1** Represent the torus with implicit equation $(r - 2)^2 + z^2 = 1$ (in cylindrical coordinates) as a parametric surface, and plot it.

**SOLUTION** A cross-section of the surface at constant $\theta$ is a circle of radius 1 centred at $r = 2$, $z = 0$. The simplest parametric representation of this has parameter $u$ as the angle at the centre from a point on the circle to the positive $r$ axis: $r = 2 + \cos u$, $z = \sin u$, where $0 \leq u \leq 2\pi$. The second parameter $v$ can be the polar coordinate $\theta$. Converting to rectangular coordinates, the parametric representation of the torus is

$$x = (2 + \cos u) \cos v, \quad y = (2 + \cos u) \sin v, \quad z = \sin u, \quad 0 \leq u \leq 2\pi,\ 0 \leq v \leq 2\pi$$

Note that this is a **closed surface**, i.e., a surface with no boundary, because $\mathbf{R}(0, \theta) = \mathbf{R}(2\pi, \theta)$ and $\mathbf{R}(u, 0) = \mathbf{R}(u, 2\pi)$.

You can plot parametric surfaces using the "`plot3d`" command. The first input is a list of expressions for the coordinates as functions of the parameters: $[x(u, v), y(u, v), z(u, v)]$. The second and third inputs give the parameter names and the intervals on which they are defined. After this, the usual "`plot3d`" options can be used.

```
> plot3d([(2+cos(u))*cos(v), (2+cos(u))*sin(v), sin(u)],
>          u = 0 .. 2*Pi, v = 0 .. 2*Pi,
>          scaling = constrained, style = PATCH);
```

I'll save our plot for later use (see Figure 9.13).

> `plot1:= ":` (Release 4) or > `plot1:= %;` (Release 5)

The lines that you see on the plot (in the "`PATCH`", "`HIDDEN`", and "`WIREFRAME`" styles) correspond to constant values of $u$ and $v$, forming a grid. Maple calculates the points for the parameter values on this grid, and joins them up to form the surface. The "`grid`" option determines the number of parameter values that are used. Thus with parameter intervals $u = 0..1$ and $v = 0..2$ and "`grid=[5,6]`", Maple calculates the points for $u = 0, 1/4, \ldots 1$ and $v = 0, 2/5, \ldots 2$. The default is "`grid=[25,25]`". As usual, larger "`grid`" numbers produce a more accurate and smoother-looking plot, but require more time and memory.

Instead of the usual Cartesian coordinates, you can specify a parametric surface using cylindrical or spherical coordinates with the "`coords`" option. For example, our torus could have been plotted as follows:

```
> plot3d([2+cos(u), v, sin(u)], u=0..2*Pi, v=0..2*Pi,
>        coords = cylindrical, scaling=constrained);
```

In cylindrical coordinates, the list to be plotted is $[r, \theta, z]$. In spherical coordinates, obtained with "`coords=spherical`", it is $[\rho, \theta, \phi]$. Note that, in contrast to the convention we have been using, $\theta$ comes before $\phi$ in Maple. ■

**EXAMPLE 2** Plot the part of the cylinder $(x-2)^2+y^2=1$ that is inside the torus of Example 1.

**SOLUTION** The natural parameters for the cylinder are $z$ and an angle $p$ measured at the centre of the cylinder, where $x = 2 + \cos p$ and $y = \sin p$. To be inside the torus we need $(r-2)^2 + z^2 < 1$, where $r = \sqrt{x^2+y^2}$. Let's see what this means in terms of $z$ and $p$. I'll use "xc", "yc", and "zc" for the coordinates of points on the cylinder, leaving "x", "y", and "z" unassigned.

```
> xc:= 2+cos(p); yc:= sin(p);
> rc:= sqrt(xc^2 + yc^2);
> eq:= (rc - 2)^2 + zc^2 = 1;
```

$$eq := \left(\sqrt{(2+\cos(p))^2+\sin(p)^2}-2\right)^2 + zc^2 = 1$$

The equation "eq" should define the boundary of our surface. We'd like to solve this for "zc". For some reason, "solve" produces an unnecessarily complicated answer here, involving "RootOf" with a fourth degree polynomial. No matter: it's easy enough to solve by hand.

```
> z0:= simplify(sqrt(1 - (rc - 2)^2));
```

$$z0 := 2\sqrt{-2-\cos(p)+\sqrt{5+4\cos(p)}}$$

There are, of course, two solutions, $z_0$ and $-z_0$. For any $p$, at $z = \pm z_0$ the equation is satisfied, which says that the corresponding point on the cylinder is also on the surface of the torus. However, because of the square root with negative quantities inside, we should worry about whether this value is real. One way to see what is happening is to consider the projection of the cylinder and the torus on the $xy$ plane. I'll use parametric curves to plot two circles forming the outline of the torus and another circle for the projection of the cylinder.

```
> plot({[cos(t), sin(t), t=0..2*Pi],
>       [3*cos(t), 3*sin(t), t=0..2*Pi],
>       [2+cos(t), sin(t), t=0..2*Pi]}, scaling=constrained);
```

All the way around the cylinder there are points on the torus (although at $p = 0$ and $\pi$ they are reduced to a single point). When $-z_0 < z < z_0$ we have a point of the cylinder inside the torus. So we want to plot the surface for $0 \le p \le 2\pi$ and $-z_0 \le z \le z_0$.

```
> plot3d([xc, yc, z],
>    p=0..2*Pi, z=-z0 .. z0,
>    orientation=[45,80],
>    scaling=constrained);
```

(Figure 9.12)

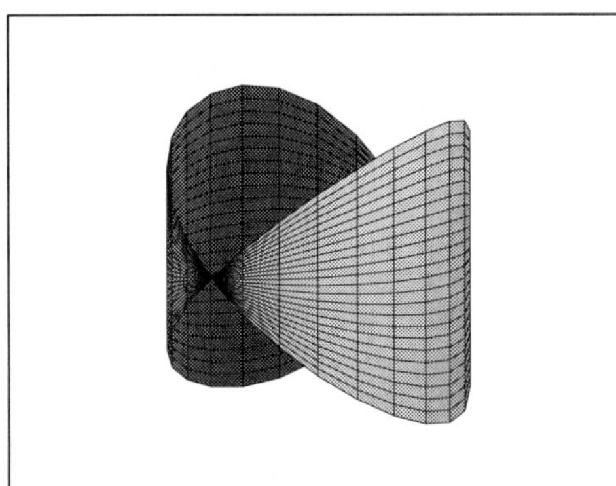

Figure 9.12. Cylinder inside torus

To see how this fits inside the torus, we might combine the plots of the two surfaces. To keep the torus from hiding the piece of cylinder inside it, we will use "style=WIREFRAME" on the torus. Now we already made a plot of the torus and saved this as "plot1", but that was in the "PATCH" style. No problem: we can simply use "subs" on the saved plot to transform its style to "WIREFRAME". I'll also use "shading=Z" because it looks better than "XYZ" for the "WIREFRAME" style.

```
> plot2:= ": (Release 4) or > plot2:= %: (Release 5)
> plots[display]({subs(PATCH=WIREFRAME,plot1), plot2},
>    orientation=[45,80], scaling=constrained, shading=Z);
```

■

## Tangent Planes and Normals

If $\mathbf{R}(u,v) = [x(u,v), y(u,v), z(u,v)]$ is a parametric surface, the derivatives $\dfrac{\partial \mathbf{R}}{\partial u}$ and $\dfrac{\partial \mathbf{R}}{\partial v}$ at a certain point are vectors tangent to the surface at that point. Assuming they are not parallel[14], the tangent plane to the surface consists of all the points $\mathbf{R} + a\dfrac{\partial \mathbf{R}}{\partial u} + b\dfrac{\partial \mathbf{R}}{\partial v}$, where $a$ and $b$ are scalars. We can consider this as a parametric representation of the tangent plane, where $a$ and $b$ are the parameters. Note that in this section we must use Cartesian coordinates, so if we have a parametric surface expressed in cylindrical or spherical coordinates we must convert these to Cartesian coordinates.

■ **EXAMPLE 3** Write a command that plots a patch of the tangent plane to the torus of Example 1 at a given point $\mathbf{R}(u_0, v_0)$.

**SOLUTION** Since we will be using vectors, we'll need the file "`dot.def`".

```
> with(linalg): read `dot.def`:
```

We define "`R`" for the torus and find the derivatives $\dfrac{\partial \mathbf{R}}{\partial u}$ and $\dfrac{\partial \mathbf{R}}{\partial v}$, which I'll call "`dRdu`" and "`dRdv`".

```
> R:= [(2 + cos(u))*cos(v), (2 + cos(u))*sin(v), sin(u)];
> dRdu:= diff(R,u);
```

$$dRdu := [-\sin(u)\cos(v),\ -\sin(u)\sin(v),\ \cos(u)]$$

```
> dRdv:= diff(R,v);
```

$$dRdv := [-(2+\cos(u))\ \sin(v),\ (2+\cos(u))\ \cos(v),\ 0]$$

Now given $u_0$ and $v_0$, we can substitute these for $u$ and $v$ in `R + a*dRdu + b*dRdv`, use "`evl`" to do the linear algebra and convert the result back to a list, and plot parametrically with parameters "a" and "b". Here is a command to do all that.

```
> tanplane:= (u0, v0) -> plot3d(subs(u=u0, v=v0,
>       evl(R + a*dRdu + b*dRdv)), a=-Pi/6 .. Pi/6,
>       b=-Pi/6 .. Pi/6, grid=[5, 5], colour=red,
>       style=PATCH);
```

The parameter intervals were picked in order to have the grid on the tangent plane correspond to the grid on the surface. Note that the linear approximation to $\mathbf{R}(u,v)$ is

$$\mathbf{R}(u_0, v_0) + (u - u_0)\frac{\partial \mathbf{R}}{\partial u}(u_0, v_0) + (v - v_0)\frac{\partial \mathbf{R}}{\partial v}(u_0, v_0)$$

which means that one unit of $u$ or $v$ corresponds to one unit of $a$ or $b$ respectively. The torus had both $u$ and $v$ go from 0 to $2\pi$, with the default grid of `[25,25]`. Thus the parameter intervals are divided into $25 - 1 = 24$ subintervals, each of length $2\pi/24 = \pi/12$. If we use a 5 by 5 grid for the tangent plane, the parameter intervals for the tangent plane will be divided into 4 subintervals. To make these subintervals the same length as the subintervals for the torus, we take the parameter interval of the tangent plane to have length $\pi/3$.

Let's try it out. The plot of the torus was already saved as "`plot1`". We need "`display`" from the "`plots`" package.

```
> with(plots, display);
> display({plot1, tanplane(0,0)}, scaling=constrained);
```

While we're at it, we can also plot an outward-pointing normal to the surface. We can use the cross product of $\dfrac{\partial \mathbf{R}}{\partial v}$ and $\dfrac{\partial \mathbf{R}}{\partial u}$ (it turns out that in this order we get an outward-pointing normal, while in the other order we'd get a normal pointing inwards). To plot it we can use the "`arrow3d`" function from the file "`arrow3d.def`", which we met in Lab 25. I'll define "`norml`"[15] to take the parameters $u_0$ and $v_0$ and draw the normal at $\mathbf{R}(u_0, v_0)$.

---

(14) If they are parallel, there may not be a tangent plane at all.

(15) We can't use the name "`normal`" because this is already a standard Maple command.

```
> read(`arrow3d.def`):
> norml:= (u0, v0) -> arrow3d(
>     subs(u=u0, v=v0, R),
>     subs(u=u0, v=v0, evl(dRdv &x dRdu)), colour=blue);
> display({plot1, tanplane(Pi/3, -Pi/6), norml(Pi/3, -Pi/6)},
>        scaling=constrained, orientation=[10, 75]);
```

(Figure 9.13) ■

## Parametrization

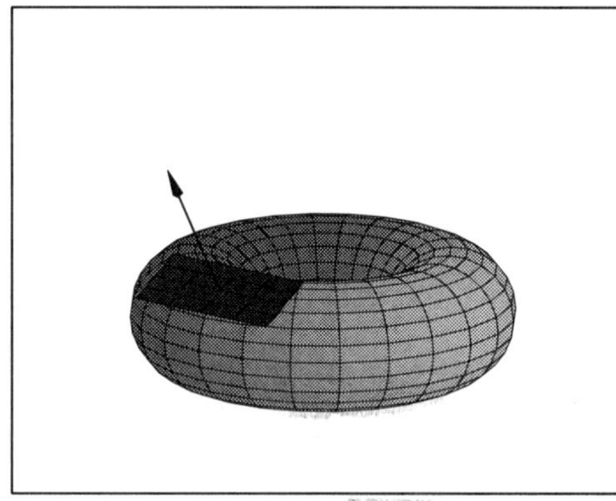

Figure 9.13. Tangent plane and normal

Given a surface (perhaps described geometrically or implicitly), how shall we represent it as a parametric surface? There are a number of considerations. We need a one-to-one[(16)] continuous function $\mathbf{R}(u, v)$ from some domain in the $uv$ plane to the surface. Ideally the domain should be a rectangle $a \le u \le b$, $c \le v \le d$. We would like the range of this function to be the whole surface, although if necessary we might construct the surface from several different "patches". For plotting purposes, it is best if the cells into which Maple divides the surface are not too long or wide. While we can ensure this by increasing the "grid" numbers, that can use up a lot of time and memory. To keep the cells small at a fixed "grid" size, we should ensure that $\partial\mathbf{R}/\partial u$ and $\partial\mathbf{R}/\partial v$ are not too large. Finally, we would usually prefer to avoid very complicated functions. This is not as important with Maple as it would be for calculations done by hand (or perhaps I should say that the standard for "very complicated" is different).

**■ EXAMPLE 4** Find a parametric representation of the surface $x^4+y^4+z^4+(x+y+z)xyz = 1$, and plot it.

**SOLUTION** Our first thought might be to use $x$ and $y$ as the parameters. We would then need to solve a fourth-degree equation to obtain $z$. Maple can solve such equations, but the result is usually very complicated.

```
> eqn:= x^4 + y^4 + z^4 + (x+y+z)*x*y*z = 1;
> S:= solve(eqn, z);
```

$$S := \mathrm{RootOf}\left(\_Z^4 + xy\,\_Z^2 + \left(x^2y + xy^2\right)\,\_Z + x^4 + y^4 - 1\right)$$

```
> zs:= allvalues(S);
```

Not only is the result complicated, but it is also difficult to determine which, if any, of the four roots are real. We would certainly need several patches, and the domains would be complicated.

Fortunately, another approach is more promising. Notice that when the left side of the equation is expanded out, all terms have total degree 4 in $x$, $y$, and $z$. If we write $x$, $y$, and $z$ in spherical coordinates, the equation will have the form $\rho^4 f(\phi,\theta) = 1$. Therefore we can write $\rho = 1/f(\phi,\theta)^{1/4}$ (when $f(\phi,\theta) > 0$), i.e., use $\phi$ and $\theta$ as the parameters.

```
> f:= unapply(subs(x=sin(phi)*cos(theta), y=sin(phi)*sin(theta),
>               z=cos(phi), op(1,eqn)), (phi,theta));
```

$$f := (\phi,\theta) \to \sin(\phi)^4\cos(\theta)^4 + \sin(\phi)^4\sin(\theta)^4 + \cos(\phi)^4$$
$$+ (\sin(\phi)\cos(\theta) + \sin(\phi)\sin(\theta) + \cos(\phi))\sin(\phi)^2\cos(\theta)\sin(\theta)\cos(\phi)$$

(16) It need not be one-to-one on the boundary of the domain.

Now we must ask, when is $f(\phi, \theta) > 0$? Is this true for all $\phi$ and $\theta$? We might try plotting.

```
> plot3d(f(phi,theta), phi=0..Pi, theta=0..2*Pi,
>       orientation = [0, 90], axes=boxed);
```

The minimum value of $f(\phi, \theta)$ seems to be approximately 0.2. If this is the case, our domain is all of $0 \le \phi \le \pi$, $0 \le \theta \le 2\pi$. The function $\mathbf{R}$ will be one-to-one except on the boundary, where we have the usual ambiguities of spherical coordinates: $\mathbf{R}(0, \theta)$ and $\mathbf{R}(\pi, \theta)$ do not depend on $\theta$, and $\mathbf{R}(\phi, 0) = \mathbf{R}(\phi, 2\pi)$. Our surface is a closed surface. Let's plot it[(17)].

```
> plot3d([f(phi,theta)^(-1/4), theta, phi],
>        phi = 0 .. Pi, theta = 0 .. 2*Pi, coords=spherical,
>        scaling = constrained); (Figure 9.14)
```

■

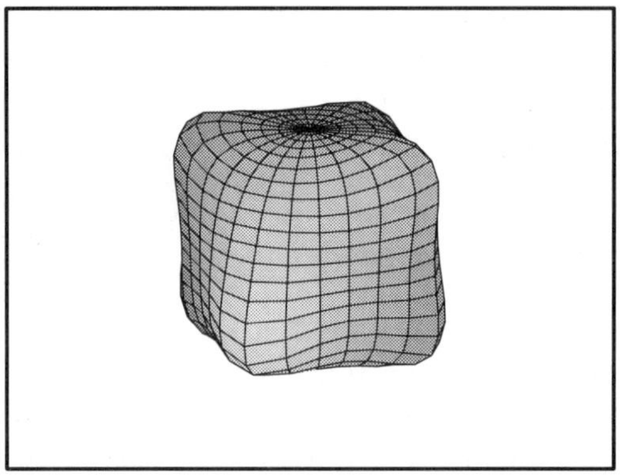

Figure 9.14. $x^4+y^4+z^4+(x+y+z)xyz = 1$

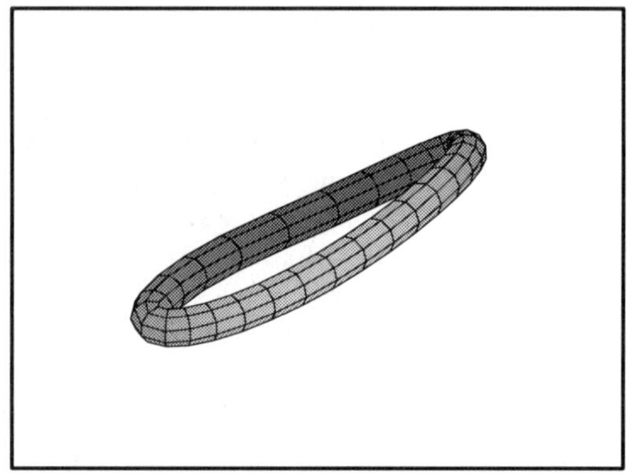

Figure 9.15. Tube around curve

■ **EXAMPLE 5** Parametrize a tube of radius 1 centred on the parametric curve $\mathbf{R} = \mathbf{R}_c(t)$ given by $x = t(t^2 - \pi^2)$, $y = 3\cos(t)$, $z = 3\sin(t)$ for $-\pi \le t \le \pi$.

**SOLUTION** Note that the "plots" package's "tubeplot" command can be used to plot such a tube. However, we want to consider it as an ordinary parametric surface.

The tube consists of those points whose distance from the curve is 1. Thus for each such point $P$ there is some point $\mathbf{R}_c(t)$ on the curve (only one such point, unless the tube has "kinks" or self-intersections) with distance 1 from $P$, and the distances from $P$ to all other points on the curve are greater than 1. This means that the curve is tangent at $\mathbf{R}_c(t)$ to the sphere of radius 1 centred at $P$, i.e., the tangent vector to the curve at $\mathbf{R}_c(t)$ is perpendicular to the vector from $\mathbf{R}_c(t)$ to $P$.

Now it seems reasonable to let $t$ be one of the two parameters for the tube. The discussion above implies that the points of constant $t$ on the tube form a circle of radius 1 centred at $\mathbf{R}_c(t)$ in the plane perpendicular to the tangent vector $\mathbf{T}(t)$ to the curve at $\mathbf{R}_c(t)$. Our second parameter $u$ should then describe a position on this circle. An angle, ranging from, say, 0 to $2\pi$, would be a reasonable choice for $u$. If $\mathbf{A}(t)$ and $\mathbf{B}(t)$ are unit vectors, perpendicular to each other and to $\mathbf{T}(t)$, then we can write parametric equations for the surface as $\mathbf{R} = \mathbf{R}_c(t) + \cos(u)\mathbf{A}(t) + \sin(u)\mathbf{B}(t)$.

The remaining question is how to choose $\mathbf{A}(t)$ and $\mathbf{B}(t)$. One possibility is to use the unit normal and binormal to the curve (assuming these are not zero). Another, somewhat simpler, possibility is the following. First choose a constant vector $\mathbf{C}$. Take the projection perpendicular to $\mathbf{T}(t)$, and normalize this to get $\mathbf{A}(t)$. Then take the cross product of this with $\mathbf{T}(t)$ to get $\mathbf{B}(t)$. This will work as long as $\mathbf{T}(t)$ is never parallel to $\mathbf{C}$. We should expect that to be true, barring an exceptional coincidence, if $\mathbf{C}$ is chosen "randomly".

```
> Rc:= [ t * (t^2-Pi^2), 3*cos(t), 3*sin(t) ];
> Vc:= diff(Rc, t);
> C:= [ 3.14, 1.59, 2.65 ];
```

(17) We could also do this with "sphereplot" in the plots package.

```
> read(`dot.def`):
> A:= unit(C - (C &. Vc) * Vc / (Vc &. Vc));
> B:= unit(A &x Vc);
> R:= evl(Rc + cos(u) * A + sin(u) * B);
> plot3d(R, t = -Pi .. Pi, u = 0 .. 2*Pi, grid = [30, 10],
>        scaling = constrained); (Figure 9.15)
```
■

## EXERCISES

**1.** Represent as a parametric surface, and plot:

**(a)** The surface obtained by rotating the curve $y = \sin x$, $0 \le x \le \pi$, about the $x$ axis.

**(b)** The surface obtained by rotating the same curve about the $y$ axis.

**(c)** The surface obtained by rotating the cardioid $r = 1 - \sin\theta$ about the $y$ axis.

**2.** Where do each of the surfaces of Exercise 1 have normal vector $[1, 1, 1]$? Plot each surface that does have this normal vector, together with a patch of the tangent plane at such a point.

**3.** Plot the part of the torus inside the cylinder of Example 2, as well as the part of the cylinder inside the torus (together constituting the surface of the intersection of the cylinder and the torus).

**4.** **(a)** Plot the parametric surface $x = u^2 - v^2$, $y = u^2 + v^2$, $z = (u - v)^2$ for $-1 \le u \le 1$, $0 \le v \le 1$.

**(b)** The function defining this surface is not one-to-one on the boundary of its domain. Explain. When do points on the boundary of the domain correspond to the same points on the surface?

**(c)** Is the surface smooth at the point $[1/4, 1/4, 1/4]$? If so, what is a normal vector to the surface there?

**(d)** Is the surface smooth at the point $[0, 0, 0]$? Why?

**5.** **(a)** Show that an ellipsoid $x^2/a^2 + y^2/b^2 + z^2/c^2 = 1$ can be parametrized as $x = a \sin\phi \cos\theta$, $y = b \sin\phi \sin\theta$, $z = c \cos\phi$ for $0 \le \phi \le \pi$, $0 \le \theta \le 2\pi$. Note that $\theta$ and $\phi$ are **not** spherical coordinates for the point they represent on the ellipsoid.

**(b)** Let $a = 1$, $b = 2$, and $c = 3$. Find (using numerical approximation, and the parametrization of (a)) the closest and farthest points to $[1, 1, 1]$ on the ellipsoid.

**(c)** Plot the ellipsoid together with a sphere centred at $[1, 1, 1]$ that is tangent to it. Use an orientation so that the point of tangency will not be hidden by the sphere and the ellipsoid.

**6.** **(a)** What quadric surface can be (partially) parametrized by $x = a \sinh u \cos v$, $y = b \sinh u \sin v$, $z = c \cosh u$? *Hint:* find a linear combination of $x^2$, $y^2$, and $z^2$ that is constant.

**(b)** Why did I have to say "partially"? How can you obtain the whole surface?

**(c)** With $a = 1$, $b = 2$, and $c = 3$, plot the part of the surface for $-10 \le z \le 10$.

**(d)** Show that $x = a \cosh s \sinh t$, $y = b \sinh s$, $z = c \cosh s \cosh t$ is another parametrization of the same surface as in (a). Use this (modified as necessary) to plot the same part of the surface as in (c). Which parametrization is better for this purpose? Why?

**7.** A **Möbius strip** is a non-orientable (one-sided) surface, obtained by twisting a strip of paper and pasting the ends together.

**(a)** Begin with a flat strip $2\pi$ units long and 1 unit wide, centred on the $y$ axis. Give it a half twist. Represent the result as a parametric surface. Plot it.

**(b)** Add a suitable constant to the $x$ coordinate from (a), and regard the coordinates as cylindrical coordinates. Plot the result.

**(c)** Produce an animation of a normal to the surface travelling around the Möbius strip and coming back to the start pointing in the opposite direction to how it started out.

**(d)** Another unusual property of the Möbius strip is that if you split it in half lengthwise, you still have one strip. To see how this works, start with the plot of a Möbius strip as in (b), where the parameter $v$ goes across the strip: say $v = 0$ is the centre of the strip and $v = \pm.5$ the edge. Then plot the same parametric function for $-.6 \le v \le -.1$ and $.1 \le v \le .6$, etc. Combine several of these in an animation of the strip in the process of splitting. Is the split strip orientable? To see, produce an animation of a normal to the surface travelling around it and coming back to the start. Note that to come back to the start on the split strip corresponds to going twice around the original Möbius strip.

**8.** The patch of tangent plane that we plot with the "`tanplane`" procedure is in the shape of a parallelogram with edges parallel to $d\mathbf{R}/du$ and $d\mathbf{R}/dv$ (where $u$ and $v$ are the parameters). Write a procedure that plots a square patch of tangent plane. Write another that plots a circular disk-shaped patch. Test them out on the surface $[\cos u \cos v,\ \sin u \sin v,\ v]$, $0 \le u \le 2\pi$, $0 \le v \le \pi/2$.

**9.** The equation

$$z^2 - \frac{2z(2w^2-1)}{\sqrt{1-w^2}w} + r^2 = 4$$

where $0 < w = 1 - 2\theta/\pi < 1$ (using cylindrical coordinates) represents a force surface for a certain magnetic field[18]. Note that the cross-section for constant $w$ is a semicircle in $r$ and $z$. Plot the surface parametrically, using $w$ as one parameter and the angle at the centre of the semicircle as the other. Don't go too close to $w = 0$ and $w = 1$ where the denominator of the second term is 0.

# LAB 28 LINE AND SURFACE INTEGRALS

**Goals:** *To use Maple to set up and calculate the integrals of scalar functions over parametric curves and surfaces, of the tangential components of vector fields over curves, and of the normal components of vector fields over surfaces*

## Line Integrals

There are two main types of line integral: the line integral of a scalar field along a curve, $\int_C f\,ds$, and the line integral of the tangential component of a vector field along a curve, $\int_C \mathbf{F} \bullet d\mathbf{r} = \int_C \mathbf{F} \bullet \mathbf{T}\,ds$. Maple has a "`Lineint`" command in the "`student`" package that does the line integral of a two-dimensional scalar field, but it will be more instructive to construct such commands ourselves. The commands I've written are in the file "`vecints.def`", which should be available to you[19].

---

[18] This example is due to Rafał Abłamowicz of Gannon University.

[19] Note that this file automatically reads "`dot.def`", which should also be available.

First, consider the line integral $\int_C f\,ds$ of a scalar field $f$ along a curve $C$. We want a command "`slineint`" that takes three inputs: a scalar field, a parametric curve, and a parameter interval. The scalar field will be an expression in the coordinate variables, usually $x$, $y$, and (for three dimensions) $z$. The parametric curve will be written as a list of equations expressing each coordinate variable as a function of the parameter, e.g., "`[x = t, y = t^2]`". This list can be substituted into the scalar field using "`subs`" to obtain the value of the scalar field at a point on the curve. The parameter interval will be written as an equation, e.g., "`t = a..b`". The left side of this equation will identify the name of the parameter. Thus to integrate $f(x, y)$ over the parametric curve $x = g(t)$, $y = h(t)$ for $t$ from $a$ to $b$, we would enter the command as

```
> slineint(f(x,y), [x=g(t), y=h(t)], t=a..b);
```

and the result should be

$$\int_a^b f(g(t), h(t)) \sqrt{\left(\frac{dg}{dt}\right)^2 + \left(\frac{dh}{dt}\right)^2}\, dt$$

The command should thus be of the form

```
> slineint:= (F,C,R) -> int(subs(C, F) * L, R);
```

where "`L`" will be the factor $\sqrt{\left(\frac{dg}{dt}\right)^2 + \left(\frac{dh}{dt}\right)^2}$. To obtain this, we will first extract the right sides of all the equations in the list "`C`", which can most conveniently be done using "`map`" with the "`rhs`" function. Secondly, we differentiate with respect to the parameter (which is `op(1,R)`)[20]. That will produce the velocity vector of the parametric curve (thinking of the parameter as "time"). Since we'll need that vector again, we'll define it as a function "`veloc(C,R)`". Then we take the length of `veloc(C,R)`, using the "`len`" function that we included in "`dot.def`" (see Lab 14).

```
> read `dot.def`:
> veloc:= (C,R) ->  diff(map(rhs,C), op(1,R));
> slineint:= (F,C,R) -> int(subs(C, F) * len(veloc(C,R)), R);
```

Now let's try it out, first in two dimensions with everything symbolic, then with a specific example in three dimensions.

```
> slineint(f(x,y), [x=g(t), y=h(t)], t=a..b);
```

$$\int_a^b f(g(t), h(t)) \sqrt{\left(\frac{\partial}{\partial t} g(t)\right)^2 + \left(\frac{\partial}{\partial t} h(t)\right)^2}\, dt$$

```
> slineint(x - y*z, [x=sin(t), y=cos(t), z=t], t=0..Pi);
```

$$4\sqrt{2}$$

It may also be useful to have an "inert" version of the command that will set up the integral but not evaluate it. This will be the same as "`slineint`", but using "`Int`" instead of "`int`". I'll call it "`Slineint`". Here's a trick: rather than retype the definition, we can simply change "`int`" to "`Int`" using "`subs`"[21].

---

[20] By using "`op(1,R)`" rather than "`lhs(R)`", we also allow for an indefinite line integral where "`R`" is simply a variable.

[21] We use it on "`eval(slineint)`" rather than just "`slineint`", since otherwise "`subs`" would just see the name "`slineint`" rather than the definition.

```
> Slineint:= subs(int=Int, eval(slineint));
> S:= Slineint(x - y*z, [x=sin(t), y=cos(t), z=t], t=0..Pi);
```

$$S := \int_0^{\pi} (\sin(t) - \cos(t)t)\sqrt{\cos(t)^2 + \sin(t)^2 + 1}\,dt$$

This integrand can be simplified.

```
> S:= simplify(S);
```

$$S := \int_0^{\pi} \sqrt{2}\sin(t) - \sqrt{2}\cos(t)t\,dt$$

To evaluate the integral, we can use "`value`".

```
> value(S);
```

$$4\sqrt{2}$$

Next we'll try a command "`vlineint(F,C,R)`" for the line integral of the tangential component of a vector field. This time the first input "`F`" is a vector field. Again we'll substitute "`C`" into "`F`", to evaluate the vector field at a point on the curve. Then we take the dot product of this vector with `veloc(C,R)`, and integrate that with respect to "`R`".

```
> vlineint:= (F,C,R) -> int(subs(C, F) &. veloc(C, R), R);
```

One extra technical point: the field "`F`" might sometimes be a Maple "`vector`" rather than a list, in which case the "`subs`" won't work directly. However, it will work if we first apply "`evl`" to change "`F`" into a list. So here is the new "`vlineint`":

```
> vlineint:= (F,C,R) -> int(subs(C, evl(F)) &. veloc(C,R), R);
```

Again it may be useful to have an inert form "`Vlineint`" that uses "`Int`" instead of "`int`".

```
> Vlineint:= subs(int=Int,eval(vlineint));
```

And now let's try out "`vlineint`". Here is the line integral of the vector field $\mathbf{F}(x, y) = [y^2, xy]$ along the curve $y = x^2$ from $(0, 0)$ to $(1, 1)$. We can use $x$ as the parameter.

```
> vlineint([y^2, x*y], [x = x, y = x^2], x = 0..1);
```

$$\frac{3}{5}$$

## Surface Integrals

Just as with line integrals, there are two main types of surface integrals: the integral of a scalar field $\iint_S f\,dS$, and the integral of the normal component of a vector field $\iint_S \mathbf{F} \bullet d\mathbf{S}$ (the flux of the vector field across the surface). We will define Maple commands "`surfint`" and "`flux`" to do these.

We begin with the scalar case. The "`surfint`" command should take four inputs: a scalar field, a parametric surface, and the intervals for the two parameters. The scalar field will be an expression in the coordinate variables, usually $x$, $y$, and $z$. The parametric surface will be given as a list of equations expressing these coordinate variables in terms of the parameters (e.g., `[x = u^2, y = v^2, z = u*v]`). The parameter intervals will be written as equations such as `u=2..3` and `v=1..u`. The left sides of these will identify the parameters. The endpoints for the second parameter are allowed to depend on the first parameter, but not vice versa. Thus to integrate $f(x, y, z)$ over the surface $x = g(u, v)$, $y = h(u, v)$, $z = k(u, v)$ for $a \le u \le b$ and $c \le v \le d$ we would enter

```
> surfint(f(x,y,z), [x=g(u,v), y=h(u,v), z=k(u,v)],
>          u=a..b, v=c..d);
```

The command should be of the form

```
> surfint:= (f, S, R1, R2) ->
>            int(int(subs(S,f) * aelt(S,R1,R2), R2), R1);
```

where "aelt(S,R1,R2)" is the factor $\left|\frac{\partial \mathbf{R}}{\partial u} \times \frac{\partial \mathbf{R}}{\partial v}\right|$ that is multiplied by $du\,dv$ to get the area element. We can use the same "veloc" that we used in line integrals to obtain $\frac{\partial \mathbf{R}}{\partial u}$ and $\frac{\partial \mathbf{R}}{\partial v}$.

```
> aelt:= (S, R1, R2) -> len(veloc(S,R1) &x veloc(S,R2));
```

**■ EXAMPLE 1** Find the integral of $x^2$ over the torus of Example 1 of Lab 27.

**SOLUTION** We define the parametric equations of the torus, and use "surfint".

```
> torus:= [x = (2+cos(u))*cos(v), y = (2+cos(u))*sin(v),
>             z = sin(u)];
> surfint(x^2, torus, u = 0 .. 2*Pi, v = 0 .. 2*Pi);
```

$$22\pi^2$$

What were the integrals that Maple performed to get this result? We could use an inert form of "surfint".

```
> Surfint:= subs(int=Int,eval(surfint));
> S:= Surfint(x^2, torus, u=0 .. 2*Pi, v=0 .. 2*Pi);
```

$$S := \int_0^{2\pi}\int_0^{2\pi} (2+\cos(u))^2\cos(v)^2\mathrm{sqrt}\Big(\cos(u)^2(2+\cos(u))^2\cos(v)^2 + \cos(u)^2(2+\cos(u))^2\sin(v)^2 + (-\sin(u)\cos(v)^2(2+\cos(u)) - \sin(u)\sin(v)^2(2+\cos(u)))^2\Big)\,dv\,du$$

This can be simplified. For example, the identity $\cos^2 v + \sin^2 v = 1$ will certainly be helpful.

```
> S:= simplify(S);
```

$$S := \int_0^{2\pi}\int_0^{2\pi} 8\cos(v)^2\mathrm{csgn}(2+\cos(u)) + 12\cos(v)^2\mathrm{csgn}(2+\cos(u))\cos(u) + 6\cos(v)^2\mathrm{csgn}(2+\cos(u))\cos(u)^2 + \cos(v)^2\mathrm{csgn}(2+\cos(u))\cos(u)^3\,dv\,du$$

The "csgn$(2+\cos(u))$" factors come from uncertainty over the sign of the square root. Maple does not realize that $2+\cos(u)$ is always positive(22).

To evaluate the simplified integral, we can use "value".

```
> value(S);
```

$$22\pi^2$$

■

Now we will try a command "flux" that will integrate the normal component of a vector field **F** over our surface. The vector area element is $d\mathbf{S} = \left(\frac{\partial \mathbf{R}}{\partial u} \times \frac{\partial \mathbf{R}}{\partial v}\right) du\,dv$. We'll also define an "inert" version.

```
> flux:= (F, S, R1, R2) -> int(int(subs(S, evl(F)) &.
>             (veloc(S,R1) &x veloc(S,R2)), R2), R1);
> Flux:= subs(int=Int, eval(flux));
```

We'll have to be careful about which of the two normal directions to a surface we want. The one that "flux" uses is the direction of $\frac{\partial \mathbf{R}}{\partial u} \times \frac{\partial \mathbf{R}}{\partial v}$ where $u$ is the first parameter and $v$ is the second. According to the right-hand rule, if you start with the fingers of your right hand straight and pointing in the direction of increasing $u$, then curve them in the direction of increasing $v$, your thumb will point in the direction of this normal. If you want the other normal, use either `-flux(F,S,R1,R2)` or `flux(F,S,R2,R1)`.

(22) Recall that Maple uses complex numbers, not just real numbers, and $2+\cos(u)$ is not necessarily positive when $u$ is complex.

**EXAMPLE 2** Find the flux of $\mathbf{R} = [x, y, z]$ outward through the torus (Example 1 of Lab 27).

**SOLUTION** As noted in the section "Tangent Planes and Normals" of Lab 27, the normal $\dfrac{\partial \mathbf{R}}{\partial u} \times \dfrac{\partial \mathbf{R}}{\partial v}$ points inwards in this case, so I'll put a negative sign in front of the flux.

```
> - flux([x,y,z], torus, u=0..2*Pi, v=0..2*Pi);
```

$$12\pi^2$$

## EXERCISES

1. Evaluate $\int_C (x + y^2 + z^3)\, ds$ and $\int_C (z^3\, dx + y^2\, dy + x\, dz)$ where $C$ is the helix $x = \cos t$, $y = \sin t$, $z = t$, $0 \le t \le 2\pi$.

2. Evaluate $\iint_S z\, dS$ and $\iint_S [x, y, z] \cdot d\mathbf{S}$, where $S$ is the conical surface $x = u \cos v$, $y = u \sin v$, $z = 1 - u$ for $0 \le u \le 1$ and $0 \le v \le 2\pi$.

3. Find the line integral $\int_C \mathbf{F} \bullet d\mathbf{r}$, where $\mathbf{F} = [yz^2, x^2 z, xy^2]$ and $C$ is the curve where the surfaces $x^2 + y^2 = 1$ and $z = x^2$ intersect, counterclockwise when viewed from above.

4. The time a particle takes to traverse a curve $C$ is $\int_C (1/v)\, ds$ where $v$ is the speed (considered as a function of position on the curve). For an object sliding without friction in a constant gravitational field of magnitude $g$, $v = \sqrt{2g(y_0 - y)}$ where $y$ is the vertical coordinate and $y_0$ is the level at which the speed would be 0 (in particular, if the object starts at rest, $y_0$ is the initial $y$ value).
   - **(a)** Consider the (inverted) cycloid $x = a(t - \sin t)$, $y = a(-1 + \cos t)$ where $a$ is a positive parameter. Calculate the time for an object to reach the bottom of the cycloid if it starts from rest at any given point on the cycloid. The answer turns out to be constant: this says that the cycloid is a "tautochrone".
   - **(b)** The cycloid is also a "brachistochrone": of all curves from point $A$ to point $B$, where $B$ is below but not directly below $A$, a cycloid is the one on which a sliding object, starting at rest at $A$, will reach $B$ the quickest. We can't prove that here, but we can try some examples. Let $A = [0, 0]$ and $B = [2, -1]$. Find $a$ for which the cycloid passes through $B$, and the time for the object to reach $B$ along this cycloid. Compare to the least time for circular arcs passing through $A$ and $B$.

5. Find the area of the surface $x = u \cos v$, $y = u \sin v$, $z = v$ for $0 \le u \le 1$, $0 \le v \le 2\pi$, and the integral of the normal component of $[x, y, z]$ over it (taking the upward-pointing normal).

6. Let $\mathbf{F}$ be the vector field $[-y, x]/(x^2 + y^2)$. Let $C(a)$ be the path in the $xy$ plane consisting of the straight-line segment from $[-1, 0]$ to $[\cos a, \sin a]$ followed by the straight-line segment from there to $[1, 0]$. Calculate $\int_{C(a)} \mathbf{F} \bullet d\mathbf{r}$. Plot this as a function of $a$ in the interval $(-\pi, \pi)$. Be careful about values of $a$ for which the integral is undefined. Explain the results. *Hint:* we saw this vector field in the section "Potentials for Fields" of Lab 26.

7. A thin spherical shell of radius $a$ has uniform density $\sigma$. Find the gravitational potential at a point $P$ at distance $r$ from the centre of the shell. Note that the gravitational potential at $P$ due to an "infinitesimal" element of the shell is $G\sigma\, dA/d$ where $G$ is the gravitational constant, $dA$ is the area of the element, and $d$ is the distance from it to $P$. This can be integrated over the sphere to obtain the actual potential. Assume that $P$ is on the positive $z$ axis, and use spherical coordinates on the shell. The result will be different depending on whether $P$ is inside or outside the shell.

8. **(a)** Find the element of area for a surface $\phi = f(\rho, \theta)$ in spherical coordinates.
   - **(b)** Use this to find the flux of $[x, y, z]$ upward through the part of the surface $z = \dfrac{x^2 + y^2}{x^2 + 2y^2}$ between $\rho = 1$ and $\rho = 2$.

# LAB 29 VECTOR ANALYSIS

**Goals:** *To use Maple in calculations involving the gradient, curl, and divergence operators, including the theorems of Gauss, Green, and Stokes, and the computation of vector potentials.*

## Div, Grad, and Curl

The three main operators of vector analysis are the gradient, divergence, and curl. The Maple commands for these are in the "`linalg`" package. As we have seen in Lab 19, "`grad(f, [x,y,z])`" finds the gradient grad $f$ (sometimes written $\nabla f$) of the scalar expression $f$, where the coordinate variables are $x$, $y$, and $z$. Similarly, "`diverge(F, [x,y,z])`"[23] finds the divergence, $\text{div}\,\mathbf{F} = \nabla \bullet \mathbf{F}$, of a vector (or list) expression $\mathbf{F}$, and "`curl(F, [x,y,z])`" finds its curl, $\text{curl}\,\mathbf{F} = \nabla \times \mathbf{F}$. The results of "`grad`" and "`curl`" are vectors.

At various points in this lab we are going to use some of the commands found in the files "`dot.def`" and "`vecints.def`", so we may as well load them now, as well as the "`linalg`" package. Moreover, we will always be using "`[x,y,z]`" as the variable list for "`grad`", "`diverge`", and "`curl`", so to save a few keystrokes let's assign this to the name "`xyz`".

```
> with(linalg): read `dot.def`; read `vecints.def`;
> xyz:= [x,y,z];
```

Here are some examples of the use of "`grad`", "`diverge`", and "`curl`".

```
> G:= grad(x + y^2 + z^3, xyz);
```

$$G := \left[1,\, 2y,\, 3z^2\right]$$

```
> d:= diverge(G, xyz);
```

$$d := 2 + 6z$$

```
> curl(d*G, xyz);
```

$$[-12y,\, 6,\, 0]$$

Curiously, the last calculation doesn't work if done with "`diverge`".

```
> diverge(d*G, xyz);
Error (in diverge) wrong number or type of arguments
```

It seems that "`diverge`" expects an actual list rather than a scalar multiple of a list. No problem: just use "`evalm`" (or "`evl`" from "`dot.def`") to multiply the scalar and the vector before applying "`diverge`".

```
> diverge(evalm(d*G), xyz);
```

$$4 + 12z + 18z^2 + 6(2 + 6z)z$$

Another useful operator is the Laplacian $\nabla^2 = (\partial^2/\partial x^2) + (\partial^2/\partial y^2) + (\partial^2/\partial z^2)$. The command "`laplacian(f, [x,y,z])`" finds the Laplacian of the scalar expression $f$. This is the same as div (grad $f$).

```
> laplacian(x + y^2 + z^3, xyz);
```

$$2 + 6z$$

---

(23) I suppose the reason Maple doesn't use the standard mathematical notation "`div`" is that "`div`" is used in many computer languages for integer division (although Maple uses "`iquo`" for that).

The Laplacian is sometimes applied to a vector field instead of a scalar field: each component of $\nabla^2\mathbf{F}$ is the Laplacian of the corresponding component of $\mathbf{F}$. The "`laplacian`" command doesn't handle vector fields directly, but you can use "`map`" to apply it to each component.

```
> map(laplacian, G, xyz);
```

$$[0, 0, 6]$$

## Physical Interpretation of Curl

One way of interpreting the curl of a vector field is the following. Consider the field $\mathbf{V}$ as representing the velocity vector field of a fluid. Place a small paddle wheel in the fluid at a point $P$. Then the rate at which the paddle wheel rotates is proportional to the component of the curl of $\mathbf{V}$ at $P$ in the direction of the paddle wheel's axis. Let us see how this works.

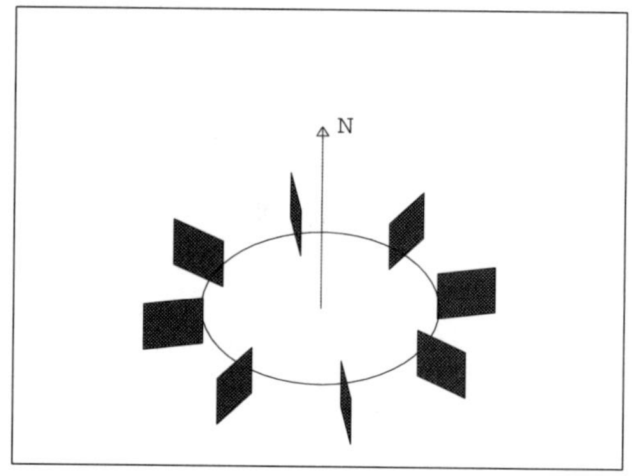


Figure 9.16. Paddle wheel

To model the paddle wheel, we consider a small circle $C$ of radius $r$, centred at $P$ (Figure 9.16). The blades of the wheel are on the circle. Left to itself, a blade would travel along with the fluid, but being constrained to travel on the circle, its velocity would be the vector projection of $\mathbf{V}$ on the tangent $\mathbf{T}$ to the circle. The angular speed about the centre of the circle would be $\mathbf{V} \bullet \mathbf{T}/r$. However, the whole paddle wheel must turn as a unit, so these must be averaged over the circumference of the circle. Thus the angular speed of the paddle wheel should be $\dfrac{1}{2\pi r}\displaystyle\int_C \dfrac{\mathbf{V} \bullet \mathbf{T}}{r}\,ds = \dfrac{1}{2\pi r^2}\displaystyle\int_C \mathbf{V} \bullet d\mathbf{r}$. I claim that in the limit as $r \to 0+$, this is proportional to the component of the curl of $\mathbf{V}$ at $P$ in the direction $\mathbf{N}$ normal to the plane of the circle. The coefficient of proportionality turns out to be $1/2$, i.e.,

$$\lim_{r\to 0+} \frac{1}{\pi r^2}\int_C \mathbf{V} \bullet d\mathbf{r} = \mathbf{N} \bullet (\operatorname{curl}\mathbf{V})(P)$$

where the orientation of $C$ is induced by $\mathbf{N}$ (i.e., $C$ goes counterclockwise as seen from a viewpoint in the direction of $\mathbf{N}$).

We'll calculate both sides of this equation using Maple, and see that they are equal. The components of the vector field $\mathbf{V}$ will be arbitrary functions[24] of $x$, $y$, and $z$. The point $P$ will also have arbitrary coordinates. A parametric representation of the circle is $\mathbf{R} = P + r\cos(t)\mathbf{A} + r\sin(t)\mathbf{B}$, where $\mathbf{A}$ and $\mathbf{B}$ are orthogonal unit vectors with $\mathbf{A}\times\mathbf{B} = \mathbf{N}$. Actually I won't tell Maple the conditions that $\mathbf{A}$ and $\mathbf{B}$ are orthogonal unit vectors, but make them arbitrary vectors.

```
> V:= [v1(x,y,z), v2(x,y,z), v3(x,y,z)]; P := [p1,p2,p3];
> A:= [a1,a2,a3]; B:= [b1,b2,b3];
> C:= evl(P + r * cos(t) * A + r * sin(t) * B);
```

$$C := [p1 + r\cos(t)\,a1 + r\sin(t)\,b1,\ p2 + r\cos(t)\,a2 + r\sin(t)\,b2,\ p3 + r\cos(t)\,a3 + r\sin(t)\,b3]$$

For use with the line-integral commands, we need this in the form "`[x=..., y=..., z=...]`".

```
> C:= [x=C[1], y=C[2], z=C[3]];
```

Now we can find the circulation of $\mathbf{V}$ around the circle. Since there is no hope of actually performing an integration (as the components of $\mathbf{V}$ are arbitrary functions), we may as well use the inert form "`Vlineint`" of the line integral.

```
> J:= Vlineint(V, C, t=0..2*Pi);
```

(24) They are assumed, of course, to be differentiable.

The most straightforward way to proceed would be to divide this by $\pi r^2$ and use "`limit`". However, this uses a significant amount of computer time. It seems somewhat quicker to use "`taylor`" instead. Since we will divide the result by $r^2$ and expect a limit, the Taylor series of $J$ should be taken to order 3.

```
> S:= taylor(J, r=0, 3);
```

$$S := (-a3\, D_1(v3)(p1, p2, p3)\, b1\, \pi + b1\, D_2(v1)(p1, p2, p3)\, a2\, \pi + b1\, D_3(v1)(p1, p2, p3)\, a3\, \pi + b2\, D_1(v2)(p1, p2, p3)\, a1\, \pi + b2\, D_3(v2)(p1, p2, p3)\, a3\, \pi - a3\, D_2(v3)(p1, p2, p3)\, b2\, \pi + b3\, D_2(v3)(p1, p2, p3)\, a2\, \pi - a1\, D_2(v1)(p1, p2, p3)\, b2\, \pi - a1\, D_3(v1)(p1, p2, p3)\, b3\, \pi - a2\, D_1(v2)(p1, p2, p3)\, b1\, \pi - a2\, D_3(v2)(p1, p2, p3)\, b3\, \pi + b3\, D_1(v3)(p1, p2, p3)\, a1\, \pi)\, r^2 + O(r^3)$$

We use "`convert(...,polynom)`" to get rid of the $O(r^3)$ term, and divide by $\pi r^2$ to obtain the limit.

```
> Left:= convert(S,polynom)/(Pi*r^2);
```

Now let's work on the right side of our equation. We calculate the curl of **V**.

```
> C:= curl(V, xyz);
```

$$C := \left[\frac{\partial}{\partial y} v3(x, y, z) - \frac{\partial}{\partial z} v2(x, y, z), \frac{\partial}{\partial z} v1(x, y, z) - \frac{\partial}{\partial x} v3(x, y, z), \frac{\partial}{\partial x} v2(x, y, z) - \frac{\partial}{\partial y} v1(x, y, z)\right]$$

Now we take the dot product with the normal $\mathbf{N} = \mathbf{A} \times \mathbf{B}$.

```
> R:= C &. (A &x B);
```

$$R := \left(\frac{\partial}{\partial y} v3(x, y, z) - \frac{\partial}{\partial z} v2(x, y, z)\right)(a2\, b3 - a3\, b2) + \left(\frac{\partial}{\partial z} v1(x, y, z) - \frac{\partial}{\partial x} v3(x, y, z)\right)(a3\, b1 - a1\, b3) + \left(\frac{\partial}{\partial x} v2(x, y, z) - \frac{\partial}{\partial y} v1(x, y, z)\right)(a1\, b2 - a2\, b1)$$

This uses the "`diff`" form of derivatives rather than the "`D`" form that we see on our expression for the left side. The "`diff`" form is less convenient when we want to evaluate the derivatives at a particular point $P$. Fortunately, we can use "`convert`" to get the "`D`" form. Then we use "`subs`" to evaluate it at $P$.

```
> Right:= subs(x=p1, y=p2, z=p3, convert(R, D));
```

$$(D_2(v3)(p1, p2, p3) - D_3(v2)(p1, p2, p3))(a2\, b3 - a3\, b2) + (D_3(v1)(x, y, z) - D_1(v3)(x, y, z))(a3\, b1 - a1\, b3) + (D_1(v2)(x, y, z) - D_2(v1)(x, y, z))(a1\, b2 - a2\, b1)$$

Finally, let's verify that the left and right sides are the same. Both involve similar-looking terms, but the right has products of differences that must be expanded.

```
> expand(Left - Right);
```

$$0$$

## Identities

There are many identities involving "div", "grad", and "curl". They are based on straightforward, if sometimes tedious, calculations that are easy to verify using Maple. Maple can even be useful in discovering such identities.

■ **EXAMPLE 1** Find an identity involving $\nabla \bullet (f\nabla g - g\nabla f)$ where $f$ and $g$ are scalar fields.

**SOLUTION**

```
> S:= diverge(evl(f(x,y,z) * grad(g(x,y,z), xyz)
>           - g(x,y,z)*grad(f(x,y,z), xyz)), xyz);
```

$$S := f(x,y,z)\frac{\partial^2}{\partial x^2}g(x,y,z) - g(x,y,z)\frac{\partial^2}{\partial x^2}f(x,y,z) + f(x,y,z)\frac{\partial^2}{\partial y^2}g(x,y,z)$$
$$- g(x,y,z)\frac{\partial^2}{\partial y^2}f(x,y,z) + f(x,y,z)\frac{\partial^2}{\partial z^2}g(x,y,z) - g(x,y,z)\frac{\partial^2}{\partial z^2}f(x,y,z)$$

You should recognize the terms of the Laplacians of $f$ and $g$ in this expression. To disentangle it, we can collect the terms in $f(x,y,z)$ and $g(x,y,z)$.

```
> collect(S,[f(x,y,z), g(x,y,z)]);
```

$$\left(\left(\frac{\partial^2}{\partial x^2}\, g(x,y,z)\right) + \left(\frac{\partial^2}{\partial y^2}g(x,y,z)\right) + \left(\frac{\partial^2}{\partial z^2}g(x,y,z)\right)\right) f(x,y,z)$$
$$+\left(-\left(\frac{\partial^2}{\partial x^2}f(x,y,z)\right) - \left(\frac{\partial^2}{\partial y^2}f(x,y,z)\right) - \left(\frac{\partial^2}{\partial z^2}f(x,y,z)\right)\right) g(x,y,z)$$

Thus the answer is $\nabla \bullet (f\nabla g - g\nabla f) = f\nabla^2 g - g\nabla^2 f$. ■

## Theorems of Gauss, Green, and Stokes

Among the most important results of vector calculus are Gauss's Theorem (also known as the Divergence Theorem), Green's Theorem, and Stokes's Theorem. Let's try a typical example of Gauss's Theorem, computing both sides of the equation using Maple and checking that they are equal.

■ **EXAMPLE 2** Verify Gauss's Theorem for the vector field $\mathbf{F} = [xy^2, yz^2, zx^2]$ and the region $R$ shown in Figure 9.17, which is bounded by $z = x^2$ and $z = 1 - y^2$.

**SOLUTION** The theorem says

$$\iiint_R \operatorname{div}\mathbf{F}\, dV = \iint_S \mathbf{F} \bullet d\mathbf{S}$$

where $S$ is the boundary of $R$. Note that the surfaces $z = x^2$ and $z = 1-y^2$ intersect on the circle $x^2+y^2 = 1$, so the bounded region between the two surfaces has $x^2 + y^2 \leq 1$. In this region, $x^2 \leq 1 - y^2$. Thus the region $R$ is defined by $x^2 \leq z \leq 1 - y^2$ for $x^2 + y^2 \leq 1$. Here is the left side of the equation.

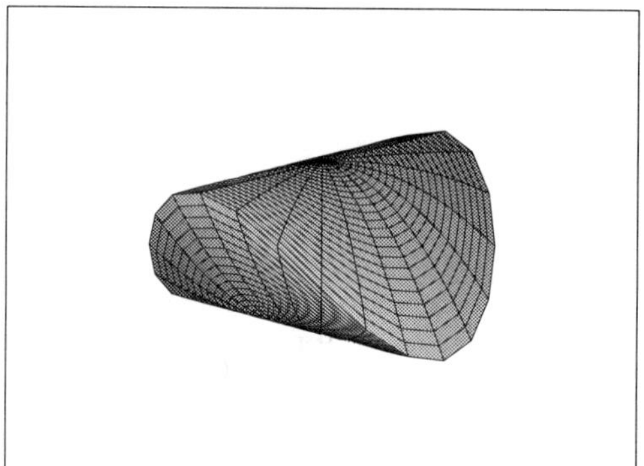

Figure 9.17. Region R

```
> F:= [x*y^2, y*z^2, z*x^2];
> int(int(int(diverge(F, xyz), z = x^2 .. 1-y^2),
>          y = -sqrt(1-x^2) .. sqrt(1-x^2)), x = -1..1);
```

$$\frac{31}{96}\pi$$

For the right side, we need to integrate over the boundary of $R$. This consists of two pieces: the upper boundary with $z = 1 - y^2$ and the lower boundary with $z = x^2$, both for $x^2 + y^2 \leq 1$. We can use $x$ and $y$ as parameters. Note that, using $x$ and $y$ in that order, the normal points upwards, which is the correct orientation for the upper boundary but not for the lower boundary. Therefore we will use a negative sign in front of "`flux`" for the lower boundary.

```
> flux(F, [x = u, y = v, z = 1-v^2], u = -1 .. 1,
>         v = -sqrt(1-u^2) .. sqrt(1-u^2))
> - flux(F, [x = u, y = v, z = u^2], u = -1 .. 1,
>         v = -sqrt(1-u^2) .. sqrt(1-u^2));
```

As expected, the result is the same as for the left side. ■

The main importance of these three theorems is theoretical, but they can also be used in practical computation with Maple. Of course, we don't usually want to calculate the same thing in two different ways. In an application, we may have to perform an integration, and we would like to do it as quickly and accurately as possible. Suppose, for example, we need to calculate a line integral $\int_C \mathbf{F} \bullet d\mathbf{r}$ where $C$ is a closed curve in space. Stokes's Theorem offers the alternative of computing $\iint_S (\text{curl}\,\mathbf{F}) \bullet d\mathbf{S}$ where $S$ is a surface with boundary $C$. There are a number of considerations:

(1) Which integrand is simpler? In some cases, $\mathbf{F} \bullet d\mathbf{r}$ or $(\text{curl}\,\mathbf{F}) \bullet d\mathbf{S})$ is 0, which of course decides the issue. If both are nonzero, the main question of interest is whether the integral can be done in closed form or will require numerical integration.

(2) If both integrals must be done numerically, it is usually much faster to do a line integral rather than a surface integral.

(3) Is it easier to parametrize the curve or the surface? In some cases, one of the integrals can be done in one piece, while the other requires several different parametrizations for different parts. Other things being equal, we would prefer to do it in one piece. Note that if we have a parametrization for the surface, say $R = R(u, v)$ for $a \le u \le b$, $c \le v \le d$, then we can represent the curve in at most four pieces: $R(t, c)$ for $t = a$ to $b$, $R(b, t)$ for $t = c$ to $d$, $R(t, d)$ for $t = b$ to $a$, and $R(a, t)$ for $t = d$ to $c$. It is also possible, given a parametric description of a closed curve, to obtain an oriented parametric surface (but possibly self-intersecting) whose boundary is the curve. So this item is usually not a very important consideration.

The conclusion is that, unless $(\text{curl}\,\mathbf{F}) \bullet d\mathbf{S}$ is much simpler than $\mathbf{F} \bullet d\mathbf{r}$, it is usually not worth the trouble to use Stokes's Theorem to calculate a line integral. The other direction, going from a surface integral to a line integral, is more likely to be useful because of (2).

■ **EXAMPLE 3** Calculate $\iint_S (\text{curl}\,\mathbf{F}) \bullet d\mathbf{S}$ where $\mathbf{F} = [\sin z^3, \sin x^3, \sin y^3]$ and $S$ is the paraboloid $z = x^2 + y^2$ for $x^2 + y^2 \le 1$, with upward-pointing normal.

**SOLUTION** Using the direct method, we might parametrize the paraboloid using cylindrical coordinates.

```
> F:= [sin(z^3), sin(x^3), sin(y^3)];
> S:= [x=r*cos(theta), y=r*sin(theta), z=r^2];
> A:= Flux(curl(F, xyz), S, r=0 .. 1, theta=0 .. 2*Pi);
```

$$A := \int_0^1 \int_0^{2\pi} -6\cos\left(r^3 \sin(\theta)^3\right) r^4 \sin(\theta)^2 \cos(\theta) - 6\cos\left(r^6\right) r^6 \sin(\theta) + 3\cos\left(r^3\cos(\theta)^3\right) r^2 \cos(\theta)^2 \left(\cos(\theta)^2 r + \sin(\theta)^2 r\right)\, d\theta\, dr$$

Maple cannot do this in closed form, and would spend a long time in an unsuccessful attempt to do it. Moreover, since it is a double integral, "`evalf`" is also time-consuming here.

Using Stokes's Theorem, we have only a single integral to do. The boundary of the surface is the circle $x^2 + y^2 = 1$ with $z = 1$.

```
> C:= subs(r=1, S);
> B:= Vlineint(F, C, theta=0 .. 2*Pi);
```

$$\int_0^{2\pi} -\sin(1)\sin(\theta) + \sin\left(\cos(\theta)^3\right)\cos(\theta)\, d\theta$$

This still can't be done in closed form, but "evalf" handles it much faster.

```
> evalf(B);
```

$$2.108561075$$

■

## Vector Potentials

A **vector potential** for a vector field **F** is another vector field **V** such that $\mathbf{F} = \operatorname{curl}\mathbf{V}$. One motivation for this might be the desire to use Stokes's Theorem to reduce a surface integral to a line integral. In order to do this for $\int_S \mathbf{F} \bullet d\mathbf{S}$, we must be able to represent **F** as the curl of something. Actually, there is a much more important motivation: the vector potential turns out to be very useful in electromagnetism (where **F** is the magnetic field). The basic existence result is the following:

> If **F** is a smooth vector field on a domain $D$ such that
>
> (1) every closed surface in $D$ is the boundary of a domain contained in $D$ (i.e., $D$ has no "holes"), and
>
> (2) $\operatorname{div}\mathbf{F} = 0$ everywhere on $D$
>
> then **F** has a vector potential.

A vector field satisfying (2) is said to be **solenoidal** on $D$.

Maple can find vector potentials with the "vecpotent" command in the "linalg" package. This is similar in style to the "potential" command for finding scalar potentials, which we met in Lab 26. Thus `vecpotent(F, [x,y,z], 'V')` is supposed to return the value "true" if **F** has a vector potential or "false" if it does not, and in the "true" case it assigns a vector potential to the variable "V". Note that "V" will be a Maple vector, not a list.

```
> vecpotent([x^2, -x*y, -x*z], xyz, 'V');
```

true

```
> evl(V);
```

$$[-xyz, -x^2z, 0]$$

Vector potentials are highly nonunique. The one returned by "vecpotent" always has its third component 0 (this is known as a **gauge condition**).

The "true" or "false" response depends only on whether or not $\operatorname{div}\mathbf{F} = 0$. There is no attention to domains. Unfortunately, in Release 4 "vecpotent" has the same type of bug as "potential": if "`diverge(F, [x,y,z])`" produces something that is not explicitly 0, even if it can be simplified to 0, "vecpotent" will answer "false"[(25)].

```
> F:= [-x,y*cos(x)^2, z*sin(x)^2];
> diverge(F, xyz);
```

$$-1 + \cos(x)^2 + \sin(x)^2$$

```
> vecpotent(F, xyz, 'V');
```

false

Well, let's treat this as an opportunity rather than a misfortune, and find a vector potential for this one ourselves. Just for variety, we'll choose a different gauge condition: $V_1 = 0$. Here's a vector field with first component 0 and the others unspecified functions of $x$, $y$, and $z$.

```
> V:= [0, V2(x,y,z), V3(x,y,z)];
```

Let's see its curl.

---

(25) The bug is corrected in Release 5.

```
> CV:= evl(curl(V,xyz));
```

$$\left[\left(\frac{\partial}{\partial y}\mathrm{V3}(x,y,z)\right)-\left(\frac{\partial}{\partial z}\mathrm{V2}(x,y,z)\right),\ -\left(\frac{\partial}{\partial x}\mathrm{V3}(x,y,z)\right),\ \left(\frac{\partial}{\partial x}\mathrm{V2}(x,y,z)\right)\right]$$

According to this, we should get $V_2$ (up to a function of $y$ and $z$) by integrating $F_3$ with respect to $x$, and $V_3$ (up to a function of $y$ and $z$) by integrating $-F_2$.

```
> V2:= unapply(int(F[3], x) + g(y,z), (x,y,z));
```

$$V2 := (x,y,z) \to \left(-\frac{1}{2}\cos(x)\sin(x)+\frac{1}{2}x\right)z+g(y,z)$$

```
> V3:= unapply(-int(F[2], x) + h(y,z), (x,y,z));
```

$$V3 := (x,y,z) \to -\left(\frac{1}{2}\cos(x)\sin(x)+\frac{1}{2}x\right)y+h(y,z)$$

Let's see what this gives us for curl **V**.

```
> CV;
```

$$\left[-x+\left(\frac{\partial}{\partial y}h(y,z)\right)-\left(\frac{\partial}{\partial z}g(y,z)\right),\ \left(-\frac{1}{2}\sin(x)^2+\frac{1}{2}\cos(x)^2+\frac{1}{2}\right)y,\right.$$
$$\left.\left(\frac{1}{2}\sin(x)^2-\frac{1}{2}\cos(x)^2+\frac{1}{2}\right)z\right]$$

The second and third components are correct (after a bit of trigonometric simplification). For the first component, we just need $\dfrac{\partial h}{\partial y}-\dfrac{\partial g}{\partial z}=0$. One possibility that will work is $g=0$ and $h=0$. Here, then, is one possibility for our vector potential.

```
> eval(subs(g = 0, h = 0, V));
```

$$\left[0,\ \left(-\frac{1}{2}\cos(x)\sin(x)+\frac{1}{2}x\right)z,\ -\left(\frac{1}{2}\cos(x)\sin(x)+\frac{1}{2}x\right)y\right]$$

We assigned values to "V2" and "V3", so remember to "unassign" them before trying other examples.

```
> V2:= 'V2'; V3:= 'V3';
```

■

## EXERCISES

**1.** Calculate the following with Maple, where $\mathbf{F}=[x^2y,\ -y^2z, z^2x]$ and $f=xyz$.

**(a)** div $(f\mathbf{F})-\mathbf{F}\bullet \text{grad}\, f$ **(b)** curl (grad $f\times\mathbf{F}$)

**(c)** $\nabla^2(\mathbf{F}\bullet\nabla^2\mathbf{F})$ **(d)** $(\mathbf{F}\bullet\nabla)\mathbf{F}$

*Hint:* in (d), define a command that takes a scalar field $g$ to $\mathbf{F}\bullet\nabla g$, and use "map".

**2.** If **F** represents a velocity field of a fluid, div **F** can be thought of as the rate of expansion of the fluid per unit time. Thus if $(\text{div}\,\mathbf{F})(P)=3$ and we take a small region $R$ around $P$ of volume $V$, there will be a net outflow from $R$ of approximately $3V$ cubic length-units per time-unit. To justify this statement, consider a sphere $S$ of radius $r$ centred at an arbitrary point $P$ and a vector field **F** whose components are arbitrary functions of $x$, $y$ and $z$. Show that

$$\lim_{r\to 0+}\frac{3}{4\pi r^3}\iint_S \mathbf{F}\bullet d\mathbf{S}=(\text{div}\,\mathbf{F})(P)$$

**3.** Verify the following identities, where $f$ and $g$ are scalar fields and **F** is a vector field.

**(a)** curl $(f\,\text{grad}\,g+g\,\text{grad}\,f)=0$ **(b)** div $(\mathbf{F}\times\text{grad}\,f)=(\text{grad}\,f)\bullet(\text{curl}\,\mathbf{F})$

**4.** Illustrate Green's Theorem by calculating $\displaystyle\iint_R\left(\frac{\partial F_2}{\partial x}-\frac{\partial F_1}{\partial y}\right)dA$ and $\displaystyle\int_C F_1\,dx+F_2\,dy$ where $R$ is the region of the plane bounded by $y=x^2$ and $y=\sqrt{x}$, $C$ its positively oriented boundary, and $\mathbf{F}=[x+y,\ xy]$.

**5.** To illustrate Stokes's Theorem, let $S$ be the part of the surface $z = 1 - y^2$ for $x^2 + y^2 \leq 1$, with normal pointing upwards, and let $\mathbf{F} = [x, xy, xyz]$.

**(a)** Calculate $\iint_S (\text{curl}\,\mathbf{F}) \bullet d\mathbf{S}$ and $\int_C \mathbf{F} \bullet d\mathbf{r}$ where $C$ is the boundary of $S$, and note that they are the same.

**(b)** Find another surface with the same boundary, and calculate the flux of curl $\mathbf{F}$ through that surface.

**6.** Let $C$ be the closed curve consisting of the straight line from the origin to $[1, 0, 1]$ followed by the curve of intersection of the cone $z = \sqrt{x^2 + y^2}$ with the sphere $(x - 1)^2 + y^2 + z^2 = 1$ for $y \geq 0$. Let $\mathbf{F} = [\sin z^2, \sin y^2, \sin x^2]$.

**(a)** Calculate $\int_C \mathbf{F} \bullet d\mathbf{r}$.

**(b)** What integral would have to be done in order to use Stokes's Theorem for this calculation?

**(c)** Would it be better to use the line integral of (a) or the result of (b)?

**7. (a)** Find a vector potential $\mathbf{V}$ for the vector field $\mathbf{F} = [z^2, x^2, y^2]$, subject to the gauge condition $V_2 = 0$.

**(b)** Compare $\mathbf{V}$ to the vector potential $\mathbf{W}$ produced by "`vecpotent`". Find a scalar field $f$ such that $\mathbf{V} - \mathbf{W} = \text{grad}\, f$. Why must such an $f$ exist?

**(c)** Suppose another vector potential for $\mathbf{F}$, $\mathbf{U} = \mathbf{W} + \text{grad}\, g$, satisfies the "Coulomb gauge" condition div $\mathbf{U} = 0$. What equation must $g$ satisfy?

**(d)** Find a solution to the equation of (c) of the form $g(x, y, z) = P(x)z$.

**8.** Find $F_3(x, y, z)$ so that the vector field $\mathbf{F} = [xz, yz, F_3(x, y, z)]$ is both conservative and solenoidal.

**9.** This exercise shows the need for a condition on the domain in the existence theorem for vector potentials. Consider $\mathbf{F} = [x, y, z]/(x^2 + y^2 + z^2)^{3/2}$.

**(a)** Show that div $\mathbf{F} = 0$ everywhere that $\mathbf{F}$ is defined. Where is that?

**(b)** Find a vector potential $\mathbf{V}$ for $\mathbf{F}$ subject to the gauge condition $V_3 = 0$. Where is $\mathbf{V}$ defined?

**(c)** Show that the natural domain of $\mathbf{F}$ does not satisfy the requirements of the theorem.

**10.** Consider the vector field $\mathbf{F} = [ax + by, cx + dy]$ in the plane, where $a$, $b$, $c$, and $d$ are arbitrary constants.

**(a)** Choose values of the constants for which $\mathbf{F}$ is conservative but not solenoidal, and so that $\mathbf{F} \bullet \mathbf{r}$ is always positive. Plot $\mathbf{F}$ together with a circle centred at the origin, and note that the component of $\mathbf{F}$ in the direction of the outward-pointing normal to the circle is always positive.

**(b)** Choose values of the constants for which $\mathbf{F}$ is solenoidal but not conservative, and so that $\mathbf{F} \bullet [-y, x]$ is always positive. Plot $\mathbf{F}$ together with a circle centred at the origin, and note that the component of $\mathbf{F}$ in the direction of the counterclockwise tangent to the circle is always positive.

### CHAPTER 10

# Differential Equations

## LAB 30 ORDINARY DIFFERENTIAL EQUATIONS

**Goals:** *To look at some of the algebraic, graphical, and numerical aspects of differential equations using Maple.*

The study of differential equations is a major field of mathematics, and very important for applications. In one lab we can only hope for a small glimpse of this subject and of Maple's extensive facilities for working in it. There are three basic aspects to Maple's treatment of the subject: algebraic (finding formulas for solutions), graphical (graphing direction fields and solutions), and numerical (finding numerical values for solutions where there may not be a formula available).

An **ordinary differential equation** is an equation involving an unknown function (the **dependent variable**) and its derivatives with respect to one variable (the **independent variable**). For example,

$$\frac{d^2y}{dx^2} - \frac{dy}{dx} - 2y = x$$

is an ordinary differential equation with dependent variable $y$ and independent variable $x$. We can represent it to Maple in either "`diff`" or "`D`" form.

```
> eq1:= diff(y(x),x$2) - diff(y(x),x) - 2 * y(x) = x;
> eq2:= (D@@2)(y)(x) - D(y)(x) - 2 * y(x) = x;
```

In both forms, you need the "`(x)`". A differential equation for Maple is an equation between two expressions, not two functions.

## Solving Differential Equations

The main command for solving differential equations is "`dsolve`". In the simplest form of the command, the inputs are the differential equation and the name of the unknown function (given in the form "`y(x)`" to show what is the independent variable as well as the dependent variable).

```
> S:= dsolve(eq1, y(x));
```

$$y(x) = \frac{1}{4} - \frac{1}{2}x + \_C1\,e^{(-x)} + \_C2\,e^{(2x)}$$

Note the arbitrary constants "$\_C1$" and "$\_C2$" in this solution. In general, the solution of an $n$th **order** differential equation (one where the highest derivative appearing is the $n$th) has $n$ arbitrary constants.

Let's check that this solution really satisfies the differential equation. We define "`Y`" as a function whose value at $x$ is the right side of the answer[(1)].

```
> Y:= unapply(rhs(S), x);
```

Now let's take another look at the equations, with this for $y$.

```
> eval(subs(y=Y,eq1));
```

$$x = x$$

(1) You could assign this to "`y`", eliminating the need for "`eval`" and "`subs`" in the next command, but you must then remember to "unassign" `y` before writing another equation with this variable.

(and the same for "eq2"). This confirms that the solution Maple found was valid.

Since the general solution contains arbitrary constants, in order to specify a particular solution we need some extra conditions. Usually these extra conditions are in the form of **initial conditions**: for an $n$-th order equation the initial conditions specify the values of $y$ and its first $n-1$ derivatives at one point $x = x_0$. Thus a typical set of initial conditions for our equation might be

$$y(1) = 2, \quad y'(1) = 3$$

Since we already have the general solution $Y(x)$, we could simply look at these initial conditions with $y = Y$ as equations in $\_C1$ and $\_C2$, and solve these.

```
> solve({Y(1)=2, D(Y)(1)=3});
```

$$\left\{\_C1 = \frac{1}{3}\mathrm{e},\ \_C2 = \frac{23}{12}\frac{1}{(\mathrm{e})^2}\right\}$$

However, if we are only interested in one set of initial conditions and not in the general solution, we can usually obtain the solution for those initial conditions in one step, by making the first input to "dsolve" a set consisting of the differential equation together with the initial conditions (2).

```
> dsolve({eq1, y(1)=2, D(y)(1)=3}, y(x));
```

$$y(x) = -\frac{1}{2}x + \frac{1}{4} + \frac{23\,\mathrm{e}^{(2x)}}{12\,(\mathrm{e})^2} + \frac{1}{3}\mathrm{e}\mathrm{e}^{(-x)}$$

Sometimes "dsolve" presents its solution in implicit form. This happens more often in Release 4 than in Release 5. An example is the differential equation $\dfrac{dy}{dx} = 2xy^2$.

```
> dsolve(D(y)(x) = 2*x*y(x)^2, y(x));
```

$$\frac{1}{y(x)} = -x^2 + \_C1 \qquad \text{(Release 4)}$$

$$y(x) = -\frac{1}{x^2 - \_C1} \qquad \text{(Release 5)}$$

The Release 4 result could easily be solved to obtain $y(x)$ as an explicit function of $x$.

```
> solve(", y(x));
```

$$-\frac{1}{x^2 - \_C1}$$

We could also obtain this in one step, using the "explicit" option in "dsolve".

```
> dsolve(D(y)(x) = 2*x*y(x)^2, y(x), explicit);
```

$$y(x) = -\frac{1}{x^2 - \_C1}$$

With this option, Maple will present the solution in explicit form if possible. Sometimes, however, it is not possible, as with the differential equation $(x\cos y + \sin x)\dfrac{dy}{dx} + \sin y + y\cos x = 0$.

```
> dsolve((x*cos(y(x))+sin(x))*D(y)(x) + sin(y(x))+y(x)*cos(x)
>          = 0, y(x));
```

$$\_C1 + \sin(y(x))x + \sin(x)y(x) = 0$$

There's no hope of solving that one for $y(x)$.

```
> dsolve((x*cos(y(x))+sin(x))*D(y)(x) + sin(y(x))+y(x)*cos(x)
>          = 0, y(x), explicit);
```

$$y(x) = \mathrm{RootOf}(\_Z\sin(x) + \_C1 + x\sin(\_Z))$$

---

(2) This might not return an implicit solution, nor a particular solution that is not a special case of the general solution (see Exercise 2).

This "RootOf" is just another way of stating the implicit solution.. Sometimes Maple is unable to come up with a solution, perhaps because the general solution can't be expressed in "closed form". That's probably the case for the equation $(x+y)\dfrac{dy}{dx} = (x+1)^2 - y^2 - 1$ (although there is one particular solution in closed form).

```
> dsolve((x+y(x))*D(y)(x) = (x+1)^2 - y(x)^2 - 1, y(x));
```

## Direction Fields

The **direction field** provides the best intuitive idea of what is really going on in a first-order differential equation. It can also tell us a lot about the qualitative behaviour of the solutions. Consider an equation $y' = f(x,y)$. What this says is that at each point $(x,y)$, there is a number $f(x,y)$ that is the slope of our curve $y = y(x)$ if it happens to pass through that point. You can imagine it as a little road sign at each point saying "go this way". To represent it visually, we can have Maple draw arrows representing the directions for certain sample points in the plane. The "DEplot" command in the "DEtools" package will do this for us.

In its simplest version, "DEplot" takes four inputs. The first is a first-order differential equation, in its usual form for Maple. The second, just as in "dsolve", gives the name of the unknown function and the independent variable. The third and fourth are ranges for the independent and dependent variables. After these, other options may be specified: besides the usual options for "plot", useful options are "dirgrid" and "arrows" (which we met in connection with the "fieldplot" command in Lab 26).

**EXAMPLE 1** Study the qualitative behaviour of the solutions of the differential equation $\dfrac{dy}{dx} = x^2 - y^2$.

**SOLUTION** We begin by defining the differential equation and plotting its direction field using "DEplot". Since we'll want to do more with this plot, we'll save it in a variable.

```
> with(DEtools, DEplot);
> de:= D(y)(x) = x^2 - y(x)^2;
> plot1:= DEplot(de, y(x), x=-4..4, y=-3..3, arrows=THIN,
>                scaling=constrained):
> plot1;
```

Near the left and right of the screen, the arrows go upwards nearly vertically. At the top and bottom of the middle section, they go nearly vertically downwards. In an "X" shaped region centred at the origin, they are nearly horizontal. This is what you should expect, as it reflects the behaviour of $x^2 - y^2$: positive when $|y| < |x|$, negative when $|y| > |x|$, and zero when $y = \pm x$.

Imagine starting at the point $(-2, 0)$ and trying to follow the arrows, as a solution of the differential equation must do. Since the arrows go upwards, you move up and to the right. According to the slopes of the arrows in the picture, your path is at first quite steep, but then becomes gentler until at some point it is horizontal. Judging by eye, this point could be somewhere near $(-1.3, 1.3)$. After that, the slopes appear to be negative and fairly gentle for a while. You cross the positive $y$ axis, maybe near $y = .7$, with a gentle negative slope. A little while later you reach another point where the arrow is horizontal, maybe near $(.5, .5)$, and come into a region where the slopes are positive. You continue upward and leave the plotting window at the top right.

Another way to study the direction field is to plot some **isoclines**, which are the curves $f(x,y)$ =constant. We can use "contourplot" for this. Alternatively, we could simply use "plot" with formulas of the form $y = g(x)$ for the isoclines, if such formulas are available. The most important value of $f(x,y)$ to plot is 0, which tells us where the arrows are horizontal. I'll also include $f(x,y) = 1$ and $f(x,y) = -1$, which in this case turn out to be very significant. The "contourplot" command may not produce a very good picture for the case $f(x,y) = 0$ (which is the two crossed lines $y = \pm x$)[3], but we can improve matters by using the "grid" option: if both "grid" numbers are odd, the origin is one of the grid points and the picture should look better. I'll combine the plot of the isoclines with the direction field plot we saved before, using "display". Recall that "contourplot" and "display" are in the "plots" package.

```
> with(plots):
> plot2:= contourplot(rhs(de), x = -4 .. 4, y = -3 .. 3,
>                   contours = [-1,0,1], grid = [25,25]):
> display({plot1,plot2}, scaling=constrained, axes=BOXED);
```
(Figure 10.1)

The isoclines divide up the plane into regions, in each of which the slopes of the arrows are between certain bounds. In our picture there are eight regions, which I have labeled as A to H in Figure 10.2. Thus in region B the slopes are always between $-1$ and 0.

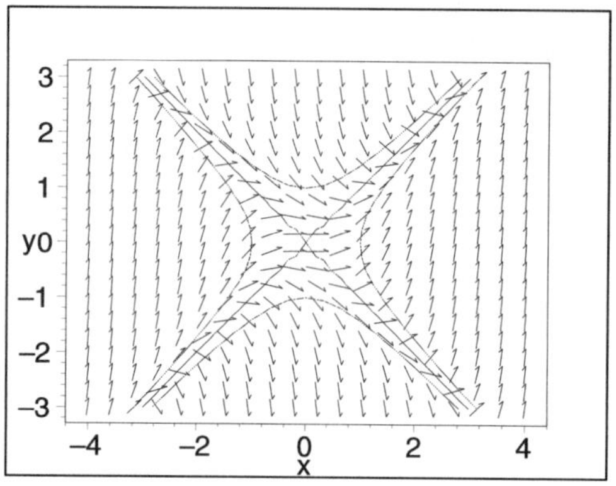


Figure 10.1. Arrows and isoclines

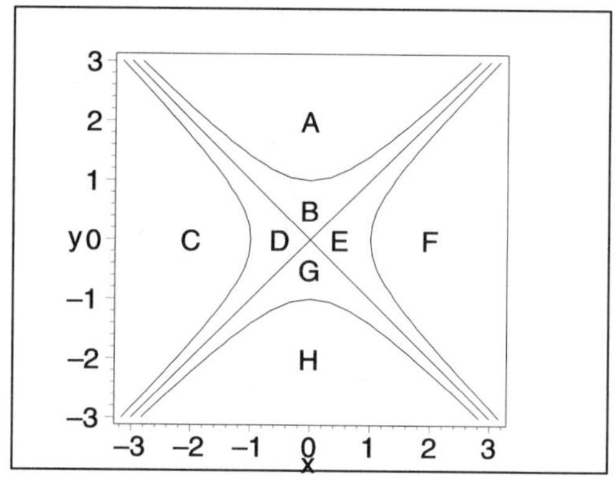


Figure 10.2. Isoclines and regions

The basic principles of interpreting the direction field and isoclines are as follows:

(1) Through every point where $f(x, y)$ is continuously differentiable there is a unique solution curve, whose slope always "follows the arrows".

(2) Solution curves can't stop, except by either going off to infinity or encountering a point where $f(x, y)$ is not continuous.

(3) Solution curves can't meet or cross, except at a point where $f(x, y)$ is not continuously differentiable.

(4) Consider a curve $y = g(x)$ for $a \leq x \leq b$. If $g'(x) > f(x, g(x))$ for $a \leq x \leq b$, then solution curves can pass (left to right) from the region above the curve to the region below, but not vice versa. If $g'(x) < f(x, g(x))$, then this is reversed.

To illustrate principle (4), consider the part of the hyperbola $x^2 - y^2 = -1$ that forms the boundary between regions A and B. This being an isocline $f(x, y) = -1$, all the arrows on it have slope $-1$. However, the slope of this curve is always greater than $-1$, so the arrows at points on the curve point into region B. The consequence is that a solution curve that starts out in region B can never enter region A.

---

(3) This is typical for "contourplot" or "implicitplot" when dealing with self-intersecting curves.

Let's look again at the solution for initial condition $(-2, 0)$. This initial point is in region C, where slopes are greater than 1. Moving to the right, as long as it is in region C the slope of the solution curve must be greater than 1. Obviously at some point it must hit the boundary between C and D, and continue into region D where the slope is between 0 and 1. After crossing D, it must hit the boundary between D and B[(4)] with slope 0. In region B, the slope is between 0 and $-1$. As mentioned above, the solution curve can't enter A, and of course it can't re-enter D[(5)]. Clearly it must then pass into E, again with slope 0 at the boundary. In E the slope is between 0 and 1. There is a somewhat subtle argument to show that the solution must enter F (rather than just continuing within the part of E that is between B and F)[(6)]. In F the solution has a slope greater than 1, and continues within F forever[(7)].

We can also follow the solution curve backwards (right to left) from the point $(-2, 0)$. The boundary between C and D for $y < 0$ has slope greater than 1, so solution curves can't go (forwards) from C to D. Therefore the curve stays within region C for all $x < -2$.

To confirm our analysis, it is helpful to plot some actual[(8)] solution curves. Together with the arrows[(9)], the "`DEplot`" command will plot solution curves for one or more initial conditions. Between the $x$ range and the $y$ range we put a set of initial conditions. Each initial condition is itself a list, with just one entry (which is an equation): `[y(x0)=y0]`. Here is the solution curve through our initial point $(-2, 0)$, together with the direction field and isoclines. Note that it agrees with our qualitative results.

```
> plot3:= DEplot(de, y(x), x=-4 .. 4, {[y(-2)=0]}, y=-3 .. 3):
> display({plot2, plot3}, scaling=constrained, axes=BOXED);
```

Sometimes the "`DEplot`" command may run into trouble when solutions "blow up", going off to infinity in the $y$ direction at a finite value of $x$. It still tries to calculate the solution even after it leaves the plotting region (which is reasonable, since in some cases it might return to the region later). But when $y$ values get extremely large (to the point where the computer can't represent them as floating-point numbers in the usual way) you may encounter errors: "`Over or underflow in conversion to double`". That in itself is not so bad, but in some cases strange lines may appear on the plots. You can cure the problem by using the option "`obsrange=true`", which causes "`DEplot`" to stop calculating the solution as soon as it leaves the plotting region. Another way to handle such problems is to modify the differential equation so that such enormous $y$ values won't arise, but without affecting it for more reasonable $y$ values. Thus in Example 1 we could replace the differential equation "`de`" by

```
> de1:= D(y)(x) = (x^2 - y(x)^2) * Heaviside(10^3 - abs(y(x)));
```

For $|y| < 1000$ this is the same as the original equation, but $y'$ becomes 0 for $|y| > 1000$. Thus if the solution ever leaves the interval $-1000 \le y \le 1000$, it will become "stuck". This method has the advantage of allowing solutions to go slightly outside the plotting region and return.

## Direction Fields for Systems

A differential equation or system of differential equations in which the independent variable does not appear explicitly is said to be **autonomous**. The direction field approach can be applied to autonomous

---

(4) Of course not G, since after leaving the initial point $y$ has been positive.

(5) The boundary between D and B has slope $-1$ and is an isocline $f(x, y) = 0$, so the second part of principle (4) says curves can go from D to B but not vice versa.

(6) Consider a point $(x_1, y_1)$ on the solution curve within E, and draw a straight line L through $(x_1, y_1)$ with a slope of 1. The solution, having a slope between 0 and 1 in E, must stay below L while it is in E and to the right of $x_1$. But the line L eventually hits region F, so by staying below L and above $y = y_1$ the solution is forced to hit F as well.

(7) It can't go back into E by principle (4), since the boundary between E and F for $y > 0$ has slope greater than 1.

(8) Or rather, approximate.

(9) If you don't want the arrows, you can use the option "`arrows=NONE`".

systems of two differential equations, i.e., systems of the form

$$\frac{dx}{dt} = f(x, y), \qquad \frac{dy}{dt} = g(x, y)$$

Here we are interested in the trajectories of the system, which are the paths of the solutions $(x(t), y(t))$ considered as parametric curves in the plane. These are the same as the field lines of the vector field $[f(x,y), g(x,y)]$, which we studied in Lab 26 using the "`DEplot`" command from the "`DEtools`" package(10).

**■ EXAMPLE 2** Study the qualitative behaviour of the trajectories for the second-order differential equation $\frac{d^2x}{dt^2} = x - x^2\frac{dx}{dt}$.

**SOLUTION** A second-order differential equation can be considered as a system of two first-order differential equations by introducing a new dependent variable which is the derivative of the original dependent variable. Thus our equation is equivalent to the system

$$\frac{dx}{dt} = v, \qquad \frac{dv}{dt} = x - x^2 v$$

At first we'll just plot arrows, and not trajectories. The "`DEplot`" command accepts a list of equations as its first input. On the right sides of these equations (i.e., the terms that are not derivatives) you can use "`x`" and "`v`" instead of "`x(t)`" and "`v(t)`"(11). The second input is a list of the two dependent variables, which must be given in the form "`[x(t), v(t)]`". The third input is an interval for the independent variable $t$ (which is a bit strange, since this doesn't really play any role when not plotting trajectories). Then come the intervals for the dependent variables, and any other options.

```
> de:= [D(x)(t) = v, D(v)(t) = x - x^2*v];
> DEplot(de, [x(t),v(t)], t = 0 .. 1, x = -3 .. 3, v = -2 .. 2);
```

The interpretation of the direction field is similar to the case of a single differential equation, with a few exceptions. First, while previously they always went from left to right, here the arrows can point in any direction. Second, trajectories can appear to stop (and perhaps meet other trajectories) at a critical point (a point where the vector field has the value $[0, 0]$). What actually happens in such a case is that the solution approaches the critical point in the limit $t \to \infty$ or $t \to -\infty$, but never actually reaches there at any finite $t$.

For our example there seem to be four main classes of trajectories: they can enter the picture either from the top left or the bottom right, and in either case can leave either to the left or the right. Examples of each of these should be the trajectories through the points $(1, 0)$, $(-1, 0)$, $(0, 1)$, and $(0, -1)$. As we saw in Lab 26, the "`DEplot`" command will draw the trajectories through specified initial points. The initial conditions can be given as a set of lists, each of the form $[x(t_0) = x_0,\ y(t_0) = y_0]$. I needed to experiment a little to get an interval for $t$ so that the trajectories extended to the edges of the plotting window. In order to avoid giving the trajectory a "segmented" appearance (and also to make it more accurate), we should decrease the step size from its default value (which is $1/20$ the length of the $t$ interval).

```
> plot1:= DEplot(de, [x(t), v(t)], t=-1 .. 5,
>        {[x(0)=1, v(0)=0], [x(0)=-1, v(0)=0],
>         [x(0)=0, v(0)=1], [x(0)=0, v(0)=-1]},
>        x = -3 .. 3, v = -2 .. 2, stepsize = .1):
> plot1;
```

Two trajectories may appear to meet in this picture near $(-2.1, -.5)$ and near $(2.1, .5)$, but that is impossible. Actually, they come very close to each other but can never touch.

---

(10) I'm assuming "`plots`" and "`DEtools`" are still loaded into Maple, as they were used in the last section. If not, load them now.

(11) This is an exception to the rule that the dependent variables are always written as functions of the independent variables.

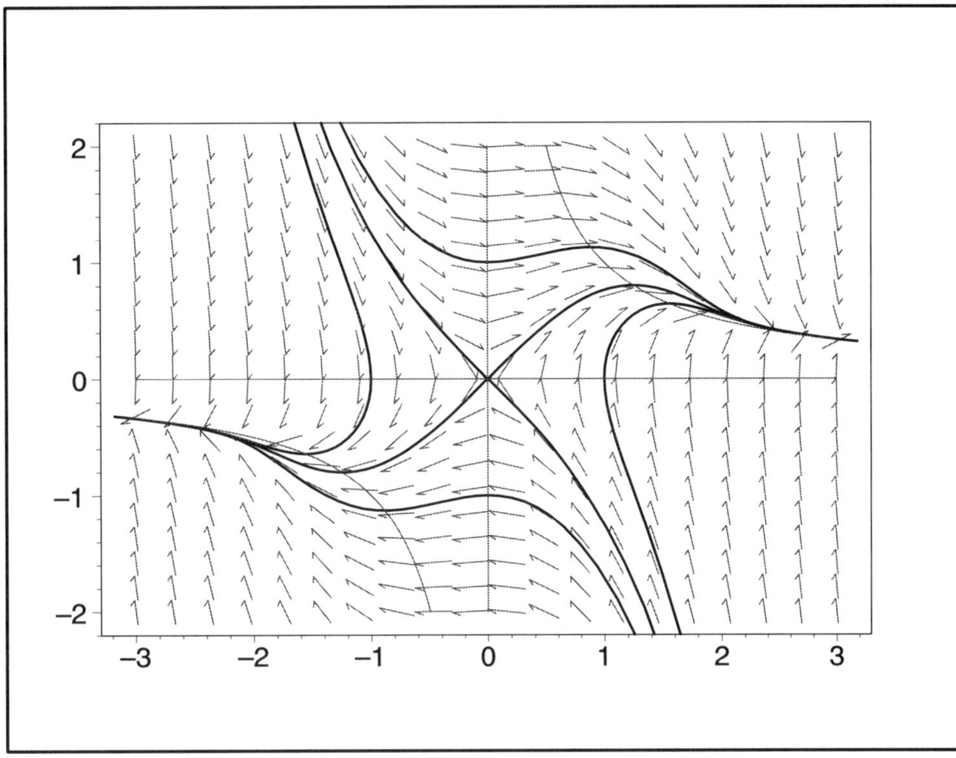


Figure 10.3. Phase portrait

Isoclines are also useful in studying these systems. An isocline in this case is a curve on which the ratio of the two components of $F$ is constant. Particularly important are the isoclines $F_1 = 0$ (where our trajectories have horizontal tangents) and $F_2 = 0$ (where they have vertical tangents). In our case $F_1 = 0$ on the $v$ axis, while $F_2 = 0$ on the $x$ axis and the hyperbola $v = 1/x$. To get them all in one command, I'll use parametric curves. I'll make them blue so (on a colour monitor) we won't confuse them with the trajectories.

```
> plot2:= plot({[0, v, v=-2 .. 2], [x, 0, x=-3 .. 3],
>         [x, 1/x, x=-3 .. -1/2], [x, 1/x, x=1/2 .. 3]},
>         colour=blue):
> display({plot1, plot2}, axes=BOXED);
```

What can happen to a trajectory that starts out in the second quadrant of the $xv$ plane ($x < 0, v > 0$)? The arrows there are pointing down and to the right. It may leave the quadrant across the $v$ axis (with horizontal tangent) or the $x$ axis (with vertical tangent). If it crosses the $v$ axis, it first enters a region where the arrows point up and to the right. It is easy to see that such a trajectory must cross the hyperbola $v = 1/x$ in the first quadrant. Once across the hyperbola, it is in a region where the arrows point down and to the right. However, the trajectory can't cross the hyperbola again, so it must continue moving down and to the right, (slightly) above the hyperbola, with $x \to \infty$ as $t \to \infty$. Similarly, a trajectory that crosses from the second quadrant to the third will end up below the hyperbola, moving up and to the left, with $x \to -\infty$ as $t \to \infty$.

Whenever two neighbouring classes of trajectories have different behaviours, the boundary between them is a trajectory (called a **separatrix**) that may have different behaviour than either of them. In this case, between the trajectories that go from the second quadrant to the first and those that go to the third there is a separatrix that approaches the critical point $(0, 0)$ as $t \to \infty$. There are three other separatrices, all approaching $(0, 0)$ as $t \to \infty$ or $t \to -\infty$. We can approximate the separatrices by taking initial points close to the origin. It's necessary to take a larger $t$ interval, or the trajectories will stop in the middle of the plotting window (the solutions spend a lot of time near the critical point).

Since we've already seen the arrows in the last plot, I'll suppress them in this one using the option "arrows=NONE".

```
> plot3:= DEplot(de, [x(t), v(t)], t = -6 .. 9,
>      {[x(0)=.01, v(0)=0], [x(0)=-.01, v(0)=0]},
>      x = -3 .. 3, v = -2 .. 2,  stepsize = .1, arrows=NONE):
```

Here, then, is the complete "phase portrait" of our system.

```
> display({plot1, plot2, plot3}, scaling = constrained,
>          axes = BOXED); (Figure 10.3)
```

■

## Numerical Methods

The "solution curves" plotted by "DEplot" are actually numerical approximations to the solution. Maple can find such an approximate solution even when "dsolve" can't find an exact solution. Assuming the approximation is sufficiently accurate, this is quite adequate for many purposes.

The simplest numerical method for differential equations is the **Euler method**. For an equation $y' = f(x, y)$ with initial condition $(x_0, y_0)$, it consists of the iteration

$$x_{n+1} = x_n + h, \qquad y_{n+1} = y_n + hf(x_n, y_n)$$

where $h$ is the **step size**. Then $y_n$ is taken as the approximate value of $y$ for $x = x_n$. This is intuitively appealing because it corresponds closely to the basic idea of the direction field: the solution takes "infinitesimal steps" in the direction of the arrows, the Euler approximation takes finite steps in the direction of the arrows.

The "dsolve" command produces numerical approximations with the option "numeric". Use the "method=..." option to specify which numerical method to use: the Euler method is "method=classical[foreuler]". Use the "stepsize=..." option to specify the step size. The result of the command is a function, which can be evaluated at any value of $x$ to produce the approximation for $y(x)$. This actually returns a list of the form $[x = \ldots,\ y(x) = \ldots]$(12).

■ **EXAMPLE 3** Approximate the value at $x = 1$ of the solution of the initial value problem $y' = y^2/(1 + x^2)$, $y(0) = 1$, using the Euler method with various step sizes, and study the accuracy of the results.

**SOLUTION** The initial value problem consists of the differential equation and the initial value. Since we'll use this same initial value problem several times, it's convenient to save it in a variable.

```
> ivp:= { D(y)(x) = y(x)^2/(1+x^2), y(0) = 1 };
```

Now to solve the problem using the Euler method with step size, say, $1/10$:

```
> S:= dsolve(ivp, y(x), numeric, method=classical[foreuler],
>            stepsize=0.1);
```

$$S := \textbf{proc}(x\_classical)\ \ldots\ \textbf{end}$$

Then to evaluate this at $x = 1$:

```
> S(1);
```

$$[x = 1, y(x) = 3.528515847536992]$$

We can use "subs" to extract the value of $y$.

```
> subs(S(1), y(x));
```

$$3.528515847536992$$

(12) In the case of a system rather than a single equation, the list includes values of all the dependent variables; in the case of a higher-order equation or system, it also includes derivatives of the dependent variable(s).

For each different step size we use, we need to call `dsolve` again. The step size $h$ should be chosen so that an integer number of steps of size $h$ takes us from the initial $x = 0$ to the final $x = 1$. Thus $h$ should be the reciprocal of an integer (which would be the number of steps). I'll use a loop to get values for $h = 1/10, 1/20, 1/40, 1/80$ and $1/160$, and call the results $V_{10}, \ldots V_{160}$.

```
> for nn from 0 to 4 do
>     S:= dsolve(ivp, y(x), numeric, method=classical[foreuler],
>               stepsize=1/(10*2^nn)):
>     V[10*2^nn]:= subs(S(1), y(x)):
> od:
> V[10], V[20], V[40], V[80], V[160];
```

$$3.528515847536992, 3.947517155376690, 4.248394789544967,$$
$$4.436333382753202, 4.542944559146185$$

Let's plot $V_h$ as a function of $h$.

```
> plot([seq([1/(10*2^n), V[10*2^n]], n = 0..4)], style=point);
```

It appears that the points lie on a smooth curve, which is fairly close to a straight line.

To know how accurate the results are, we need to know the correct answer. Fortunately, Maple can find a closed-form solution for this differential equation. Of course, most of the time when we need a numerical approximation it is because there is no closed-form solution.

```
> sol:= dsolve({D(y)(x)=f(x,y(x)), y(0)=1},y(x));
```

$$sol := y(x) = -\frac{1}{\arctan(x) - 1}$$

```
> exact:= eval(subs(x=1, rhs(sol)));
```

$$exact := -\frac{1}{\frac{1}{4}\pi - 1}$$

```
> evalf(exact);
```

$$4.659792369$$

Here, then, are the errors in the Euler-method approximations.

```
> seq(evalf(exact - V[10*2^n]), n = 0 .. 4);
```

$$1.131276521, .712275214, .411397579, .223458986, .116847810,$$

Going to a smaller step size, it seems, does reduce the error. Moreover, dividing the step size by 2 divides the error by about 2. This is in accord with the theoretical result that the cumulative error in the Euler method is approximately proportional to the step size. ■

The Euler method is generally not suitable if a very accurate result is desired: because the error is approximately proportional to the step size, a very large number of steps would be needed. Fortunately, better methods are available. The **Improved Euler method** or **Heun method**(13) consists of the iteration

$$x_{n+1} = x_n + h, \quad y_{n+1} = y_n + \frac{h}{2}\left(f(x_n, y_n) + f\left(x_n + h, y_n + hf(x_n, y_n)\right)\right)$$

Our third numerical method is the **fourth-order Runge-Kutta method**, which is the most popular of a whole family of Runge-Kutta methods(14). It is defined by a rather complicated iteration:

---

(13) Euler was one of the greatest mathematicians of all time, so don't think that the relatively obscure Heun was an "Improved Euler".

(14) The Improved Euler method is actually the second-order Runge-Kutta method.

$$x_{n+1} = x_n + h, \quad y_{n+1} = y_n + \frac{h}{6}\left(m_1 + 2m_2 + 2m_3 + m_4\right)$$

where

$$m_1 = f(x_n, y_n), \qquad m_2 = f\left(x_n + \frac{h}{2}, y_n + \frac{h}{2}m_1\right)$$
$$m_3 = f\left(x_n + \frac{h}{2}, y_n + \frac{h}{2}m_2\right), \qquad m_4 = f(x_n + h, y_n + hm_3)$$

Maple uses the Improved Euler method if you use "`method=classical[heunform]`" in the "`dsolve`" command, and fourth-order Runge-Kutta if you use "`method=classical[rk4]`". The other options are the same as for the Euler method.

**EXAMPLE 4** Repeat the last example with the Improved Euler and fourth-order Runge-Kutta methods.

**SOLUTION** Here are the results for Improved Euler:

```
> for nn from 0 to 4 do
>    S:= dsolve(ivp, y(x), numeric, method=classical[heunform],
>              stepsize=1/(10*2^nn)):
>    V[10*2^nn]:= subs(S(1), y(x)):
> od:
> V[10], V[20], V[40], V[80], V[160];
```

$$4.499965636521797, 4.614286333056052, 4.647754248867632,$$
$$4.656705801189754, 4.659011587271492$$

Here are the errors:

```
>  seq(evalf(exact - V[10*2^n]), n = 0 .. 4);
```

$$.159826732, .045506036, .012038120, .003086568, .000780782$$

This is indeed an improvement on the Euler results. Not only are the errors smaller, but they become smaller more rapidly: dividing the step size by 2 divides the error by about 4. This is in accord with the theoretical result that the error in the Improved Euler method is approximately proportional to the square of the step size. Next we try fourth-order Runge-Kutta.

```
> for nn from 0 to 4 do
>    S:= dsolve(ivp, y(x), numeric, method=classical[rk4],
>              stepsize=1/(10*2^nn)):
>    V[10*2^nn]:= subs(S(1), y(x)):
> od:
> V[10], V[20], V[40], V[80], V[160];
```

$$4.659299608226976, 4.659758238839909, 4.659790124501190,$$
$$4.659792222753928, 4.659792357243614$$

Even for $h = .1$ this is fairly accurate. Here are the errors:

```
>  seq(evalf(exact - V[10*2^n]), n = 0 .. 4);
```

$$.000492761, .000034130, .2244\,10^{-5}, .146\,10^{-6}, .12\,10^{-7}$$

The theory says that (as the "fourth-order" in its name implies) the error in fourth-order Runge-Kutta is approximately proportional to the fourth power of the step size, so dividing the step size by 2 should divide the error by about 16. Here are the actual ratios we obtained (I'll need to use a larger setting of "`Digits`" here to reduce the effect of roundoff error):

```
> seq(evalf((exact - V[10*2^n])/(exact-V[10*2^(n+1)]), 16),
>       n = 0 .. 3);
```

$$14.43874607711071, 15.22308667261268, 15.61468241074125, 15.80858474898295$$

The fourth-order Runge-Kutta method, though it works fairly well for many purposes, is far from being the last word in numerical methods. The default method used by "`dsolve(...,  numeric)`" if no "`method`" is specified, is "`rkf45`" which is a combination of fourth-order and fifth-order Runge-Kutta methods with a step size that is adjusted automatically during the course of the computation.

```
> S:= dsolve(ivp, y(x), numeric);
```

$$S := \mathbf{proc}(rkf45\_x) \ldots \mathbf{end}$$

```
> V:= subs(S(1), y(x));
```

$$V := 4.659792311390192$$

```
> evalf(exact-V, 16);
```

$$.54935295\,10^{-7}$$

## A Chaotic System

Autonomous systems of three or more differential equations can have a much more varied behaviour than those with two equations. In particular, they can exhibit **chaos** and **strange attractors**. One of the first such systems to be studied was the Lorenz system, introduced by the meteorologist Edward N. Lorenz in 1963. This arises from a model of convection in a layer of fluid that is heated from below. The equations are

$$\frac{dx}{dt} = -\sigma x + \sigma y$$
$$\frac{dy}{dt} = rx - y - xz$$
$$\frac{dz}{dt} = -bz + xy$$

where $\sigma$, $r$, and $b$ are positive parameters. The most commonly used parameter values are $\sigma = 10$, $r = 28$, and $b = 8/3$.

```
> Lorenz:= { D(x)(t)  = -sigma*x(t) + sigma*y(t),
>            D(y)(t)  = r*x(t) - y(t) - x(t)*z(t),
>            D(z)(t)  = -b*z(t) + x(t)*y(t)};
> sigma:= 10; r:= 28; b:= 8/3;
```

We can use "`DEplot`" to plot solution curves in a system such as this one, specifying with the "`scene`" option which two variables are plotted on which axes. For example, here is $x$ as a function of $t$:

```
> with(DEtools):
> p1:= DEplot(Lorenz, [x(t),y(t),z(t)], t=0..50,
>      {[x(0)=5,y(0)=10,z(0)=10]}, scene=[t,x], stepsize=.01,
>      linecolour=red, thickness=0, axes=box):
> p1;
```

The $x$ coordinate sometimes oscillates for a while near $x = -10$, the oscillations gradually growing, and then switches to an oscillation near $x = +10$. Sometimes one of these oscillations has only one or two cycles, sometimes 10 or more. There doesn't seem to be any overall pattern, only an apparently random process.

Part of the concept of chaos is sensitive dependence on initial conditions: two solution curves that start off very close together separate and eventually lose all connection to each other. Let's try another initial condition close to the first one.

```
> p2:= DEplot(Lorenz, [x(t),y(t),z(t)], t=0..50,
>     {[x(0)=5.01,y(0)=10,z(0)=10]}, scene=[t,x], stepsize=.01,
>     linecolour=black, thickness=0, axes=box):
> p2;
```

Up to about $t = 10$ the two graphs are almost indistinguishable. After that they are very different. Here are the two plotted together.

`> with(plots): display({p1,p2});` (Figure 10.4)

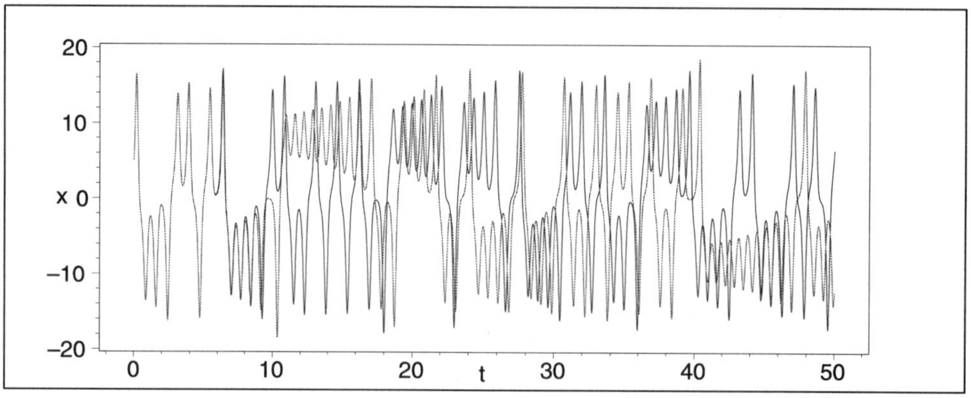


Figure 10.4. Sensitive dependence on initial conditions

Now here is our first solution curve projected in the $xz$ plane:

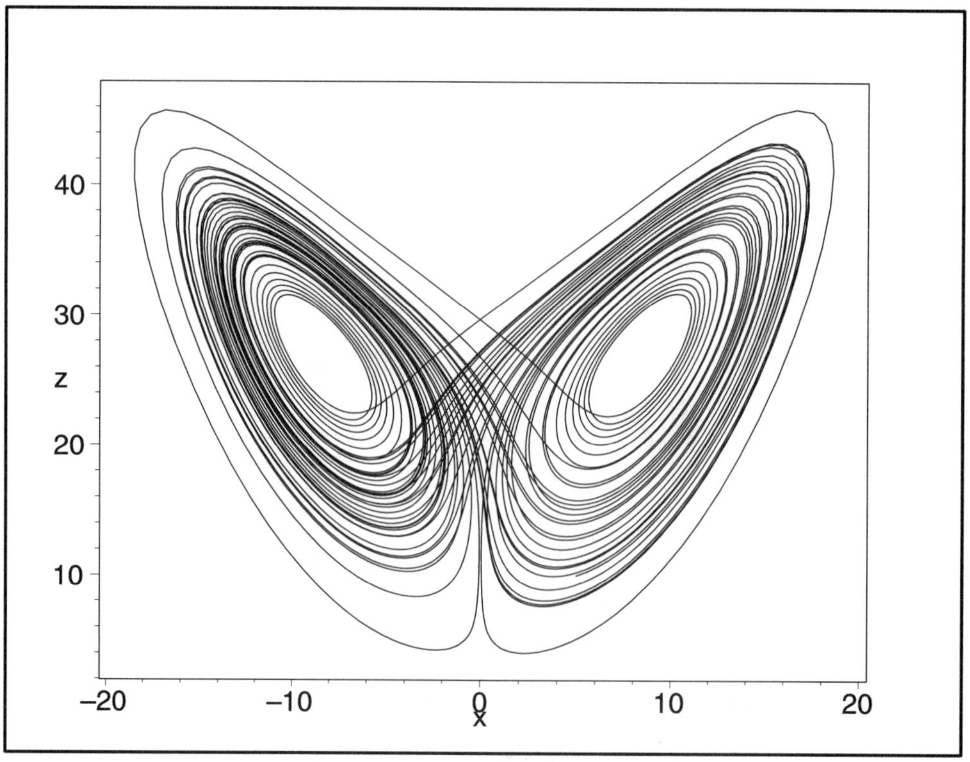


Figure 10.5. The Lorenz attractor

```
> DEplot(Lorenz, [x(t),y(t),z(t)], t=0..50,
>     {[x(0)=5,y(0)=10,z(0)=10]}, scene=[x,z], stepsize=.01,
>     linecolour=red, thickness=0, axes=box);
```
(Figure 10.5)

It's just one solution curve, but other solution curves would produce quite similar pictures. The solution curves approach a complicated set known as the **Lorenz attractor**. It might be described as two interconnected, warped phonograph disks. Each "oscillation" in $x$ corresponds to a loop around one of the disks. A two-dimensional picture hardly does justice to the attractor. You can use "DEplot3d" to draw a three-dimensional plot. This is especially good when viewed with real-time rotation.

```
> DEplot3d(Lorenz, [x(t),y(t),z(t)], t=0..50,
>     {[x(0)=5,y(0)=10,z(0)=10]}, scene=[x,y,z], stepsize=.01,
>     linecolor=black, thickness=0, axes=box);
```

## EXERCISES

**1.** Solve the following differential equations or initial value problems using Maple. Note: (a), (b) and (c) can be done by hand using "standard techniques": (a) is separable, (b) exact and (c) linear. However, (d) seems not to be so easy.

**(a)** $\dfrac{dy}{dx} = \dfrac{x(y^2+1)}{(x^2+1)(y+1)^2}$  **(b)** $x + \sin y + (x\cos y + 2y)y' = 0$

**(c)** $y' - y\cot x = \csc x,\ y(\pi/2) = 1$  **(d)** $(xy-1)y' = y^2 - 1$

**2.** Sometimes a differential equation has a particular solution that is not part of the general solution (i.e., there are no values of the parameters in the general solution that produce it), but is a limiting case of it.

**(a)** Find the general solution of $y' = 2xy^2$.

**(b)** Find the solution of this equation with the initial condition $y(0) = 0$. Why do you think "dsolve" can't find it?

**(c)** Find the general solution of $y' = -y^2 + 2y - 1 + 2x + 2x^2y - 2x^2 - x^4$.

**(d)** Find an initial condition for which "dsolve" can't find the solution. What is the solution?

**3. (a)** Find all polynomials $y(x)$ of degree 5 that satisfy the differential equation $y''y''' = y$. It turns out that these are all the solutions that are polynomials.

**(b)** Find the eighth-degree Maclaurin polynomial of the solution of this differential equation with initial conditions $y(0) = 0, y'(0) = 0, y''(0) = 1$.

**4. (a)** Plot the direction field for the differential equation $y' = xy/(y-x)$.

**(b)** What happens to a solution that hits the line $x = y$? What does "DEplot" show for such a solution?

**(c)** Consider a solution with $y(-2) = -1$. Describe its qualitative behaviour both to the right and to the left from this point.

**(d)** Do the same for $y(2) = -1$.

**5.** Study the qualitative behaviour of the trajectories for the second-order differential equations

**(a)** $\dfrac{d^2x}{dt^2} = \left(\dfrac{dx}{dt}\right)^2 + x$

**(b)** $\dfrac{d^2x}{dt^2} = -\left(\dfrac{dx}{dt}\right)^2 - x$

What are the possible behaviours of the trajectories as $t \to \infty$? *Hint:* in (b), trajectories cross the line $v = -1 + x$ for $x < 0$ in only one direction.

**6.** In this exercise we show that in Example 1, some solutions have $y \to -\infty$ at a finite value of $x$.

**(a)** Consider the curve $y = -2/(2-x)$ for $0 < x < 2$. Show that solutions of the differential equation $y' = x^2 - y^2$ can cross from the region above the curve to the region below, but not vice versa.

**(b)** Conclude that solutions of the equation with $y(0) < -1$ have $y \to -\infty$ as $x \to a-$ for some $a$ in the interval $(0, 2)$.

**(c)** How might you investigate numerically the value of $a$? *Hint:* find a differential equation for $u(x) = 1/y(x)$.

**7.** Find differential equations that have solution curves to match each of the following pictures. *Hint:* guess an equation for the isoclines $f(x, y) = 0$.

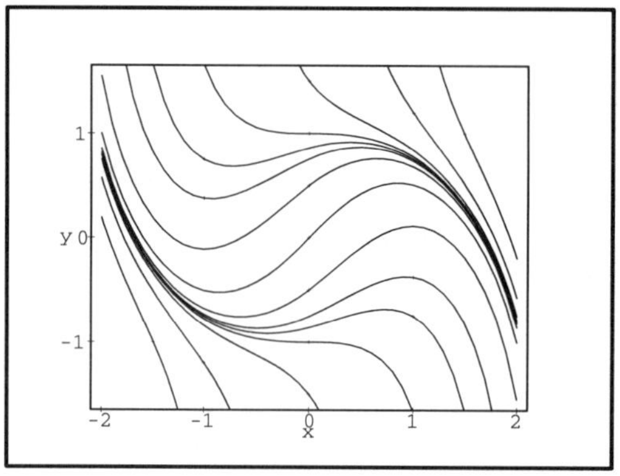


Figure 10.6. Solution curves (a)

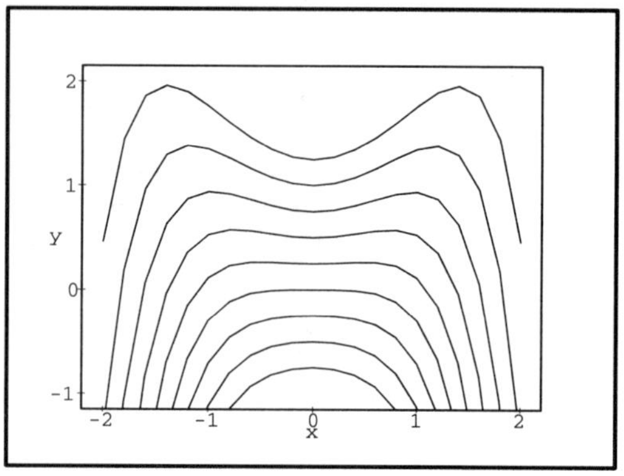


Figure 10.7. Solution curves (b)

**8.** **(a)** Find the value at $x = 2$ of the solution to the initial value problem $y' = -x/y$, $y(0) = 3$. Compare the results of the Euler, Improved Euler, and fourth-order Runge-Kutta methods with step sizes $0.1, 0.05, 0.025$, and $0.0125$.

**(b)** What step size would be necessary for the Euler method to match the accuracy of the Runge-Kutta method with step size $0.1$?

**9.** When is the slope $0$ for the solution to the initial value problem $y' = x^2 - y^2$, $y(-2) = 0$? Use "`dsolve(..., numeric)`".

**10.** Choose a nonlinear initial value problem $y' = f(x, y)$, $y(x_0) = y_0$ that Maple can solve, where $f$ depends on both $x$ and $y$. Consider the result of one step of the Euler, Improved Euler, and fourth-order Runge-Kutta methods with step size $h$ as functions $e(x_0 + h)$, $ie(x_0 + h)$, and $rk(x_0 + h)$. Expand these and the exact solution $y(x_0 + h)$ in a Taylor series about $h = 0$. To what order $O(h^p)$ do each of the approximations agree with the exact solution?

**11.** Produce a 3-dimensional animation of points moving along the Lorenz attractor. *Hint:* use "`DEplot3d`" to plot the attractor. Use "`dsolve(...,numeric)`" to produce points $P_i$ on the trajectory, and plot them using "`pointplot3d`". Let the first frame of the animation have points $P_1$, $P_{21}$, $P_{41}$, ..., the second have $P_2$, $P_{22}$, $P_{42}$, ..., etc. Then after 20 frames the animation can repeat (in "continuous-cyle" mode), with $P_1$ having "moved" to $P_{21}$.

# APPENDIX I

# Additional Files

There are a number of definitions that we find useful at various places in this book. They are available in electronic form, in files with the extension "`.def`", at the web site

`http://www.geocities.com/CapeCanaveral/Campus/7705.`

If you don't have access to the web, you can type them in from the listings below. The definitions are listed here file-by-file, in the order in which the files first appear in the book.

## CHART.DEF (Lab 3)

```
chart:= (head,vals) ->
      MATRIX([head, seq(evalf(eval(subs(head[1]=vals[i],head))),
           i=1..nops(vals))]):
```

## RIEMANN.DEF (Lab 11)

```
h:= ('b'-'a')/'n';
X:= i -> a + i*h;
between:= (x,L,R) -> 0 < evalf(x - L) and 0 < evalf(R - x);

upper:= i -> max(f(X(i - 1)), f(X(i)),
     op(map(f, select(between, cp, X(i - 1), X(i)))));

lower:= i -> min(f(X(i - 1)), f(X(i)),
     op(map(f, select(between, cp, X(i - 1), X(i)))));
rsum:= g -> h * sum('g(i)', i=1..n);
rect:= (x1,x2,y) -> [[x1, 0], [x1, y], [x2, y], [x2, 0]];

rbox:= g ->
     plot({[seq(op(rect(X(ii - 1), X(ii), g(ii))), ii = 1 .. n)],
          f(x)}, x = a .. b);
```

## DOT.DEF (Lab 14)

```
evl:= V -> convert(evalm(V), list);
&.:= (U,V) -> linalg[dotprod](U, V, orthogonal);
len:= V -> sqrt(V &. V);
unit:= V -> evl(V/sqrt(V &. V));
`&x`:= (U,V) -> convert(linalg[crossprod](U,V),list);
```

## IMPTAY.DEF (Lab 20)

```
imptaylor:= (eq, y, b, x, a, n) ->
     taylor(b +(x-a)
        * convert(taylor(RootOf(subs(y=b+(x-a)*_Z, eq)),x=a,n-1),
             polynom)
        + (x-a)^n, x-a, n);
```

## CLASSIFY.DEF (Lab 21)

```
classify:= H ->
     if linalg[definite](H,positive_def) then 'loc_min'
     elif linalg[definite](H,negative_def) then 'loc_max'
     elif linalg[definite](H,positive_semidef) or
          linalg[definite](H,negative_semidef) then 'uncertain'
     else 'saddle'
     fi;
```

## ARROW3D.DEF (Lab 25)

```
evl := t -> convert(evalm(t), list):
arrow3d := proc(base::{list,vector},vect::{list,vector,`+`})
local b1,v1,stem,r,hbase,hpoint,v2,v3,head,opts, scopts;
    b1 := evalf(evl(base));
    v1 := evalf(evl(vect));
    if not type(v1,list) or nops(b1) <> 3 or nops(v1) <> 3 then
        ERROR(`points need three dimensions`)
    fi;
    if not (type(b1,[realcons,realcons,realcons]) and
        type(v1,[realcons,realcons,realcons])) then
        ERROR(`coordinates must evaluate to real constants`)
    fi;
    opts := args[3 .. nargs];
    scopts := indets({opts}, identical('scalefactor')=realcons);
    if scopts <> {} then
      v1:= evalf(op(2,scopts[1])) * v1;
      opts:= op({opts} minus scopts);
    fi;
    hbase := evl(b1+.8*v1);
    hpoint := evl(b1+v1);
    if not has({opts},{color,colour}) then
        opts := opts,colour = red
    fi;
    if not has({opts},thickness) then
      opts:= opts, thickness=3
    fi;
    stem := plots[spacecurve]([b1,hbase],opts);
    if v1 = [0,0,0] then RETURN(stem) fi;
    r := sqrt(linalg[norm](v1,2));
    v2 := evl(linalg[crossprod]([1,0,0],v1));
    if linalg[dotprod](v2,v2) < .01*linalg[dotprod](v1,v1) then
        v2 := evl(linalg[crossprod]([0,1,0],v1))
    fi;
    v2 := evl(.05*r*v2/linalg[norm](v2,2));
    v3 := linalg[crossprod](v1,v2);
    v3 := evl(.05*r*v3/linalg[norm](v3,2));
    head := plots[polygonplot3d]([hpoint,evl(hbase+v2),
        evl(hbase+v3), evl(hbase-v2),
        evl(hbase-v3), evl(hbase+v2)]
        ,style = patchnogrid,opts);
    plots[display]({stem,head})
end;
```

## SKIER.DEF (Lab 25)

```
spheroid:= proc(c,u,v,w)
   local p,t,cc;
   if nargs = 5 then cc := args[5] else cc := tan fi;
   plot3d(evl(c+ cos(p)*u +sin(p)*(cos(t)*v + sin(t)*w)),
     p=0..Pi, t = 0 .. 2*Pi,style=patch,grid=[8,8],colour=cc)
end;
skier:= proc(R,T,C,B,U) local y1,y2,h1,h2, a, a1,a2;
  h1:= .3*U &. B; y1:= .6 - h1; h2:= -h1;
  a1:= fsolve((.4+.4*cos(a))^2+(y1-.4*sin(a))^2-.36);
  y2:= 1.2 - y1;
  a2:= fsolve((.4+.4*cos(a))^2+(y2-.4*sin(a))^2-.36);
  display({
    spheroid(R+.75*B,.5*T+.1*B,-.04*T+.2*B,.3*C),
    spheroid(R+.25*T+.2*C+.6*B,.25*T-.1*C,.02*T+.05*C,.06*B,brown),
    spheroid(R+.25*T-.2*C+.6*B,.25*T+.1*C,.02*T-.05*C,.06*B,brown),
    spheroid(R+.5*T+.8*B,.15*T,.15*C,.15*B,black),
    spheroid(R+(-.2+.2*cos(a1))*T+.15*C+(.3+h1/2+.2*sin(a1))*B,
      (.2+.2*cos(a1))*T+(h1/2-.3+.2*sin(a1))*B+.05*C,.13*C,
      (-.15*h1-.09+.06*sin(a1))*T-.06*(1+cos(a1))*B,brown),
    spheroid(R+.2*cos(a1)*T+.2*C+(h1+.2*sin(a1))*B,
      .2*cos(a1)*T+.2*sin(a1)*B,.1*C,-.05*sin(a1)*T+.05*cos(a1)*B,
      brown),
    spheroid(R+(-.2+.2*cos(a2))*T-.15*C+(.3+h2/2+.2*sin(a2))*B,
      (.2+.2*cos(a2))*T+(h2/2-.3+.2*sin(a2))*B-.05*C,.13*C,
      (-.15*h1-.09+.06*sin(a2))*T-.06*(1+cos(a2))*B,brown),
    spheroid(R+.2*cos(a2)*T-.2*C+(h2+.2*sin(a2))*B,
      .2*cos(a2)*T+.2*sin(a2)*B,-.1*C,-.05*sin(a2)*T+.05*cos(a2)*B,
      brown)})
end;
```

## VECINTS.DEF (Lab 28)

```
veloc:= (C,R) ->  diff(map(rhs,C), op(1,R));
slineint:= (F,C,R) -> int(subs(C, F) * len(veloc(C,R)), R);
Slineint:= subs(int=Int, eval(slineint));
vlineint:= (F,C,R) -> int(subs(C, evl(F)) &. veloc(C, R), R);
Vlineint:= subs(int=Int,eval(vlineint));
aelt:= (S, R1, R2) -> len(veloc(S,R1) &x veloc(S,R2));
surfint:= (f, S, R1, R2) ->
    int(int(subs(S,f) * aelt(S,R1,R2), R2), R1);
Surfint:= subs(int=Int,eval(surfint));

flux:= (F, S, R1, R2) -> int(int(subs(S, evl(F)) &.
    (veloc(S,R1) &x veloc(S,R2)), R2), R1);
Flux:= subs(int=Int, eval(flux));
```

## APPENDIX II

# Common Maple Errors

Everybody who uses Maple makes errors. There are many causes of errors, from simple typing mistakes to misunderstanding complicated syntax. Fortunately, it is rare for an error to have serious results (such as crashing the system or deleting your files). Usually the result is one of the following:

(1) Maple just sits there doing nothing

(2) Maple gives you a warning or error message

(3) Maple gives a strange-looking answer, or no answer when there should be one

(4) Everything seems normal now, but (2) or (3) may occur later in the session.

Most errors can easily be corrected once they are identified. The main difficulty is often in diagnosing the error. This appendix is intended to help with the diagnosis. It lists some of the most common errors, their symptoms, and how to correct them.

## Forgetting to save your work

You should be aware that the computer's memory is ephemeral. Anything and everything that is not saved onto a disk could disappear at any time. Sometimes this seems to happen especially when it will cause you the most inconvenience. It might be due to a power failure, or a problem with the computer system. Or you might accidentally start Maple on a calculation that seems to go on forever, and be unable to interrupt it.

**Prevention:** Get in the habit of saving your work every 15 minutes or so. See the section "Saving Your Work" in Lab 1.

## Omitting ";" or ":"

A command to Maple, or an expression you want Maple to evaluate, must end with "`;`" (if you wish to see the result) or "`:`" (if you don't). The only exceptions are "?" and "`quit`".

**Symptoms:** Maple produces another input line in the same group, and issues a warning:

```
Warning, premature end of input
```

**Correction:** Enter the "`;`" or "`:`", and press `Enter`.

## Unbalanced parentheses

Every "`(`", "`{`" or "`[`" needs a corresponding "`)`", "`}`" or "`]`".

**Symptoms:** Maple prints an error message, such as "`` `;` unexpected``" if an end parenthesis is missing, or "`` `)` unexpected``" if a beginning parenthesis is missing or if, say, '`[`' was paired with '`)`'. It leaves the cursor at the point on the line where the error became evident.

**Diagnosis:** Note that the cursor is often not at the point where the missing parenthesis should go. In a complicated expression with lots of parentheses, you can check the balance by counting: start at the beginning with a count of 0, add 1 at each "(" parenthesis and subtract 1 at each ")". If the count ever gets negative, you have a ")" with no matching "(". If the count is not 0 at the end of the expression, there is a "(" with no matching ")".

**Correction:** Edit the expression and enter it again. If you find it very difficult to spot the error because there are so many parentheses, you might try building the expression up in several steps rather than all at once. This can also improve the readability and clarity of your work. Thus instead of the almost unreadable and error-prone

```
> f((x+(b+c)^(1/2))/(q+x^(-3)));
```

you could try

```
> n:= x + (b+c)^(1/2);
> d:= q + x^(-3);
> f(n/d);
```

## Omitting "*" for multiplication

Unlike normal mathematical notation, Maple requires the use of a multiplication sign: you must use "a * b", not "a b."

**Symptoms:** Usually the error message is "missing operator or `;`". However, if the second factor is in parentheses, Maple will interpret it as evaluating a function at the value in parentheses. The result may look quite normal, and the error may not show up until later. For example:

```
> a (3 + a);
```

$$\mathrm{a}(3 + a)$$

The only clue that this is not the same as "a * (3 + a)" is that (in the standard Windows-version math style) the first "a" is in a Roman font (used for function names) while the second is in the italic font normally used for variable names.

If both factors are in parentheses, the result may look strange:

```
> (3 + a) (3 - a);
```

$$3 + \mathrm{a}(3 - a)$$

The explanation is that this is evaluating the function x -> 3 + a(x) at $x = 3 - a$.

**Correction:** Put in the "*"

## Omitting "(" and ")" for a function

The input to any function or command (with very few exceptions such as "read") must be placed in parentheses "()". Thus you must write "sin(x)", not "sin x" as in ordinary written mathematics.

**Symptoms:** "missing operator or `;`".

**Correction:** Put in the parentheses.

## Invalid inputs to a function or command

Most Maple functions and commands are very particular about the number, types and order of inputs they accept.

**Symptoms:** Typical error messages are

```
Error, (in ...)  wrong number (or type) of arguments
Error, (in ...)  invalid arguments
Error, (in ...)  expecting 2 arguments, got 1
Error, ...  expects its 1st argument, x, to be of type integer,
but received ...
```

**Correction:** Look at the help for the function or command, in particular the "calling sequence" and "parameters" sections, to see the correct way to use it. Or imitate one of the examples that are given, making sure that your inputs correspond in number, types and order to those in the example. In some cases it is helpful to use the "whattype" command, which will tell you what type a certain expression has.

## Wrong case

Maple is a "case-sensitive" language: it distinguishes between uppercase letters A, B, C ... and lowercase letters a, b, c .... So, for example, Maple doesn't recognize the "Sin" function, which it considers to be different from "sin".

**Symptoms:** When Maple encounters a name it doesn't recognize (e.g., "`Sin`"), it just treats this as a new variable or function name. It doesn't usually cause an error message, but it won't return numerical values: `Sin(0)` will just return Sin(0).

**Prevention:** Be careful to type the name correctly. Consult the help system to see if your spelling is correct.

## Using "pi" instead of "Pi"

This is a special case of the previous error: "Wrong case". Maple uses "`Pi`" for the constant 3.14159.... If you use "`pi`" instead, Maple will just consider this to be a variable.

**Symptoms:** You can't obtain numerical values, e.g., `sin(pi)` just returns $\sin(\pi)$ instead of 0.

**Diagnosis:** Unfortunately, Maple's "pretty-printer" shows both `Pi` and `pi` as $\pi$. You can use "`has(`*expr*`, pi);`" if you suspect that *expr* contains `pi`. The result will be "`true`" if it does and "`false`" if it doesn't.

**Correction:** `> pi:= Pi;`

## Using "e" or "E" instead of "exp"

The base of natural logarithms, $e = 2.71828\ldots$, is not a standard constant in Maple. It was denoted by "`E`" up to Release 3, but this is not a predefined constant any more. The preferred way to refer to this number is "`exp(1)`". In typeset output $\exp(t)$ is shown as $\mathbf{e}^t$. Note that this $\mathbf{e}$ is in a different font than the ordinary letter $e$.

**Correction:** If you wish, you can assign a value to the variable `E`:

```
> E:= exp(1);
```

But it is usually better to use `exp(t)` rather than `E^t` in formulas, as this is easier for Maple to simplify.

## Using "sign" instead of "signum"

The signum or sgn function in Maple is "`signum`". There is also a "`sign`" function but this is different: it returns the sign of the leading coefficient of a polynomial. The potential for confusion is high because "`sign`" and "`signum`" both work for numbers. But "`sign(x)`", where "`x`" is a variable, will always return 1.

## Defining an expression instead of a function

An example of a function definition is "`f:= x -> x^2`". On the other hand, "`f:= x^2`" defines an expression. There's no problem if that's what you wanted, but the proper uses of expressions and functions are different.

**Symptoms:** If you try to use an expression as a function, Maple will assume that any variables used in the expression are functions. For example:

```
> f(t);
```

$$\mathrm{x}(t)^2$$

```
> D(f)(t);
```

$$2D(x)(t)\mathrm{x}(t)$$

**Correction:** `> f:= unapply(f,x);`

## Defining "f(x)" instead of "f"

In ordinary mathematics we usually speak of "the function $f(x) = \ldots$," so it is very tempting to use "`f(x):= ...`" instead of "`f:= x -> ...`" in Maple.

**Symptoms:** Everything seems fine, as long as the only input you use with $f$ is $x$. But when you try to use $f$ with some other input, Maple acts as if $f$ hasn't been defined (this is not surprising, since it hasn't). For example:

```
> f(x):= x^2 + 1;
```

$$f(x) := x^2 + 1$$

```
> f(x) - 1;
```

$$x^2$$

```
> f(3);
```

$$f(3)$$

**Correction:** `> f:= unapply(f(x),x);`

## Using "y" instead of "y(x)" in a differential equation

In differential equations, the dependent variables should be written as functions of the independent variable. Thus you should write "`D(y)(x) = y(x)`" or "`diff(y(x), x) = y(x)`", but not "`D(y) = y`" or "`diff(y, x) = y`".

**Symptoms:** Typical error messages are

```
Error, (in ODEtools/info) y(x), and, y, cannot both appear in
the given ODE.
```

```
Error, (in DEplot) Unable to isolate the derivative
```

**Prevention:** If you are going to work with a number of differential equations with the same dependent and independent variables, it may help to define and work with abbreviations such as "`yp:= D(y)(x)`".

## Using "f" instead of "f(x)" in a plot

In this book, we have almost always plotted expressions rather than functions, e.g., "`plot(f(x), x=0..1)`". It is also possible to plot a function without giving a name to the independent variable, e.g., "`plot(f, 0..1)`". However, the two methods can't be mixed: you should not use "`plot(f, x=0..1)`".

**Symptoms:** The function is plotted as if it were a very large constant.

**Prevention:** Be consistent. Always plot expressions (except in the rare cases where there is a good reason not to, as in Example 2 of Lab 25).

## Changing without recalculating

It is very convenient to move around in the worksheet and make whatever changes you want. However,

(1) Changes in an input region have no effect unless you press `Enter` with the cursor in that region or group.

(2) When you press `Enter`, only the commands in this particular input region or group are performed.

(3) Maple uses the current values of any variables involved in the calculation, not values corresponding to this position in the worksheet. For example:

```
> x:= 1:
> x^2;
```

$$1$$

```
> x:= 2:
```

If you now go back and press `Enter` on the line "`> x^2;`", the result is 4.

**Symptoms:** Changes may not appear to have taken effect, or the wrong values to have been used.

**Correction:** You can recalculate the entire worksheet at once, or a selected portion of it, by choosing "Execute" from the Edit menu. In both cases, however, recalculation is done with current values in effect. This can cause problems if you use a variable that is not supposed to have a value near the beginning of your worksheet, and then assign it a value later on. For example:

```
> x^2;
```

$$x^2$$

```
> x:= 2:
```

When the worksheet is recalculated, the first output changes to 4. To fix this, you can proceed as follows:

(1) `> restart;` (this removes all defined values, and puts Maple back in the state in which it started).

(2) Remove the "`restart`" line (because you don't want everything to be forgotten after you recalculate the worksheet).

(3) Recalculate the worksheet.

## Omitting "readlib" or "with"

Many commands and functions are not known to Maple until you tell Maple to read them in with a "`readlib`" or (if they are part of a package) a "`with`" command.

**Symptoms:** The function or command is not evaluated or executed. Maple returns an expression containing the command name.

**Diagnosis:** The help page you get with "?" and the name will tell you whether a "`readlib`" or "`with`" is needed.

**Correction:** Enter the "`readlib`" or "`with`" command and try again.

## Premature evaluation

With very few exceptions, the inputs to a function or command are evaluated before the function or command is called. This usually has the desired results, but occasionally there are problems. A typical case involves a function that requires a numerical input. For example, suppose you have defined a function by

```
> f:= x -> fsolve(y + sin(y) = x, y);
```

Now something like "`f(5)`" will work fine, but "`f(x)`" will produce an error message because "`fsolve`" can't deal with an equation containing variables other than the one being solved for. This same error message will occur if you try, e.g., "`plot(f(x),...)`".

**Symptoms:** Typical error messages are

```
Error, (in fsolve) x, is in the equation, and is not solved for
Error, (in f) cannot evaluate boolean
```

**Correction:** Enclose the "`f(x)`" in single quotes `''` to prevent it from being evaluated before being passed to "`plot`":

```
> plot('f(x)', x=-5..5);
```

## Variable has been assigned value

A name that has not been assigned a value can be used as a symbolic variable. But you may forget and use a name that does have a value.

**Symptoms:** The variable name doesn't appear in an expression where it should appear. A curve plotted using this variable appears to be constant. There may be an error message if the value is of a type that shouldn't appear in the expression you entered.

**Diagnosis:** Check whether the variable has a value:

```
> print(x);
```

This will return the current value, if there is any, or "$x$" if there is none. It is a better test than just "`> x;`", which won't warn you if `x` is a function, array or table.

**Correction:** Use a different name for the variable, or unassign the value if it is no longer needed:

```
> x:= 'x';
```

## Side effect of "for"

This is really a special case of the previous error: "Variable has been assigned value". The command

```
> for i from first to last do ... od;
```

performs a command several times, for different values of the "loop variable" `i`. However, it has a (generally unwanted) side effect: it leaves the loop variable `i` with a value, `last+1`. This will affect any later commands (other than "`seq`" or "`for`") that use `i` as a variable.

**Prevention:** Reserve repeated-letter names such as "`ii`" or "`nn`" for use of "`for`" loop variables.

## Using a formal parameter outside a function

In a function definition such as "`f:= x -> ...`", the "`x`" is called a "**formal parameter**." This formal parameter has a meaning only within the function definition. When you evaluate the function by "`f(...)`", whatever is inside the parentheses is substituted for the formal parameter in the function definition. But the formal parameter has no connection to any occurrence of the same name outside of the function definition, and no such substitution will occur there.

**Example:** `> a:= x^2:`

```
> f:= x -> x^3 - a;
```

$$f := x \to x^3 - a$$

```
> f(p);
```

$$p^3 - x^2$$

Did you expect $p^3 - p^2$? Although "`f(p)`" substitutes "`p`" for the formal parameter "`x`", the definition of "`a`" uses the global variable "`x`", not the formal parameter.

**Symptoms:** The function's values still contain the name of the formal parameter. If the function is used in a command that requires a numerical value, you may get error messages such as

```
Error, wrong number (or type) of parameters in function ...
```

**Correction:** Change the expression using the formal parameter into a function. In the example above,

```
> a:= x -> x^2:
> f:= x -> x^3 - a(x):
> f(p);
```

$$p^3 - p^2$$

## Forgetting "od" in a "for" command

The correct form of the "`for`" command is

> `for` *n* `from` *initial* `to` *final* `do` *commands* `od;`

The loop is ended with "`od`".

**Symptoms:** Maple expects more commands to go in the loop. It gives you another prompt and waits for more input, with the warning message

```
Warning, premature end of input
```

**Correction:** Type the "`od;`" in the proper place, and press `Enter`.

## Assigning to a standard name

Maple uses a wide variety of names for variables, functions and options. When you define a new variable or function, its name may be one that Maple uses. This could lead to trouble if some Maple code later tries to access the name, as your definition will be used instead of the original one. Alternatively, Maple may try to deal with your new object in the same way that it would deal with the original one (e.g., if the name is originally used for a function, apply a differentiation formula for that function).

**Symptoms:** Sometimes there is an error message such as

```
Error, attempting to assign to `D` which is protected
```

Otherwise, symptoms can be quite varied. For example, suppose you assign a value to "`left`", and later try a definite integral. The result is

```
Error, (in limit) invalid directional argument
```

This is because the code for definite integrals calls "`limit`" with options "`left`" and "`right`" to evaluate the antiderivative at the endpoints of the interval. This won't work if "`left`" or "`right`" has a value.

**Prevention:** Sometimes you can use the help facility to tell whether Maple uses a name. To check on "`foo`", just enter "`?foo`". Or you could try "Full Text Search" in the Help menu: this may catch cases where the name is an option to a command or function, rather than a command or function itself. However, for common words it may find a lot of irrelevant pages. If Maple doesn't come up with a help page, it's probably safe to use the name. It's a good idea to check on any name that is an English or mathematical word, or a likely abbreviation for one. You should also avoid using names that start with an underscore "_", since Maple often uses these for internal variables (and may not mention them in the help system).

If a name is used by Maple, but only in a context that you are sure will not come up in your session, you may be able to use it. However, this is a dangerous practice, and I must recommend against it.

## Using exact arithmetic in complicated expressions

Strictly speaking, this isn't an error, but it is a source of difficulties. Maple uses exact arithmetic. Rational numbers are expressed as fractions, and irrational numbers are expressed using symbols. If you ask Maple for, say, "`(1 + 1/1000)^1000`", it will quite happily give you the answer as a rational number. Both numerator and denominator will have 3001 digits. This one will probably not be too much of a problem (except that it won't all fit on your screen at one time), but you can easily construct similar examples that will cause Maple to use up all available memory, or take a very long time to compute. Yet probably all you wanted was a few decimal places of the answer. You could use "`evalf((1+1/1000)^1000)`", but that would still compute the exact rational number before taking the numerical approximation.

**Solution:** Putting a decimal point in a number causes calculations involving that number and rational numbers to be carried out in floating point, i.e., using decimals to a certain number of significant digits (given by the variable "`Digits`"). So you could use "`1.001^1000`" or even "`(1+1./1000)^1000`". Note, however, that expressions such as "`1.0/Pi`" or "`1.0*sqrt(2)`" (involving a decimal and an irrational number) are not automatically calculated in floating point: you could use "`1/evalf(Pi)`" or "`sqrt(2.)`" instead.

## Inappropriate use of decimals

This is the reverse of the previous error: "Using exact arithmetic in complicated expressions". In some cases, you need to use exact arithmetic rather than decimal approximations. For example, the limit of an indeterminate form such as $\lim_{x \to b} f(x)/(x-b)$ will probably give you an incorrect result if you use a decimal approximation for $b$ rather than the exact value.

```
> b:= evalf(Pi/2): limit(cos(x)/(x-b), x = b);
```

$$undefined$$

Also, decimals in exponents should be avoided, especially with commands such as "`solve`" and "`int`" that perform extensive symbolic computations.

```
> int((x+3)*(x+1)^(-1/3.0)*(x+2)^(-4/3.0), x);
```

$$\int \frac{x+3}{(x+1)^{.3333333333}(x+2)^{1.333333333}}\,dx$$

**Correction:** These examples work correctly if you replace the decimals by exact values.

```
> b:= Pi/2: limit(cos(x)/(x-b), x = b);
>  int((x+3)*(x+1)^(-1/3)*(x+2)^(-4/3), x);
```

## Insufficient digits

When doing numerical work with floating-point numbers, you should be aware of the effects that roundoff error can have on the accuracy of your calculations. This can be particularly bad when you subtract two numbers that are very close to each other, or add a small number to a large number.

**Symptoms:** Especially in the case of subtracting numbers that are close to each other, Maple may report the result with fewer digits than usual. Answers of 0 are particularly suspect. In other cases, you may get inaccurate results with very little indication of it.

**Diagnosis:** Try the calculation with a larger setting for "`Digits`". If more than one or two of the digits of the original answer change, this could be an indication of a problem.

**Treatment:** Set "`Digits`" to a larger number, say 15 instead of the default 10. Try to arrange your calculations to avoid those that are particularly susceptible to roundoff error.

## Fractional powers of negative numbers

Maple's mathematics is largely based on complex numbers rather than real numbers. When $x$ is negative, it gives a complex value rather than a real value to an odd root of $x$ (e.g., $x^{(1/3)}$).

**Symptoms:** The value returned has an $I$ in it ($I$ is Maple's notation for $i = \sqrt{-1}$). In a plot, this results in missing portions of the plot.

**Treatment:** The "`surd`" function provides "elementary" $n$-th roots. Thus if $x$ is a negative number and $n$ is odd, "`surd(x,n)`" returns the negative (real) $n$-th root of $x$.

## Plotting with non-numerical parameters

In a plot command such as "`plot(f(x), x = 0 .. 1);`", Maple needs to obtain numerical values for the expression `f(x)` at each value of `x`. If `f(x)` contains parameters which haven't been given numerical values, this can't be done.

**Symptoms:** The expression can't be plotted. If there's nothing else in the plot, there is an error message:

```
Plotting error, empty plot
```

**Diagnosis:** If necessary, evaluate the expression at some numerical value of `x`. Any names that appear in the result have not been given numerical values.

**Correction:** Choose numerical values for the parameters.

## Plotting with singularities

An expression $f(x)$ you plot may be undefined for some $x$ values in the interval you specify. It is usually not a problem if "`plot`" tries to evaluate the expression at a point where it is undefined (the result is simply a small "hole" in the graph), but difficulties can arise near to a singularity.

**Symptoms:** The graph may involve very large $y$ values, or have an unexpected "spike" or an almost-vertical line segment.

**Solution:** If very large $y$ values occur, specify an interval for the $y$ coordinate. Use the `discont=true` option, or break the interval up into pieces to avoid the singularity.

## Unstated assumptions

Maple generally uses complex numbers rather than just real numbers. Unless you tell it otherwise, it will assume that a symbolic parameter you use can have any complex value. This can cause problems when you enter an expression that may not be defined except for certain values of the parameter.

**Symptoms:** Maple may refuse to do a calculation that you think it should do. A typical example is an improper integral such as

```
> int(exp(-a*x^2), x=0..infinity);
```

$$\lim_{x\to\infty-} \frac{1}{2}\frac{\sqrt{\pi}\,\mathrm{erf}(\sqrt{a}\,x)}{\sqrt{a}}$$

The improper integral only converges if $a > 0$. Maple doesn't know that you want $a > 0$, so you can't expect it to make that assumption.

In other cases, Maple may produce a result that seems excessively complicated: it refuses to do some simplification (e.g., changing $\sqrt{a^2}$ to $a$) that is valid for the cases you are interested in, but may not be valid in general.

**Solution:** It may help to use the "`assume`" command. In the example above,

```
> assume(a > 0);
> int(exp(-a*x^2), x=0..infinity);
```

$$\frac{1}{2}\frac{\sqrt{\pi}}{\sqrt{a}}$$

Sometimes you can force a desired simplification by using "`simplify(..., symbolic)`".

## Circular definition

If a variable `x` has been assigned a value, there is nothing wrong with a statement such as "`> x:= x+1;`". This simply adds 1 to the value of `x`. But if `x` does not have a value, this sets up a circular definition: we are attempting to define `x` in terms of itself.

**Symptoms:** The statement produces a warning message:

```
Warning, recursive definition of name
```

If you try to evaluate `x` when it is defined this way, the result will be an error:

```
Error, too many levels of recursion
```

**Correction:** Presumably, you meant to start out with some initial value for `x`, but forgot to enter this into Maple. Assign `x` this initial value and try again.

## Asking the impossible

Some problems simply don't have solutions. Others have solutions that are extremely complicated. It is quite easy to ask Maple to do a calculation that will use up all available memory or take forever to compute.

**Symptoms:** Maple may calculate for a very long time, perhaps with the "Bytes" indicator gradually increasing. Eventually there may be a crash, or an error message such as

```
Error, object too large
```

**Diagnosis:** Try smaller problems of the same type first, and see how the time Maple takes to solve them and the size of the result increase as the problem gets larger.

**Treatment:** This can be difficult. In some cases you may be able to organize your calculation in a more efficient way, or reduce the demands on Maple by having fewer symbolic parameters. Ask yourself whether you really need this result (and what you could do with it if Maple produced an "answer" that was 100 pages long).

## *APPENDIX III*

# *Answers to Odd-Numbered Exercises*

## Chapter 1. Preliminaries

### Lab 1 (page 13)

**1.** $8x^4$

**3.** **(a)** $2,\ 1+\sqrt{2},\ 1-\sqrt{2}$

**(b)** `> plot(P, x= -1 .. 3);`

**5.** **(a)** $3^{\pi}$ **(b)** 4 **(c)** Yes. **(d)** Yes. **(e)** 45

**7.** **(a)**
```
> plot(sqrt((100-30*t)^2 +
> (70-25*t)^2), t=0..4);
```
**(b)** 10.24 metres at 3.11 seconds. **(c)** 23 m/s

**9.** **(a)** All pass through $(-1,0)$ and $(1,2)$. **(b)** 0

**(c)** $y = b + \dfrac{d-b}{c-a}(x-a) + k(x-a)(x-c)$

**11.** If "q" is defined as `1+p` when "p" is not assigned a value, a later change in `p` changes `q` too. If `p` already has a value, that value is used in the definition of `q`, and later changes in `p` don't affect `q`.

### Lab 2 (page 20)

**1.** `> f:= x -> 2*x^2 - 1; g:= x -> 2*(x-1)^2;`

**(a)** 17 **(b)** $2t^2$ **(c)** $2x^4 - 4x^2 + 2$ **(d)** $8x^4 - 1$

**3.** **(a)** $a = -1$

**(b)** $f(x-2) - 7 = \dfrac{x^4 - 4x^3 + 8x - 13}{x^2 - 2x + 2}$

**5.** **(a)** Domain: $[-1/4, \infty)$, range: $[-1/4, \infty)$

**(b)** $[-1/4, \infty)$ **(c)** $f(-1/2)$ is the complex number $\frac{1}{2} - i$, but $f(\frac{1}{2} - i) = -1/2$. **(d)** $-\infty < x < 3/4$

**7.** **(a)**
```
> with(plots):
> display({plot(f(x),x=-1..-0.001),
>  plot(f(x),x=0.001..1)});
```
**(b)** `> simplify((1 - f(x)) - f(-x));`

**9.** **(a)** $16\cos^5 x - 20\cos^3 x + 5\cos x$

**(b)** $(\cos(5x) + 5\cos(3x) + 10\cos(x))/16$

**(c)**
```
> normal(eval(subs(cos = 1/sec,
> sin = tan/sec, expand(sin(3*x)))));
```

**11.** The variable name "a" appearing in the definition of `f` is not evaluated when "`->`" defines the function, so whenever the function is called, whatever value of `a` is current at that time is used. When "`unapply`" defines `g`, `a` is evaluated, so the value 1 rather than the name "a" becomes part of the definition of `g`.

## Chapter 2. Limits and Differentiation

### Lab 3 (page 29)

**1.** **(a)** Limit is $-2$. **(b)** Limit is 1. **(c)** Limit is $-1/2$.

**(d)** Limit is 1.

**3.** **(a)** Limit is 0.8. **(b)** 0.1353

**5.** **(a)** Limit is 0. **(b)** $|f(.00004)| < .1$, $|f(3\,10^{-44})| < .01$

**(c)** Appears to be 0.

**(d)** $g(x) = 1$ for such $x$. Limit doesn't exist.

**7.** **(a)** (i) **(b)** 14, 16, never

### Lab 4 (page 35)

**1.** **(a)** Derivative is a straight line.

**(b)** Derivative is a parabola.

**(c)** Original is an even function.

**(d)** Derivative is 0 at maxima and minima of the original.

**3.** **(a)** $y = 1 - (x-1)/2$ **(b)** Does not exist.

**(c)** Does not exist. **(d)** $y = 0$ **(e)** $y = x - 1/4$

**5.** **(a)** $m < 1$ **(b)** None. **(c)** Anything except 1.

**7.** $\mathrm{D}(f)(x)\,g(x) + f(x)\,\mathrm{D}(g)(x),\ -\dfrac{\mathrm{D}(f)(x)}{f(x)^2},$

$\dfrac{\mathrm{D}(f)(x)}{g(x)} - \dfrac{f(x)\,\mathrm{D}(g)(x)}{g(x)^2},\ \mathrm{D}(f)@g\,\mathrm{D}(g)$

**9.**
```
> chart([h, dq, normal(dq)],
> [seq((-.1)^k, k=2..10)]);
```

### Lab 5 (page 41)

**1.**
```
> f:= x -> cos(x^2);
> (D@@3)(f)(x)+4*x^2*D(f)(x)+12*x*f(x);
```

**3.** At $x = 0$, $f$ is positive, $f'$ negative and $f''$ zero.

**5.** **(a)** $\dfrac{x}{\sqrt{x^2-1}},\ -\left(x^2-1\right)^{-3/2},\ \dfrac{3\,x}{\left(x^2-1\right)^{5/2}},$

$-\dfrac{12\,x^2+3}{\left(x^2-1\right)^{7/2}},\ \dfrac{15\,x\left(4\,x^2+3\right)}{\left(x^2-1\right)^{9/2}},$

$-\dfrac{360\,x^4+540\,x^2+45}{\left(x^2-1\right)^{11/2}}$

**(b)** The $n$th derivative is a polynomial in $x$ divided by $(x^2-1)^{n-1/2}$. The polynomial is even when $n$ is even and odd when $n$ is odd.

**(c)** $\dfrac{d}{dx}\dfrac{P(x)}{(x^2-1)^{n-1/2}} = \dfrac{(x^2-1)P'(x) + (1-2n)xP(x)}{(x^2-1)^{n+1/2}}$

**7.** **(a)** $(\sin(1)-1)/(\sin(1)+1)$

**(b)** $(1.726742857, .9878649628)$, $(-1.726742857, -.9878649628)$

**(c)** $(-.3185870484, 2.817354163)$, $(.3185870484, -2.817354163)$

**(d)** $(1.560208294, .9872848186)$, $(-1.560208294, -.9872848186)$

**9.** **(a)** $y = (x\cos(x))/(1-\cos(x))$ **(b)** $0 < x < 2\pi$

## Chapter 3. Applications of Differentiation

### Lab 6 (page 50)

**1.** **(a)** 12.07019698 m/s **(b)** 10.24295 m at 3.114754098 s.

**3.** 1.023551089 $\mathrm{cm}^2/\mathrm{s}$

**7.** $1/(12\pi)$ cm/s

**9.** 3.134940955

**11.** .9484026269 × .125

## Lab 7 (page 58)

**1.** `> plot({x*sin(x), x}, x = 1 .. 2);`

**3.** **(a)** 1.802001501 **(b)** $-.04 < E(1.2) < .04$

**5.** **(b)** $-1.841470985x^2 \le E(x) \le 1.301168679x^2$

**(c)** $-.02330330717 \le x \le .02330330717$

**(d)** $-1.059543185x^2 \le E(x) \le 1.301168679x^2$, $-.02772255155 \le x \le .02772255155$

**7.** 4.71238898

**9.** **(a)** $1 - x + x^3/2 = 0$

**(b)** $x_0 = -1$ or $-2$ works; $x = -1.769292354$

**11.** **(a)** $g(x) \ge r$ where $r$ is the root near 0.9.

**(b)** e.g., $0.7 \le x \le 1$

**(c)** with $0.7 \le x_0 \le 1$, $x_8$ will do.

## Lab 8 (page 63)

**1.** **(a)** Attractors $\pm 1$, repeller 0. **(b)** They converge to 1.

**3.** **(a)** Yes. **(b)** .7390851332 is an attractor.

**(c)** no; yes (period 2).

**5.** **(a)** $1 < a < 3$

**(b)** $(a + 1 \pm \sqrt{a^2 - 2a - 3})/2$ for $a < -1$ or $a > 3$.

**(c)** $1 - \sqrt{6} < a < 1$ or $3 < a < 1 + \sqrt{6}$.

**7.** **(a)** .5173448557, 1.749999237, and 3.762499695, or .7058454252, 2.254579394, and 3.709731393.

**(b)** $(f \circ f \circ f)'(p) = -4.154715070$ or 5.334715063.

**(c)** 3.831874055

## Lab 9 (page 71)

**1.** All except (b).

**3.** **(a)** $-1 \le x \le 1$

**(b)** $g(x) = -(1 + I\sqrt{3})\left(-4x + 4\sqrt{-4 + x^2}\right)^{1/3}/4$
$-(1 - I\sqrt{3})\left(-4x + 4\sqrt{-4 + x^2}\right)^{-1/3}$

**(c)** when $-1 \le x \le 1$, $f^{-1}(f(x)) = x$; when $-2 \le x < 1$ or $1 < x \le 2$, it is the number $u$ with $-1 \le u \le 1$ and $f(u) = f(x)$.

**5.** **(a)** $x = t^{1/t}$; 1; $0 < x \le e^{1/e}$

**(b)** one for $0 < x \le 1$ and $x = e^{1/e}$, two for $1 < x < e^{1/e}$.

**(c)** $e^{-1} < t < e$; $e^{-e} < x < e^{1/e}$

**(d)** $f'(.08960084093)f'(.7494512696) = .6957671687$

**7.** **(a)** approximately 158 ft/s and 37 degrees.

**(b)** approximately 107 degrees/s and 55 degrees.

**9.** **(a)** $-1 \le x < \infty$ **(b)** "`LambertW`".

**(c)** Only apply the simplification if it is known (or assumed) that $x \ge -1$.

**(d)** $t = -\text{LambertW}(-\ln(x))\ln(x)$ for $0 < x \le e^{1/e}$; $0 < t \le e$.

## Lab 10 (page 81)

**1.** **(a)** Intercepts, (0, 0), (.3819660113, 0), (2.618033989, 0), local max. (.1835034191, .08866210787), local min. (1.816496581, −2.088662109), inflections (1, −1), asymptotes $y = \pm\pi/2$.

**(b)** inflections (−.8908987181, −.6154797086), (0, 0), (.8908987181, .6154797086).

**(c)** Min. (0, 0), inflections $(\pm 1, \ln 2)$.

**(d)** Local max $(-1 - \sqrt{2}, -2 - 2\sqrt{2})$, local min $(\sqrt{2} - 1, 2\sqrt{2} - 2)$, asymptotes $y = x$ and $x = -1$.

**(e)** Minimum (1, 1), asymptote $x = 0$.

**(f)** Min. $x = -1.254773748$, max. −.5858594692, loc. min. .6042824276, loc. max. .6851293726, inflections $x = -1.495835157, -.8274343692, -.3727936452, .6453392393, 1.212335431$.

**3.** **(a)** $f'(0) = \lim_{x\to 0} \dfrac{f(x) - f(0)}{x} = 0$.

**(b)** 0: local min., ±.5128056557: local max.

**(c)** ±.2364419550

**5.** **(b)** Increasing: $-\infty \le x \le -1/3$ and $1 \le x < \infty$. Decreasing: $-1/3 \le x \le 1$. Concave up: $-\infty \le x \le -1$. Concave down: $-1 \le x \le 1$ and $1 \le x < \infty$.

**7.** **(a)** $a \le x \le \min\{f(a)/A, c\}$

# Chapter 4. Integration

## Lab 11 (page 88)

**1.** **(a)** 385 **(b)** 2701/315 **(c)** 135

**3.** **(b)** $\tan((n + 1)a)/\sin a$

**5.** **(a)** Left: 32.8125, 35.165625; middle: 35.46875, 35.4171875; right: 37.8125, 35.665625;
lower: 29.34269754, 34.81846351;
upper: 41.25220988, 36.01276474.

**(b)** 425/12

**7.** **(a)** Left=lower, right=upper. **(b)** 614

**(c)** Errors are .03078457195, −.000065655629, −.03052195382. Middle sum is much more useful.

**9.** $-(a + b)(b - a)^2/2$, $(a + b)(b - a)^2/2$, 0

## Lab 12 (page 98)

**1.** $\ln(x^4 + 1)/4$

**3.** $\dfrac{3}{2}\ln(\sqrt{x} + 3) + \dfrac{1}{2}\ln|\sqrt{x} - 1|$

**5.** $\dfrac{1}{2}\ln\left(\sqrt{1 + a^2 + \tan^2 x} + \tan x\right)$
$-\dfrac{1}{2}\ln\left(\sqrt{1 + a^2 + \tan^2 x} - \tan x\right)$
$+ a\arctan\left(\dfrac{a\tan x}{\sqrt{1 + a^2 + \tan^2 x}}\right)$

**7.** $\sqrt{2}\ln((2 - \sqrt{2})/(2 + \sqrt{2}))/4 - \sqrt{2}(\arctan(-\sqrt{2} - 1) + \arctan(-\sqrt{2} + 1))/2$

**9.** $a_n = -\dfrac{n - 1}{n + 1}a_{n-2} - \dfrac{\arctan(x)}{n + 1}(x^{n+1} + x^{n-1}) + \dfrac{x^n}{n^2 + n}$

## Lab 13 (page 110)

**1.** **(a)** .6928353604, .6937714032, .6931502307; .003118202, −.006242226, −.000030501

**(b)** 1/32, −1/16, −1/32 **(c)** 18, 25, 5

**3.** **(a)** All three errors are $O(n^{-3/2})$. **(b)** 187490

**5.** **(a)** $O(n^{-2m-2})$ for all $m$.
**(b)** Errors are $2.64\ 10^{-11}$, $-2.64\ 10^{-11}$, $4.61\ 10^{-6}$.

**7.** $x = 1/t^2 - 1$, $n = 14$ will do.

### Lab 14 (page 118)

**1.** **(b)** They are the semi-major and semi-minor axes.
**(c)** The curve for $-a$ and $b$ or $a$ and $-b$ is the same as for $a$ and $b$, but with a different parametrization (and goes clockwise instead of counterclockwise).

**3.** **(a)** $x = \dfrac{a(b^2 - m^2a^2)}{b^2 + m^2a^2}$, $y = \dfrac{2mab^2}{b^2 + m^2a^2}$

**(b)** $2ab^2 \displaystyle\int_{-\infty}^{\infty} \frac{\sqrt{m^4a^4 + (4a^4 - 2a^2b^2)m^2 + b^4}}{(b^2 + m^2a^2)^2}\, dm$

**(c)** $t = \arctan(2mba,\ b^2 - m^2a^2)$

**5.** **(a)** $a/b$ must be an integer. **(b)** 12
**(c)** $(a^2 + 3ab + 2b^2)\pi$

**7.** **(b)** $[r,\ \theta]$ and $[-r,\ \theta \pm \pi]$ correspond to the same point.

**9.** **(a)** $x = \dfrac{t\cos\alpha - (t^2 - 1/4)\sin\alpha}{\cos\alpha - 2t\sin\alpha}$,
$y = \dfrac{t^2\cos\alpha + (t/2)\sin\alpha}{\cos\alpha - 2t\sin\alpha}$
**(b)** The curve is the straight line $y = -1/4$.

## Chapter 5. Series

### Lab 15 (page 127)

**1.** Each difference is about 1/4 the previous one. Series converges, sum is about 1.2021.

**3.** **(a)** If $p > -1$, $a_k > k^{-1}(\ln k)^{-1}$ for sufficiently large $k$. If $p < q < -1$, $a_k < k^{-1}(\ln k)^q$ for sufficiently large $k$.
**(b)** Limit is $-\infty$ so series converges.
**(c)** Limit is $+\infty$ so series diverges. **(d)** $\displaystyle\sum_{k=2}^{\infty} \frac{1}{k\ \ln k}$

**5.** **(a)** No. **(b)** Yes.

**7.** Sum is $-1/20$.

**9.** **(b)** $(B+1)\ln B - B\ln(B+1) + A\ln(A+1) - (A+1)\ln A$, $(B+3)\ln(B+1) - (B+2)\ln(B+2) + (A+2)\ln(A+2) - (A+3)\ln(A+1)$. **(c)** 616

### Lab 16 (page 135)

**1.** **(a)** Interval grows by about the same amount for each new term. **(b)** Interval approaches $(-\pi/2,\ \pi/2)$. **(c)** Interval approaches $(-1, 1)$. **(d)** Interval grows indefinitely, but slower each time. **(e)** Interval approaches $(-1, 1)$.

**3.** $1 - \dfrac{1}{2}x^4 + \dfrac{1}{2}x^5 - \dfrac{7}{24}x^6 + \dfrac{1}{4}x^7 - \dfrac{5}{24}x^8 + \dfrac{3}{20}x^9$

**5.** **(a)** The curve has a vertical tangent. **(b)** $3^{3/8}2^{-1/2}$
**(c)** $x^3/2 + x^{11}/32$ **(e)** $.1492\,10^{-7}$

**7.** $1 + \dfrac{1}{6}x + \dfrac{3}{40}x^2 + \dfrac{5}{112}x^3 + \dfrac{35}{1152}x^4 + \dfrac{63}{2816}x^5$

## Chapter 6. Vectors and Geometry

### Lab 17 (page 142)

**1.** **(a)** $[28,\ -4,\ -36]$ **(b)** $[5,\ 20,\ -15]$
**(c)** $-20$ **(d)** $\pi - \arccos(\sqrt{2}/3)$

**3.** $\pi - \arccos(1/3)$

**5.** **(a)** $(-5,\ 14,\ -3)$ **(b)** $\pi/2 - \arccos(1/\sqrt{46})$
**(c)** (15.28921142, −13.39236524, 13.22750948) or (−.957596336, 10.97784638, 5.104105602).

**7.** $ri = \dfrac{1}{\sqrt{50 - 22\sqrt{5}}}$, $rc = \dfrac{1}{4}\sqrt{6}\sqrt{\dfrac{7 - 3\sqrt{5}}{9 - 4\sqrt{5}}}$

**9.** There are eight, including $(x-1)^2 + (y-1)^2 + (z-1)^2 = 7^2$ and $(x - 45.88326527)^2 + (y + 79.61991228)^2 + (z + 45.05868810)^2 = 97.85198722^2$.

### Lab 18 (page 150)

**3.** **(a)**
```
> plot3d(sqrt(x^2+y^2), x=-1..1,
> y=-sqrt(1-x^2)..sqrt(1-x^2),
> scaling=constrained);
```
**(b)**
```
> plot3d({sqrt(1-x^2-y^2),
> -sqrt(1-x^2-y^2)},
> x=-1 .. 1, y=-.99999*sqrt(1-x^2)
> .. .99999*sqrt(1-x^2),
> scaling=constrained);
```
**(c)**
```
> plot3d({3/2*sqrt(1-4*x^2-y^2),
> -3/2*sqrt(1-4*x^2-y^2)},
> x=-.49999 .. 0.49999,
> y=-.99999*sqrt(1-4*x^2) ..
> .99999*sqrt(1-4*x^2),
> scaling=constrained);
```
**(d)**
```
> plot3d({sqrt(1+x^2-y^2),
> -sqrt(1+x^2-y^2)},
> x=-1 .. 1, y=-.99999*sqrt(1+x^2)
> .. .99999*sqrt(1+x^2),
> scaling=constrained);
```

**5.**
```
> plot3d(1- abs(y)/sqrt(1-x^2),
>   x=-0.9999..0.9999,
>   y=-sqrt(1-x^2)..sqrt(1-x^2));
```

**7.** When $z = 2$, the coefficient of "kk" is 0. Replace $z - 2$ by $z^2 - 4$ or $(z - 2)x$.

**9.**
```
> contourplot((x-1)^2 + y^2-1/2)*((x+1)^2
>   + y^2 -1/2)* (x^2/10+(y+1)^2-1/2),
>   x = -2.5 .. 2.5, y = -1.8 .. 0.9,
>   axes = FRAMED, contours=[-1.2, 1],
>   filled=true, coloring=[khaki,green]);
```

## Chapter 7. Partial Differentiation

### Lab 19 (page 157)

**1.**
```
> diff(z, x,y$2), diff(z, y,x,y),
>   diff(z,y$2,x);
```

**3.** **(b)** $\dfrac{\partial^2 f}{\partial x \partial y} = -g'(x)h'(y)f(x, y)$

**5.** **(a)** [1.618033989, .6180339890], [−.6180339890, −1.618033989], [−.7937005260, .7937005260]

**7.** **(a)** $2\sin\phi\cos\theta - 4\cos\phi$
**(b)** Direction of $[\sin\phi\cos\theta,\ \sin\phi\sin\theta,\ \cos\phi]$.
**(c)** $\phi = \pi - \arctan(1/2)$, $\theta = 0$

**9.** **(a)** $[0,\ 2q(a,b),\ 2D_2(f)(a,b)q(a,b)]$ and $[2p(a,b),\ 2r(a,b)p(a,b),\ 2D_1(f)(a,b)p(a,b) + 2D_2(f)(a,b)r(a,b)p(a,b)]$.

**(b)** $r(a,b) = -\dfrac{D_2(f)(a,b)D_1(f)(a,b)}{1 + D_2(f)(a,b)^2}$,
$p(a,b) = \dfrac{3}{10}\sqrt{\dfrac{1 + D_2(f)(a,b)^2}{D_2(f)(a,b)^2 + 1 + D_1(f)(a,b)^2}}$,
$q(a,b) = \dfrac{3}{10}\dfrac{1}{\sqrt{1 + D_2(f)(a,b)^2}}$

## Lab 20 (page 166)

**1.** $-2\dfrac{x^5 - 3z^4y - 2z^3x^2 + 4x^2y^3}{(2yz + x^2)^3}$

**3.** $-(D_{1,2}(F)(x,y,z)D_3(F)(x,y,z)^2$
$- D_3(F)(x,y,z)D_{1,3}(F)(x,y,z)D_2(F)(x,y,z)$
$- D_1(F)(x,y,z)D_{2,3}(F)(x,y,z)D_3(F)(x,y,z)$
$+ D_1(F)(x,y,z)D_{3,3}(F)(x,y,z)D_2(F)(x,y,z))$
$/(D_3(F)(x,y,z)^3)$

**7. (a)** $x + 1 - 2y - \frac{1}{2}(x-1)^2 + 2(x-1)(y-1)$
$-3(y-1)^2 + \frac{1}{3}(x-1)^3 - 2(x-1)^2(y-1)$
$+5(y-1)^2(x-1) - \frac{14}{3}(y-1)^3$

## Lab 21 (page 174)

**1. (a)** saddle. **(b)** minimum.

**5. (a)** $(0,0)$ is a saddle point.
**(b)** $(n\pi, -n\pi)$ is a saddle for any integer $n$.
**(d)** $(1.813471007, -1.328121646)$ is a local minimum.
**(e)** $\partial^2 f/\partial x^2$ and $\partial^2 f/\partial y^2$ are always positive.

**7. (a)** $(.6426667844, -1.961526843)$, $(1.995014278, -.8752838891)$, $(1.213451017, .7701014499)$, $(-1.498414482, 1.896406235)$
**(b)** $f(1.213451017, .7701014499) = -.2187545609$
**(c)** $-6.459457429$ at $(.5176380902, -1.931851653)$, $9.078185377$ at $(-.719427824, 1.866125286)$.

**9.** Closest points $(\pm.8372719302, \pm.8372719302, \pm1.139726158)$ with an even number of "$-$" signs. Farthest points $(\pm1.395437717, \pm1.395437717, \pm.229681471)$ with an odd number of "$-$" signs.

## Lab 22 (page 181)

**1.** $(.6140640208, 2.137676172)$, $(1.197913901, 2.023147422)$, $(2.157979960, -.3670432295)$

**3. (a)** $\dfrac{x}{\sqrt{x^2+y^2}} + \dfrac{x-1}{\sqrt{(x-1)^2+y^2}} + \dfrac{x}{\sqrt{x^2+(y-1)^2}} + \dfrac{x-2}{\sqrt{(x-2)^2+(y-1)^2}} = 0$,
$\dfrac{y}{\sqrt{x^2+y^2}} + \dfrac{y}{\sqrt{(x-1)^2+y^2}} + \dfrac{y-1}{\sqrt{x^2+(y-1)^2}} + \dfrac{y-1}{\sqrt{(x-2)^2+(y-1)^2}} = 0$
**(b)** $(1/3, 2/3)$
**(c)** The solution is the intersection of the line segments joining $(0,0)$ to $(2,1)$ and $(1,0)$ to $(1,1)$. Any point on a line segment is a minimum of the sum of the distances to the endpoints of the segment (because a line is the shortest distance between two points). A point that is a minimum of one function and a minimum of another is a minimum of the sum of the two.

**5. (a)** $a = 9$, $b = -4$, $c = 12$
**(b)** $1 + \frac{4}{9}x - \frac{2}{9}y$ and $2 - \frac{1}{9}x + \frac{1}{18}y$
**(c)** $K > .5121969143$. It converges when the initial point is close enough to $(1,2)$.
**(d)** 63 **(e)** `newt([x, y])` $\to \dfrac{2x^3+1}{3x^2}[1,2]$ if $x \neq 0$.
**(f)** Undefined: the right and left limits are $\pm\infty$ if $x \neq 1$.

# Chapter 8. Multiple Integration

## Lab 23 (page 187)

**1. (a)** .239948660 **(b)** .01698274990

**3.**
```
> ti:= f -> int(int(int(f,
>   z=0..x*y-x^3),x=0..sqrt(y)),
>   y=0..1);
> [ti(x)/ti(1), ti(y)/ti(1),
>   ti(z)/ti(1)];
[16/35, 3/4, 32/315]
```

**5. (a)** $\displaystyle\int_0^1 \int_{\sqrt{z}}^{2-\sqrt{1-z}} f(y,z)\,dy\,dz$
**(b)** $\displaystyle\int_0^1 \left( \int_0^a \int_{2x^2-x}^{3x-2x^2} f(x,y,z)\,dy\,dx + \int_a^b \int_{x-\frac{\arccos(z)}{\pi}}^{x+\frac{\arccos(z)}{\pi}} f(x,y,z)\,dy\,dx + \int_b^1 \int_{2x^2-x}^{3x-2x^2} f(x,y,z)\,dy\,dx \right) dz$
where $a = \left(1 - \sqrt{1 - \frac{2}{\pi}\arccos(z)}\right)/2$
and $b = \left(1 + \sqrt{1 - \frac{2}{\pi}\arccos(z)}\right)/2$.

**7.** 9.414795026

## Lab 24 (page 192)

**1.** 15

**3.** $\frac{1}{6}\sqrt{2} - \frac{1}{6}\ln\left(\sqrt{2} - 1\right)$

**5.** $r = 1.982939186$, $\theta = z = 0$

**7. (b)**
```
> animate({sqrt(1-(r-2)^2),
> -sqrt(1-(r-2)^2),
> 1+r*cos(theta)},
> theta=0 .. 2*Pi, r=1 .. 3);
```
**(c)** $\arccos\left(\left(\sqrt{-3 - r^2 + 4r} - 1\right)/r\right)$
and $2\pi - \arccos\left(\left(\sqrt{-3 - r^2 + 4r} - 1\right)/r\right)$
**(d)** $2 \le r \le 3$,
$\theta = \pi \pm \arccos\left(\left(\sqrt{-3 - r^2 + 4r} - 1\right)/r\right)$.
**(e)** $14.83191477 - 2.293666170 = 12.53824860$

# Chapter 9. Surfaces, Curves, and Fields

## Lab 25 (page 204)

**1. (b)** `orientation=[45, 35.26438969]`

(c) $\kappa = \sqrt{2}/8$, $\tau = -\sqrt{6}/8$, $\mathbf{T} = [1 + \cos t - (\sqrt{6}/3)\sin t, 1 - \cos t - (\sqrt{6}/3)\sin t, 2 + (\sqrt{6}/3)\sin t]/\sqrt{8}$, $\mathbf{N} = [-\sin(t)/\sqrt{2} - \cos(t)/\sqrt{3}, \sin(t)/\sqrt{2} - \cos(t)/\sqrt{3}, \cos(t)/\sqrt{3}]$, $\mathbf{B} = [(\sqrt{6}/4)\cos(t) - \sqrt{6}/12 - \sin(t)/2, -(\sqrt{6}/4)\cos(t) - \sqrt{6}/12 - \sin(t)/2, \sin(t)/2 - \sqrt{6}/6]$

3. (a) $x = r(\theta)\cos\theta$, $y = r(\theta)\sin\theta$, $z = r(\theta)^2$ where $r(\theta) = (\cos^2\theta - \sin^2\theta)/\sin\theta$.

5. (a) $\mathbf{R} = [\cos(t) + 3\cos(2t), 3\sin(2t) - \sin(t), 1.2\sin(3t)]$, radius 0.9.

   (c) $[3\cos(2t)+\cos(3t), -3\sin(2t)+\sin(3t), 1.2\sin(5t)]$, radius 0.6.

7. (a) 6.5 m/s (b) $x = 31.67$

9. (a) $(g+gf_2^2+v^2f_{22})yp'^2+(2gf_1f_2+2f_{12}v^2)yp'+g+gf_1^2+f_{11}v^2 = 0$

## Lab 26 (page 212)

1. Rays from the origin, and circles centred at the origin.

3. True for all $n$; $r^{2-n}/(2-n)$ for $n \neq 2$, $\ln r$ for $n = 2$.

5. (c)
```
> eqn:= (x(t)^2+y(t)^2)^2
> - 2*a*x(t)*y(t) = 0;
> Fp:= subs(x=x(t), y=y(t), F);
> ae:= solve(eqn, a);
> normal(subs(diff(x(t),t)=Fp[1],
>   diff(y(t),t)=Fp[2], a=ae,
>   diff(eqn,t)));
```

7. (a) $a = -1 \pm \sqrt{2}$ (b) $b = c = 0$, $d = 2a$

9. (a) $\arctan(x^2 + y + 1)$ (b) A family of parabolas opening downwards, vertical translates of each other.

## Lab 27 (page 219)

1. (a)
```
> plot3d([u, sin(u)*cos(v),
> sin(u)*sin(v)], u=0..Pi,
> v=0..2*Pi);
```
   (b)
```
> plot3d([u*cos(v), sin(u),
> u*sin(v)], u=0..Pi,
> v=0..2*Pi);
```
   (c)
```
> plot3d([
> (1-sin(u))*cos(u)*cos(v),
> (1-sin(u))*sin(u),
> (1-sin(u))*cos(u)*sin(v)],
> u=0..2*Pi, v=0..2*Pi);
```

3.
```
> xp:= 2 + q*cos(p); yp:= q*sin(p);
> rp:= sqrt(xp^2 + yp^2);
> zp:= sqrt(1-(rp-2)^2);
> plot3:= plot3d([xp,yp,zp], q=0..1,
>   p=0..2*Pi, scaling=constrained):
> plot4:= plot3d([xp,yp,-zp],q=0..1,
>   p=0..2*Pi, scaling=constrained):
> display({plot2, plot3, plot4});
```

5. (b) [.6336697591, .3018925961, .1612127918] for $\phi$ = 1.065989495, $\theta$ = .7612528329; [−.9455967744, −.1383100445, −.05708479713] for $\phi$ = 1.742899056, $\theta$ = 3.426186771.

   (c) Use `orientation=[theta,phi]` where $.09243846805\sin\phi\cos\theta + .1761579353\sin\phi\sin\theta + .2116565761\cos\phi = 0$.

7. (a) $[\cos(u/2)v, u, \sin(u/2)v]$, $0 \le u \le 2\pi$, $-1/2 \le v \le 1/2$

   (b)
```
> plot3d([2+cos(u/2)*v, u,
> sin(u/2)*v], u=0..2*Pi,
> v=-.5..0.5, coords=cylindrical);
```

9.
```
> z0:= (2*w^2-1)/(sqrt(1-w^2)*w);
> r0:= sqrt(4+ z0^2);
> plot3d([r0*cos(t), (1-w)*Pi/2,
>   z0 + r0*sin(t)], t=0..Pi,
>   w=.1 .. 0.95, coords=cylindrical);
```

## Lab 28 (page 224)

1. $4\sqrt{2}\pi^4 + \sqrt{2}\pi$ and $8\pi^3 - 12\pi$

3. $-\pi/8$

5. Area: $(\sqrt{2} + \ln(\sqrt{2} + 1))\pi$; Flux: $\pi^2$.

7. $4\pi G\sigma a$ for $z \le a$, $4\pi G\sigma a^2/z$ for $z > a$.

## Lab 29 (page 231)

1. (a) $2x^2y^2z - 2xy^2z^2 + 2x^2yz^2$

   (b) $[-3y^2z^2 - x^3z + 3xyz^2, 3x^2z^2 + xy^3 + 3x^2yz, 3x^2y^2 - yz^3 - 3xy^2z]$ (c) $8(x^2 + y^2 + z^2)$

   (d) $[2x^3y^2 - x^2y^2z, 2y^3z^2 - xy^2z^2, 2x^2z^3 + x^2yz^2]$

3. (b)
```
> F:= [F1(x,y,z), F2(x,y,z),
>   F3(x,y,z)];
> G:= grad(f(x,y,z),xyz);
> Q:= diverge(F &x G,xyz) -
>   G &. curl(F, xyz);
> expand(Q);
```

5. (a) $-\pi/4$

7. (a) $[x^2z - y^3/3, 0, yz^2]$

   (b) $f(x, y, z) = yz^3/3$, curl $(\mathbf{V} - \mathbf{W}) = \mathbf{F} - \mathbf{F} = 0$

   (c) $\nabla^2 g = -2xz$ (d) $g = -x^3z/3$

9. (a) Everywhere except the origin.

   (b) $[yz, -xz, 0]/\left(\left(x^2 + y^2\right)\sqrt{x^2 + y^2 + z^2}\right)$, defined everywhere except the $z$ axis.

   (c) Consider a sphere centred at the origin.

# Chapter 10. Differential Equations

## Lab 30 (page 245)

1. (a) $y + \ln\left(y^2 + 1\right) - \frac{1}{2}\ln\left(x^2 + 1\right) = c$

   (b) $x^2/2 + x\sin y + y^2 = c$

   (c) $-\cos x + \sin x$ (d) $(y - x)^2 = c(y^2 - 1)$

3. (a) $y(x) = (x - c)^5/1200$

   (b) $x^2/2 + x^5/120 - x^8/4480 + O(x^9)$

5. (a) $(x, v) \to (0, 0)$, $(+\infty, +\infty)$, or $(-\infty, -\infty)$.

   (b) $(x, v) \to (-\infty, -\infty)$ or closed loop.

7. (a) $y' = 1 - x^2 - y^2$ (b) $y' = x(y - x^2)$

9. $x = -1.281419294$ and $.5298511667$.

11.
```
> att:= DEplot3d(Lorenz, [x(t),y(t),z(t)],
>   t=0..50, {[x(0)=5,y(0)=10,z(0)=10]},
>   scene=[x,y,z], stepsize=.01,
>   thickness=0, linecolour=black):
> S:= dsolve({op(Lorenz), x(0)=5, y(0)=10,
>   z(0)=10}, [x(t),y(t),z(t)], numeric);
> for i from 0 to 799 do P[i]:=
>   subs(S(i*.04), [x(t),y(t),z(t)]) od:
> for i from 0 to 19 do frame[i]:=
>   pointplot3d({seq(P[i+20*j],j=0..39)},
>   style=point,colour=red) od:
> display({att, display([seq(frame[i],
>   i=0..19)], insequence=true)});
```

## *APPENDIX IV*

# *Coordination Chart*

This appendix shows the correspondence between the labs in this manual and sections of several popular calculus texts. The correspondence is not always perfect: there may be sections of the lab containing material not covered in the text, and of course not all material from the text is referred to in the lab. If your text is not listed here, don't worry: you can probably get a fairly good idea of the correspondences with your text by comparing tables of contents.

The following texts are referred to:

**Adams:** R.A. Adams, *Calculus: A Complete Course* (4th ed.). Don Mills, Ont.: Addison Wesley Longman, 1999.

**E & P:** C.H. Edwards and D.E. Penney, *Calculus with Analytic Geometry* (4th ed.). Englewood Cliffs, NJ: Prentice Hall, 1998.

**HHG:** D. Hughes-Hallett, A.M. Gleason et al., *Calculus.* New York: John Wiley & Sons, 1994, and W.G. McCallum, D. Hughes-Hallett, A.M. Gleason et al., *Multivariable Calculus.* John Wiley & Sons, 1996.

**SHE:** G.J. Etgen, *Salas and Hille's Calculus: One and Several Variables, revised by Garret J. Etgen* (8th ed.). New York: John Wiley & Sons, 1999.

**Stewart:** J. Stewart, *Calculus* (4rd ed.). Pacific Grove, CA: Brooks/Cole Publishing, 1999.

**Stewart-E:** J. Stewart, *Calculus, Early Transcendentals* (4rd ed.). Pacific Grove, CA: Brooks/Cole Publishing, 1999.

**T & F:** G.B. Thomas and R.L. Finney, *Calculus and Analytic Geometry* (9th ed.). Reading, Mass.: Addison-Wesley Publishers, 1996.

| Lab | Adams | E & P | HHG | SHE | Stewart | Stewart-E | T & F |
|---|---|---|---|---|---|---|---|
| 1 | P.3 | 1.1 | — | 1.4 | 1.2 | 1.2 | P.4 |
| 2 | P.4-6 | 1.1-4 | 1.9-10 | 1.5-7 | 1.1-4 | 1.1-4 | P.3,5 |
| 3 | 1.2-3,5 | 2.2-3 | 2.7 | 2.1-3 | 2.2-4 | 2.2-4, 6 | 1.1-4 |
| 4 | 2.1-5 | 2.1, 3.1-4 | 2.1-3, 4.1-2,4-5 | 3.1-6 | 3.1-6 | 3.1-5 | 1.6, 2.1-2,5 |
| 5 | 2.8-10 | 3.8, 4.6, 5.2 | 2.5, 4.8, 6.4 | 3.3,7, 5.5 | 3.7-8, 4.10 | 3.6-7, 4.10 | 2.6 |
| 6 | 4.1-2,5 | 3.5-6,8 | 5.1-2, 5.5-6 | 3.8, 4.3-5 | 3.9, 4.1,7 | 3.10, 4.1,6 | 2.7, 3.1,3,6 |
| 7 | 4.6-7 | 3.9, 4.2 | 4.9, 5.7 | 3.9 | 3.10, 4.9 | 3.11, 4.9 | 3.7-8 |
| 8 | 4.6 | 3.9, 11.2 | App. A | 3.9 | 12.1 | 11.1 | — |
| 9 | 3.1-3,5 | 7.1-4, 8.2 | 1.5-7,10 | 7.1-5,7 | 7.1-5 | 1.5-6, App. D | 6.1-3, 6.8-9 |
| 10 | 2.6, 4.3-4 | 4.3-7 | 5.1-3 | 4.2,6-8 | 4.2-6 | 4.2-5 | 3.2,4-5 |
| 11 | 5.1-5 | 5.3-6 | 3.1-4 | 5.1-4 | 5.1-4 | 5.1-4 | 4.4-7 |
| 12 | 5.6, 6.1-4 | 9.1-7 | 7.2-4 | 8.2-6 | 5.5, 8.1-6 | 5.5, 7.1-6 | 7.1-5 |

| Lab | Adams | E & P | HHG | SHE | Stewart | Stewart-E | T & F |
|---|---|---|---|---|---|---|---|
| 13 | 6.6-8 | 5.9 | 7.6-7 | 8.7 | 8.7-8 | 7.7-8 | 4.9, 7.6 |
| 14 | 8.2-6 | 10.2-5 | 12.1-3 | 9.3-9 | 11.1-5 | 10.1-5 | 9.4-9, 10.1-3 |
| 15 | 9.2-4 | 11.3-7 | — | 11.1-4 | 12.2-7 | 11.2-7 | 8.3-7 |
| 16 | 9.5-9 | 11.8-9 | 10.1-5 | 11.5-9 | 12.8-12 | 11.8-12 | 8.8-11 |
| 17 | 10.1-4,6 | 12.2-4 | 11.2, 12.4 | 12.1-7 | 13.1-5 | 12.1-5 | 10.2-5 |
| 18 | 12.1 | 13.2 | 11.3,4 | 14.3 | 15.1 | 14.1 | 12.1 |
| 19 | 12.3-5,7 | 13.4,7-8 | 13.1-7 | 14.4, 15.1-2,4 | 15.3-6 | 14.3-6 | 12.3-5,7 |
| 20 | 12.8-9 | — | 13.9, App. I | 15.7 | 15.4-5 | 14.4-5 | 12.5,10 |
| 21 | 13.1-3 | 13.5,9-10 | 14.1-3 | 15.5-6 | 15.7-8 | 14.7-8 | 12.8,9 |
| 22 | 13.6 | — | — | — | — | — | — |
| 23 | 14.1-2, 5 | 14.1-2,6 | 15.1-3 | 16.2-3,6-7 | 16.1-3,7 | 15.1-3,7 | 13.1,4 |
| 24 | 14.4,6 | 14.4,7 | 15.5-6,8 | 16.4,8-10 | 16.4,8-9 | 15.4,8-9 | 13.3,6 |
| 25 | 11.1-5 | 12.5-6 | 16.1-3 | 13.3-5,7 | 14.1-4 | 13.1-4 | 11.1,3 |
| 26 | 15.1-2 | 15.1 | 17.1-2 | 15.8 , 17.3 | 17.1 | 16.1 | 14.2,3 |
| 27 | 15.5 | 14.8 | 16.4 | 17.6 | 17.6 | 16.6 | 14.6 |
| 28 | 15.3-6 | 15.2-3,5 | 18.1-3, 19.1-3 | 17.1-4,6-7 | 17.2-3,7 | 16.2-3, 7 | 14.1,5 |
| 29 | 16.1-5 | 15.4,6-7 | 20.1-5 | 17.5,8-10 | 17.4-5,8-9 | 16.4-5,8-9 | 14.4,7-8 |
| 30 | 17.1-6 | 6.5, 7.6 | 9.1-4,8-9 | 18.1-4 | 10.1-7 | 9.1-7 | 6.11-12 |

## APPENDIX V

# Glossary of Maple Commands

`"` Last result (Release 4). 4, 5.

`""` Second-last result (Release 4). 5.

`"""` Third-last result (Release 4). 5.

`%` Last result (Release 5). 4, 5.

`%%` Second-last result (Release 5). 5.

`%%%` Third-last result (Release 5). 5.

`&*` Matrix multiplication sign: `M &* V` for $M\mathbf{V}$. 141.

`&*()` Identity matrix. 142.

`&.` Dot product operator: `A &. B` for $\mathbf{A} \bullet \mathbf{B}$ (in "`dot.def`"). 115, 247.

`&x` Cross product operator: `A &x B` for $\mathbf{A} \times \mathbf{B}$ (in "`dot.def`"). 139, 247.

`''` Quotes: `'x'` for unevaluated name `x`. 25, 254.

`()` Parentheses, used for grouping and for inputs to functions and commands. 14, 17, 250, 251.

`*` Multiplication sign: `a*b` for $ab$. 3, 4, 251.

`**` Power sign: `a**b`, equivalent to `a^b`, for $a^b$. 4.

`->` Arrow for function definition: `x -> x^2`. 21, 55.

`..` Dots indicating an interval: `a .. b` represents numbers from `a` to `b`. 6.

`/` Division sign: `a/b`, or directory separator: `` `c:/maple/myfile.txt` ``. 9.

`:` End of command, result not to be shown. 2, 4, 31, 250.

`:=` Assignment sign: `a := b+1` assigns value of `b+1` as value of variable `a`. 4, 5.

`;` End of command, result to be shown. 2, 4, 31, 250.

`=` Equal sign in an equation (not an assignment): `a = b`. 5.

`?` Help on following word: `?abs`. 6, 254.

`@` Composition sign: `f@g` for $f \circ g$. 17.

`@@` Iteration sign: `f@@3` for $f \circ f \circ f$, in particular `(D@@n)(f)(x)` for $n$th derivative $f^{(n)}(x)$. 36, 62

`[]` List delimiters: `[a,b,c]`. 12, 250.

`^` Power sign: `a^b` for $a^b$. 4.

``` `` ``` Back quotes: `` `string` `` for a literal string. 9.

`{}` Set delimiters: `{a,b,c}`. 11, 12, 250.

`about` Show assumptions made on a variable `x`: `about(x)`. 68.

`abs` Absolute value: `abs(x)` for $|x|$. 17, 33.

`add` Add a sequence of values: `add(f(n), n =1..5)`. 102

`additionally` Add additional assumptions on a variable `x`: `additionally(x > 3)`. 69.

`align` Option in "`textplot`" for alignment of labels: `align=ABOVE`, `BELOW`, `LEFT`, `RIGHT` or set of these. 112-113.

`ambientlight` Option in 3D plotting for non-directional light source: `ambientlight=[r,g,b]` for red, green, blue components $r$, $g$ and $b$. 145.

`and` Logical operator: `a and b` is true when both `a` and `b` are true. 169.

`angle` Angle between vectors **A** and **B**: `angle(A,B)` (in "`linalg`" package). 116.

`animate` Animation of graphs of functions: `animate(f(x,t), x=a..b, t=c..d)` (in "`plots`" package). 12, 184.

`animate3d` Animation of 3D graphs of functions: `animate3d(g(x,y,t), x=a..b, y=c..d, t=e..f)` (in "`plots`" package). 150.

`arccos` Inverse cosine function: `arccos(x)`. 69.

`arccot` Inverse cotangent function: `arccot(x)`. 69.

`arccsc` Inverse cosecant function: `arccsc(x)`. 69.

`arcsec` Inverse secant function: `arcsec(x)`. 69.

`arcsin` Inverse sine function: `arcsin(x)`. 69.

`arctan` Inverse tangent function: `arctan(x)` or `arctan(y,x)`. 69, 70, 117.

`arrow3d` Draw 3D arrow to represent vector **V**, starting at **R**: `arrow3d(R,V)` (in "`arrow3d.def`"). 195, 216, 248.

`arrows` Option in "`DEplot`", "`fieldplot`" and "`fieldplot3d`" for arrow type: `arrows=LINE`, `THIN`, `SLIM`, `THICK` or `NONE`. 206, 235.

`assign` Assign variables corresponding to equations: `assign({a=b, c=d})` assigns `a:= b` and `c:= d`. 45.

`assume` Make assumption on a variable `x`: `assume(x < 4)`. 92, 68, 258.

`asympt` Asymptotic expansion of expression as a variable $x \to \infty$: `asympt(x/(1+x^2), x)`. 125, 134.

`axes` Option in all plot commands to control type of axes: `axes=BOXED`. 26, 148.

`BesselI` Modified Bessel function of the first kind: `BesselI(v,z)` (usually written mathematically as $I_v(z)$). 135.

`between` Test whether $a$ is between $b$ and $c$: `between(a,b,c)` (in "`riemann.def`"). 86, 247.
```

`BOXED` Choice for "`axes`" in plot commands, putting axis labels outside a two- or three-dimensional box drawn around the plot: `axes=BOXED`. 148.

`changevar` Change variables in integral, sum or limit: `changevar(x=t^2, Int(f(x),x), t)` (in "`student`" package). 83, 90.

`chart` Make chart of values of expressions for given values of variable `x`: `chart([x,f(x),g(x)], [1,2,3,4])` (in "`chart.def`"). 24, 25, 29, 247.

`CIRCLE` Choice for "`symbol`" when plotting points: `symbol=CIRCLE`. 75, 112.

`classical` Choice for "`method`" in "`dsolve(..., numeric)`", for Euler, Improved Euler or Runge-Kutta methods: `method=classical[foreuler]`. 240, 242.

`classify` Classify multivariate critical point with Hessian matrix $H$ using second derivative test: `classify(H)` (in "`classify.def`"). 169, 247.

`coeff` Coefficient of $x^n$ ($x$ a variable or expression and $n$ a given integer) in polynomial or series $p$: `coeff(p,x,n)`. 125, 133.

`collect` Write expression as a sum of coefficients times powers of variable `x`: `collect((x+1)^2, x)`. 5, 228.

`color` or `colour`
Option in plot commands to control colour of curve, surface etc.: `colour=black, blue, brown` etc. 54, 84, 147, 206.

`coloring` Option in "`contourplot`" to specify colours of contours, or the regions between them: `coloring=[red, yellow]`. 148.

`colourstyle`
Option in "`densityplot`" to use colour instead of shades of gray: `colourstyle=HUE`. 149.

`combine` Combine terms into a single term when possible: `combine(Int(f(x),x)+Int(g(x),x))`. `combine(f,trig)` writes products and powers of trigonometric functions as functions of sums, differences and multiples of angles. 19, 70, 82, 90.

`completesquare`
Complete the square for a quadratic expression in variable `x`: `completesquare(x^2 + 2*x + 3, x)` (in "`student`" package). 95.

`CONTOUR` Choice for "`style`" in 3d plot commands, drawing curves of constant $z$: `style=CONTOUR`. 145.

`contourplot`
Plot level curves of an expression in two variables $x$ and $y$: `contourplot(f(x,y), x=a..b, y=c..d)` (in "`plots`" package). 147, 171, 235.

`contourplot3d`
3D plot of level curves of an expression in two variables $x$ and $y$: `contourplot3d(f(x,y), x=a..b, y=c..d)` (in "`plots`" package). 148, 196.

`contours` Option in "`contourplot`" and "`contourplot3d`" to control which levels are plotted: `contours=10` or `contours=[-2, .5]`. 148, 167.

`convert` Convert an expression to a different type: `convert(R, parfrac, x)` for partial fractions where `R` is a rational function in variable `x`, `convert(S, polynom)` for polynomial where `S` is a series, `convert(V, list)` where `V` is a vector. 94, 97, 115, 129, 227.

`coords` Option in "`plot`" and "`plot3d`" for non-Cartesian coordinates: `coords=polar, cylindrical` or `spherical`. 117, 214.

`cos` Cosine function: `cos(x)`. 19.

`cot` Cotangent function: `cot(x)`. 19.

`crossprod` Cross product of 3D vectors: `crossprod(A,B)` (in "`linalg`" package). 139. See also "`&x`".

`csc` Cosecant function: `csc(x)`. 19.

`csgn` Signum function: `csgn(x)`. 17, 68, 91.

`curl` Curl of 3D vector field **V** with respect to coordinate variables `x, y, z`: `curl(V, [x,y,z])` (in "`linalg`" package). 225.

`cylinderplot`
3D plot of surface $r = f(\theta, z)$ in cylindrical coordinates: `cylinderplot(f(theta,z), theta=a..b, z=c..d)` (in "`plots`" package). 190, 193.

`CYLINDRICAL`
Choice for "`coords`" in 3D plotting functions: `coords=CYLINDRICAL`. 214.

`D` Differentiation operator for functions: `D(f)`. 32, 33, 36, 152, 195, 199, 227, 233. See also "`diff`".

`definite` Test whether square matrix $M$ is (positive or negative) definite or semidefinite: `definite(M,t)` where `t` is `negative_def`, `negative_semidef`, `positive_def` or `positive_semidef` (in "`linalg`" package). 169.

`denom` Denominator of fraction (after applying "`normal`" if necessary): `denom(a/b)`. 39.

`densityplot` Density plot of expression in two variables $x$ and $y$: `densityplot(f(x,y), x=a..b, y=c..d)` (in "`plots`" package). 148.

`DEplot` Plot direction field and/or trajectories for differential equation or system of differential equations: `DEplot(de, y(x), x=a..b, {[y(x0)=y0]}, y=c..d)` or `DEplot(system, [x(t),y(t)], t=a..b, {[x(t0)=x0, y(t0)=y0]}, x=c..d, y=e..f)` (in "`DEtools`" package). 207, 235, 237, 238, 240, 243, 245.

`DEplot3d` Plot trajectories in 3 dimensions for system of differential equations: `DEplot3d(system, [x(t), y(t), z(t)], t=a..b, {[x(t0)=x0, y(t0)=y0,z(t0)=z0]}, scene=[x,y,z])`. 245.

`det` Determinant of square matrix $M$: `det(M)` (in "`linalg`" package). 142, 189.

`DEtools` Package containing commands dealing with differential equations, including "`DEplot`" and "`DEplot3d`": `with(DEtools)`. 207, 235, 238.

`DIAMOND` Choice for "`symbol`" when plotting points: `symbol=DIAMOND`. 75.

`diff` Derivative or partial derivative of expression $f$ with respect to variable $x$: `diff(f,x)`. 32-34, 36, 152, 195, 227, 233. See also "`D`".

`Digits` Variable: the number of digits used in floating-point numbers. 4, 25, 77, 257.

`dilog` Dilogarithm function, defined as $\text{dilog}(x) = \int_1^x \frac{\ln t}{1-t}dt$: `dilog(x)`. 187.

`dirgrid` Option in "`DEplot`" to specify dimensions of the rectangular grid of direction field arrows: `dirgrid=[m,n]` for $m$ arrows in the $x$ direction and $n$ in the $y$ direction. 235.

`discont=true` Option in "`plot`" to avoid vertical segments where the function plotted is discontinuous. 18, 257.

`display` Show several 2D or 3D plot structures together, or as an animation: `display({plot1, plot2})` or `display([plot1, plot2], insequence=true)` (in "`plots`" package). 12, 18, 32, 56, 57, 130, 146, 150, 184, 196.

`diverge` Divergence of vector field **V** with respect to coordinate variables `x`, `y`, `z`: `diverge(V,[x,y,z])` (in "`linalg`" package). 225.

`do` Introduces commands to be done in a "`for`" loop:
```
for ii from 1 to 5 do
    f[ii]:= ii * f[ii-1] od.
```
55.

`dotprod` Dot product of vectors **A** and **B**: `dotprod(A,B, orthogonal)` (in "`linalg`" package). 115. See also "`&.`".

`dsolve` Solve ordinary differential equation or system: `dsolve(de, y(x))` for general solution, `dsolve({de, y(x0)=y0}, y(x))` with initial condition. 233, 234, 240, 242, 243.

`eval` Evaluate an expression: `eval(f)`. 41, 169.

`evalf` Evaluate an expression numerically: `evalf(f)`, or `evalf(f,d)` to use `d` digits. 4, 41, 55, 66, 84, 98, 109, 120, 164, 186.

`evalm` Evaluate an expression involving matrices and/or vectors: `evalm(F)`. 115, 141, 225.

`evalr` Evaluate an expression involving intervals: `evalr(sin(INTERVAL(a..b)))` (requires `readlib(evalr)`). 53, 58.

`evl` Evaluate an expression involving vectors and convert the result to a list: `evl(f)` (in "`dot.def`"). 115, 209m 225, 247.

`exp` Exponential function: `exp(x)`. 68.

`expand` Expand an expression. Expands products or powers of sums, and expresses trigonometric functions of sums or multiples of angles in terms of functions of the individual angles: `expand((x+y)^3 + sin(x+y))`. 5, 19, 70, 82, 90.

`explicit` Option in "`dsolve`" to find explicit rather than implicit expression for the solution (if possible). 234.

`false` Logical constant (opposite of "`true`"). 169.

`fieldplot` Plot a vector field in 2 dimensions using arrows: `fieldplot([f(x,y), g(x,y)], x=a..b, y=c..d)` (in "`plots`" package). 206.

`fieldplot3d` Plot a 3D vector field using arrows: `fieldplot3d([f(x,y,z), g(x,y,z), h(x,y,z)], x=x1..x2, y=y1..y2, z=z1..z2)` (in "`plots`" package). 206.

`filled = true` Option in "`contourplot`" to fill in regions between contours with various colours. 148

`flux` Integrate normal component of 3D vector field over surface: `flux([f(x,y,z), g(x,y,z), h(x,y,z)], [x=p(u,v), y=q(u,v), z=r(u,v)], u=a..b, v=c..d)` (in "`vecints.def`"). 222, 223, 249.

`Flux` Inert version of "`flux`" (in "`vecints.def`"). 223, 249.

`for` Do one or more commands in a loop, with index variable taking a range of integer values: `for ii from 1 to 5 do f[ii]:= ii * f[ii-1] od`. 55, 255.

`foreuler` Choice for "`method`" in "`dsolve(..., numeric)`", for Euler method: `method=classical[foreuler]`. 240.

`FRAMED` Choice for "`axes`" in plot commands, to draw axes along the bottom and left edges of a 2D plot, or three edges of a 3D plot: `axes=FRAMED`. 26, 148.

`frames` Option in "`animate`" and "`animate3d`" to specify number of pictures to produce: `frames=10)`. 12, 150.

`fsolve` Solve an equation or set of equations numerically: `fsolve(a(x)=b(x))` if $x$ is the only variable, `fsolve(a(x)=b(x), x)` solves for $x$, `fsolve(a(x)=b(x), x, x=c..d)` looks for a solution in the interval $(c, d)$, `fsolve({a(x,y)=b(x,y), c(x,y)=d(x,y)}, {x,y})` solves two equations in two unknowns $x$ and $y$. 5, 6, 39, 54, 65, 66, 73, 77, 202, 254.

`fulldigits` Option in "`fsolve`" to perform all intermediate calculations with "`Digits`" digits. 77.

`grad` Gradient of a vector field: `grad([f(x,y),g(x,y)], [x,y])` where $x$ and $y$ are the coordinate variables (in "`linalg`" package). 154, 225.

`grid` Option in several plotting commands to specify dimensions of the rectangular grid of points where the functions will be evaluated: `grid=[m,n])` to use $m$ values of $x$ and $n$ values of $y$. 67, 145, 149, 206, 214, 235.

`hessian` Hessian matrix of an expression in several variables: `hessian(f(x,y,z), [x,y,z])` (in "`linalg`" package). 168.

`heunform` Choice for "`method`" in "`dsolve(..., numeric)`", for Improved Euler method: `method=classical[heunform]`. 242.

`HIDDEN` Choice for "`style`" in 3D plot commands, for grid lines drawn on an opaque surface. `style=HIDDEN`. 145, 149, 214.

`ifactor` Prime factorization of an integer: `ifactor(12345)`. 13.

`implicitplot` Plot one or more curves given by implicit equations in two variables: `implicitplot(f(x,y)=g(x,y), x=a..b, y=c..d)` (in "`plots`" package). 10, 12, 37, 67, 171.

`implicitplot3d` Plot one or more surfaces given by implicit equation in three variables: `implicitplot3d(f(x,y,z) = g(x,y,z), x=a..b, y=c..d, z=e..f)` (in "`plots`" package). 149.

`imptaylor` $n$ terms of Taylor series for $y(x)$ in powers of $x - a$ satisfying implicit equation $F(x, y) = c$ and $y(a) = b$: `imptaylor(F(x,y)=c, y,b,x,a,n)` (in "`imptay.def`"). 165, 247.

`infolevel` Cause information to be displayed when commands are performed: `infolevel[all]:= 5`. 109.

`insequence=true` Option in "`display`" to produce an animation from a list of plots. Note that this must come before any other options. `display(listofplots, insequence = true)`. 12, 57, 150.

`int` Definite or indefinite integration: `int(f(x), x)` or `int(f(x), x=a..b)`. 85, 90, 98, 183, 186, 257.

`Int` Inert form of "`int`". 90, 98, 186.

`INTERVAL` Used in "`evalr`" input and output to denote an interval: `INTERVAL(a..b)` for $[a, b]$. 53.

`intparts` Integrate by parts, $\int u\,dv$ becomes $uv - \int v\,du$: `intparts(Int(x^2*sin(x),x), x^2)` (in "`student`" package). 92.

`jacobian` Jacobian matrix of a list of expressions (vector field): `jacobian([f(x,y), g(x,y)], [x,y])` (in "`linalg`" package). 161, 189.

`labels` Option in plotting commands for text of labels on axes: `labels=[h,v]` to put `h` on $x$ axis and `v` on $y$ axis. 184.

`LambertW` Lambert's W function, satisfying $W(z)\exp(W(z)) = z$: `LambertW(z)`. 72.

`laplacian` Laplacian of expression in several variables: `laplacian(f(x,y), [x,y])` (in "`linalg`" package). 225.

`left` Option in "`limit`" for one-sided limit $x \to a-$: `limit(f(x), x=a, left)`. 28, 256.

`leftbox` Plot $f(x)$ and approximating rectangles for $\int_a^b f(x)\,dx$ using left sum with $n$ intervals: `leftbox(f(x), x=a..b, n)` (in "`student`" package). 83.

`leftsum` Unevaluated left sum for approximating $\int_a^b f(x)\,dx$ using $n$ intervals: `leftsum(f(x), x=a..b, n)` (in "`student`" package). 83, 99.

`len` Length of a vector:. `len([a,b,c])` (in "`dot.def`"). 115, 247.

`lhs` Left side of an equation: `lhs(a = b)`. 202.

`light` Option in 3D plot commands for a light source: `light=[phi,theta,r,g,b]` for direction angles `phi` and `theta` (degrees) and red, green, blue intensities `r, g, b`. 145.

`lightmodel` Option in 3D plot commands for predefined lighting options: `lightmodel=light4`. 145.

`limit` Limit of an expression $f(x)$ as $x \to a$: `limit(f(x), x=a)`. 22, 28, 256.

`linalg` Package containing commands for working with vectors and matrices: `with(linalg)`. 115, 139, 141, 142, 154, 161, 168, 169, 189, 209, 225, 230.

`LINE` Choice for "`arrows`" in "`DEplot`", "`fieldplot`" and "`fieldplot3d`", drawing them as straight lines: `arrows=LINE`. 206.

`linecolour` Option in "`DEplot`" to specify colour of solution curves: `linecolour=red`. 207.

`linestyle` Option in curve and line plotting commands to draw solid or various types of dashed and dotted lines: `linestyle=1` (solid), `2`, etc. 54.

`linsolve` Solve system of linear equations in matrix form: `linsolve(A, B)` solves $A\mathbf{X} = \mathbf{B}$ where $A$ is a matrix and $\mathbf{B}$ a vector (in "`linalg`" package). 142.

`ln` Natural logarithm function: `ln(x)`. 68.

`log` General logarithm function: `log[b](x)` is logarithm of $x$ to base $b$, `log(x)` is the natural logarithm. 68.

`lower` Used in "`rsum`" and "`rbox`" to produce lower Riemann sum: `lower(i)` is minimum value on $i$th subinterval (in "`riemann.def`"). 87, 247.

`map` Apply a function $f$ to each operand of an expression, such as a list or equation: `map(f, [a,b,c])` produces $[f(a), f(b), f(c)]$
`map(f, a = b, c)` uses $c$ as second input to $f$, producing $f(a, c) = f(b, c)$. 86, 118, 170, 225.

`matrix` Produce a matrix from a list of lists: `matrix([[a,b],[c,d]])` produces $\begin{pmatrix} a & b \\ c & d \end{pmatrix}$ (in "`linalg`" package). 141.

`max` Find the maximum of two or more numerical expressions: `max(E, Pi, 3)`. 86.

`method` Option in "`dsolve(...,numeric)`" to specify which numerical method to use.
`dsolve({de, y(x0)=y0}, y(x), numeric, method = classical[heunform])`. 240, 242, 243.

`middlebox` Plot $f(x)$ and approximating rectangles for $\int_a^b f(x)\,dx$ using middle sum with $n$ intervals: `middlebox(f(x), x=a..b, n)` (in "`student`" package). 83.

`middlesum` Unevaluated middle sum for approximating $\int_a^b f(x)\,dx$ using $n$ intervals: `middlesum(f(x), x=a..b, n)` (in "`student`" package). 83, 99.

`min` Find the minimum of two or more numerical expressions: `min(E, Pi, 3)`. 86.

`mtaylor` Taylor series in several variables: `mtaylor(f(x,y), [x=a, y=b], n)` for series in powers of $x - a$ and $y - b$ to truncation order $n$ (requires `readlib(mtaylor)`). 162, 176.

`negative_def`
`negative_semidef`
Options for "`definite`", to test for negative (semi)definite matrices: `definite(M,negative_def)`. 169.

`NONE` Choice for "`axes`" in plot commands, "`arrows`" in "`DEplot`", "`shading`" in 3D plot commands: `arrows=NONE`. 145, 148, 207, 237.

`NORMAL` Choice for "`axes`" in plot commands, to draw axes going through the origin (if that is in the plotted region). `axes=NORMAL`. 148.

`normal` Simplify expression by using common denominator and cancelling common factors of numerator and denominator: `normal(x/y + y/x)`. 19, 35, 159, 210.

`not` Logical operator: `not a` is true when `a` is false. 169.

`numer` Numerator of a fraction (after applying "`normal`" if necessary): `numer(a/b)`. 39.

`numeric` Option in "`dsolve`" to use numerical methods: `dsolve({de, y(x0)=y0}, y(x), numeric)`. 240, 243.

`numpoints` Option in plot commands for minimum number of points plotted: `numpoints=25`. 112.

`obsrange` Option in "`DEplot`" to stop calculating the solution as soon as it leaves the plotting region: `obsrange=true`. 237.

`od` Marks the end of a "`for`" loop:
`for ii from 1 to 5 do f[ii]:= ii; od`. 55, 255.

`op` Extract an operand or operands from an expression:
`op(2, a+b+c)` produces `b`
`op([a,b,c])` produces `a,b,c`. 86, 95, 172, 178.

`Open` In "`assume`", specifies an endpoint of a "`RealRange`" that is not included in the interval: `assume(a, RealRange(Open(0),1))`. 68.

`or` Logical operator: `a or b` is true when a or b or both are true. 169.

`Order` Variable: the default truncation order for "`taylor`" and "`mtaylor`". 129, 162.

`orientation` Option in all 3D plotting commands for viewpoint direction: `orientation=[theta, phi]` views the plot from a point at longitude `theta` degrees and colatitude `phi` degrees. 144, 196.

`orthogonal` Option in "`dotprod`", avoids taking complex conjugates of the second input: `dotprod(A, B, orthogonal)`. 115.

`parfrac` Option in "`convert`" to convert a rational function to partial fractions: `convert(f(x)/g(x), parfrac, x)`. 94.

`PATCH` Choice for "`style`" in 3D plotting commands, for shaded or coloured surface with grid lines. `style=PATCH`. 145, 150, 214.

`PATCHCONTOUR` Choice for "`style`" in 3D plotting commands, for shaded or coloured surface with curves of constant $z$. `style=PATCHCONTOUR`. 145, 150.

`PATCHNOGRID` Choice for "`style`" in 3D plotting commands, for shaded or coloured surface without grid lines. `style=PATCHNOGRID`. 145, 149, 150.

`Pi` The ratio of circumference to diameter of a circle, $\pi = 3.14159....$ 13, 252.

`piecewise` Defines a function using different formulas on different intervals: `piecewise(x<0, f(x), g(x))` is $f(x)$ when $x < 0$, $g(x)$ otherwise. 18.

`plot` Make a 2-dimensional plot of one or more curves:
`plot(f(x), x=a..b)` for $y = f(x)$
`plot({f(x), g(x)}, x=a..b)` for two curves
`plot(f(theta), theta=a..b, coords=polar)` for $r = f(\theta)$ in polar coordinates
`plot([x(t), y(t), t=a..b])` for a parametric curve.
`plot([[x1,y1],[x2,y2],[x3,y3]])` for straight lines joining points. 6, 10–12, 18, 56, 73, 111, 116, 117, 198, 254, 257.

`plot3d` Make a 3D plot of one or more surfaces:
`plot3d(f(x,y), x=a..b, y=c..d)` for $z = f(x, y)$
`plot3d([x(u,v), y(u,v), z(u,v)], u=a..b, v=c..d)` for a parametric surface. 144, 191, 214.

`plots` Package containing various plotting functions: `with(plots)`. 10, 12, 19, 32, 37, 112, 144, 199-195, 199, 206, 217, 236.

`plottools` Package containing various tools for plotting, including "`transform`": `with(plottools)`. 68.

`POINT` Choice for "`style`" in plots, to plot individual points rather than curves, and choice for "`symbol`" to plot individual dots rather than symbols: `style=POINT, symbol=POINT`. 60, 75, 112, 145, 180.

`pointplot3d` Plot a list of points in 3D: `pointplot3d(L)` (in "`plots`" package). 246.

`positive_def`
`positive_semidef` Options for "`definite`", to test for positive (semi)definite matrices: `definite(M,positive_def)`. 169.

`potential` Check whether vector field $\mathbf{F}(x, y, z)$ has a scalar potential, and if so compute a potential: `potential(F(x,y,z), [x,y,z],'V')` returns `true` if potential exists and assigns potential to "`V`", otherwise returns `false` (in "`linalg`" package). 209, 210.

`print` Print the value of any expression: `print(x)`. 254.

`quit` End the Maple session: `quit`. 3.

`quo` Divide polynomial $p(x)$ by polynomial $q(x)$: `quo(p(x), q(x), x)` returns the quotient, `quo(p(x), q(x), x, 'r')` also assigns the remainder to variable "`r`". 74.

`radius` Option in "`tubeplot`" to specify radius of tube (can be a constant or a function of the parameter): `radius=r(t)`. 195.

`rbox` Plot a sequence of rectangles for a Riemann sum for $\int_a^b f(x)\,dx$: `rbox(upper)` for upper sum, `rbox(lower)` for lower sum (in "`riemann.def`"). 87, 247.

`Re` Real part of a complex expression: `Re(z)`. 71.

`read` Read the contents of a file (either a text file or an "`.m`" file) into the session: `` read `myfile.def` ``. 9.

`readlib` Read a file from the Maple library to define a certain name: `readlib(mtaylor)`. 9, 254.

`real` In "`assume`", indicates that a variable is real: `assume(x, real)`. 68.

`RealRange` In "`assume`", indicates an interval in which a variable's values are assumed to lie: `assume(x, RealRange(a,b))`. 68.

`rect` Make a list of corners of a rectangle, for use by "rbox": `rect(x1, x2, y)` for lower left corner $[x_1, 0]$ and upper right $[x_2, y]$ (in "riemann.def"). 87, 247.

`restart` Restart Maple as if in a new session: `restart`. 9, 30, 254.

`rhs` Right side of an equation: `rhs(a=b)`. 202, 221.

`right` Option in "limit" for one-sided limit $x \to a+$: `limit(f(x), x=a, right)`. 28, 256.

`rightbox` Plot $f(x)$ and approximating rectangles for $\int_a^b f(x)\,dx$ using right sum with $n$ intervals: `rightbox(f(x), x=a..b, n)` (in "student" package). 83.

`rightsum` Unevaluated right sum for approximating $\int_a^b f(x)\,dx$ using $n$ intervals: `rightsum(f(x), x=a..b, n)` (in "student" package). 83, 99.

`rk4` Choice for "method" in "dsolve(..., numeric)", for fourth-order Runge-Kutta method: `method=classical[rk4]`. 242.

`rkf45` Default choice for "method" in "dsolve(..., numeric)", for an adaptive Runge-Kutta method: `method=rkf45`. 243.

`RootOf` Used to represent roots of an equation: `RootOf(p(x))` if "x" is "_Z" or the only variable present, `RootOf(p(x),x)` to use variable "x", `RootOf(p(x),x,a)` for the root near $x = a$. 5, 66, 73, 163, 172.

`rsum` Calculate a Riemann sum for $\int_a^b f(x)\,dx$: `rsum(upper)` for upper sum, `rsum(lower)` for lower sum (in "riemann.def"). 87, 247.

`save` Save function definitions and variable values in a file: `save f, g, 'myfile.def'`, or `save 'myfile.m'` to save all defined values. 8, 9.

`scaling` Option in plot commands to make equal scales on the axes: `scaling = constrained`. 11, 52, 145.

`scene` Option in "DEplot" or "DEplot3d" to specify which variables are plotted on which axes. `scene = [t,x]`. 243.

`sec` Secant function: `sec(x)`. 19, 21.

`select` Extract the operands of an expression (typically a list) satisfying a condition: `select(between, [1,3,4,6], 2, 5)` returns `[3,4]`. 86.

`seq` Produce sequence of expressions, one for each value of a loop variable: `seq(k^2, k=2..4)`. 11, 12, 60, 61.

`shading` Option in 3D plotting commands for colouring or shading curves or surfaces: `shading = XYZ`. Also in "leftbox" etc. for colours of rectangles: `shading=red`. 84, 145, 150, 194.

`signum` Signum or sgn function: `signum(x)`. 17, 33, 252.

`simplify` Make an expression simpler, using algebraic cancellation, trigonometric identities, etc.: `simplify(expr)`. `simplify(expr, symbolic)` uses simplification rules such as $\sqrt{x^2} = x$ without worrying about signs. `simplify(expr, trig)` uses trigonometric identities but does not try to simplify square roots. 5, 19, 68, 92, 108, 210.

`simpson` Unevaluated Simpson's Rule approximation for $\int_a^b f(x)\,dx$ using $n$ (an even number) intervals: `simpson(f(x), x=a..b, n)` (in "student" package). 102.

`sin` Sine function: `sin(x)`. 19.

`SLIM` Choice for "arrows" in "DEplot", "fieldplot" and "fieldplot3d", drawing lines with triangular heads. 206.

`slineint` Line integral $\int_C f\,ds$ of a scalar field along a parametric curve: `slineint(f(x,y), [x=g(t), y=h(t)], t=a..b)` (in "vecints.def"). 220, 249.

`Slineint` Inert form of "slineint" (in "vecints.def"). 221, 249.

`solve` Solve an equation or system of equations for a variable or variables: `solve(a(x)=b(x))` (where there is just one variable $x$) `solve(a(x)=b(x), x)` (solve for $x$) `solve({a(x,y)=b(x,y), c(x,y)=d(x,y)}, {x,y})` (solve system for $x$ and $y$). 5, 13, 31, 38, 65, 73, 257.

`spacecurve` Plot one or more curves in 3D: `spacecurve([x(t),y(t),z(t)], t=a..b)` (in "plots" package). 194.

`sphereplot` Plot one or more surfaces in spherical coordinates: `sphereplot(f(phi,theta), theta=a..b, phi=c..d)` (in "plots" package). 193, 217.

`sqrt` Square root function: `sqrt(x)`. 13, 17.

`stepsize` Option in "DEplot" and "dsolve(..., numeric, method=classical[...])" for interval between values of the independent variable at which the approximate solution is calculated: `stepsize=0.1`. 208, 240.

`student` Package containing commands for manipulating derivatives, sums and integrals: `with(student)`. 82-84, 90, 92, 99, 100, 102, 220.

`style` Option in plotting commands for the way surfaces are shown, or to plot points: `style=PATCH`. 60, 112, 145, 149, 150, 214.

`subs` Substitute equations into expression, replacing each occurrence of a left side of an equation by the right side: `subs(a=b, c=d, f(a,c))`. 15, 21, 38, 41, 73, 215, 221, 240.

`sum` Sum of an expression, where an index variable takes a range of integer values: `sum(f(i), i=a..b)`. 13, 82, 102, 120.

`Sum` Inert version of "sum". 121.

`surd` "Elementary" $n$-th root of an expression, where odd roots of negative numbers are negative: `surd(x, n)`. 257.

`surfint` Integrate a scalar field over a parametric surface: `surfint(f(x,y,z), [x=p(u,v), y=q(u,v),z=r(u,v)], u=a..b, v=c..d)` (in "`vecints.def`"). 222, 249.

`Surfint` Inert form of "`surfint`" (in "`vecints.def`"). 222, 249.

`symbol` Option for plotting with "`style=POINT`" to specify what is drawn at each plotted point: `symbol=CIRCLE`. 75, 112, 180.

`symbolic` Option in "`simplify`" to allow indiscriminate use of the simplification $\sqrt{x^2} = x$: `simplify(expr, symbolic)`. 92, 108.

`tan` Tangent function: `tan(x)`. 19, 21.

`taylor` Taylor series expansion of expression about $x = a$ to order $O(x^n)$: `taylor(f(x), x=a, n)`. 129, 163, 226.

`textplot` Place one or more text labels in a two-dimensional plot: `textplot([x, y, 'Hello'])` (in "`plots`" package). 32, 112.

`textplot3d` Place one or more text labels in a 3D plot: `textplot3d([x, y, z, 'Hello'])` (in "`plots`" package). 199.

`THICK` Choice for "`arrows`" in "`DEplot`", "`fieldplot`" and "`fieldplot3d`": `arrows=THICK`. 206.

`thickness` Option in plotting commands for thickness of lines: `thickness=0` (default), `1`, `2` or `3` (thickest). 171, 195, 207.

`THIN` Choice for "`arrows`" in "`DEplot`", "`fieldplot`" and "`fieldplot3d`", drawing lines with one barb: `arrows=THIN`. 206.

`transform` Generates a function to transform a plot by applying a function to each point: `transform((x,y) -> [y,x])(pl)` (in "`plottools`" package). 68.

`transpose` Transpose a matrix: `transpose(M)`. 141.

`trapezoid` Unevaluated Trapezoid Rule approximation for $\int_a^b f(x)\,dx$ using $n$ intervals: `trapezoid(f(x), x=a..b, n)` (in "`student`" package). 100.

`trig` Option in "`combine`" and "`simplify`" to apply trigonometric identities: `combine(expr, trig)` writes products and powers of trigonometric functions as functions of sums, differences and multiples of angles, `simplify(expr, trig)` uses $\sin^2 x = 1 - \cos^2 x$. 70, 202.

`true` Logical constant (opposite of "`false`"). 169.

`tubeplot` Plot one or more tubes about parametric 3D curves: `tubeplot([x(t),y(t),z(t)], t=a..b, radius = r(t))` (in "`plots`" package). 195, 218.

`tubepoints` Option in "`tubeplot`" for number of points around the tube: `tubepoints=n`. 195.

`unapply` Make a function from an expression involving a variable: `unapply(expr, x)`. 15, 21, 33, 55.

`unit` Unit vector (list) in the same direction as a given vector: `unit(V)` (in "`dots.def`"). 115, 247.

`upper` Used in "`rsum`" and "`rbox`" to produce upper Riemann sum: `upper(i)` is maximum value on $i$th subinterval (in "`riemann.def`"). 86, 247.

`value` Perform inert commands: `value(Int(f(x),x))`. 83, 84, 221, 223.

`vecpotent` Check whether vector field $\mathbf{F}(x, y, z)$ has a vector potential, and if so compute one: `vecpotent(F(x,y,z), [x,y,z], 'V')` returns `true` if vector potential exists and assigns vector potential to "V", otherwise returns `false` (in "`linalg`" package). 230.

`vlineint` Line integral $\int_C \mathbf{F} \bullet d\mathbf{r}$ of tangential component of a vector field along a parametric curve: `vlineint([F1(x,y),F2(x,y)], [x=g(t), y=h(t)], t=a..b)` (in "`vecints.def`"). 222, 249.

`Vlineint` Inert form of "`vlineint`" (in "`vecints.def`"). 222, 226, 249.

`whattype` Return the type of any expression: `whattype(expr)`. 251.

`WIREFRAME` Choice for "`style`" in 3D plot commands to plot grid lines in transparent surface: `style=WIREFRAME`. 145, 149, 214.

`with` Define some or all commands from a library package: `with(linalg)` or `with(linalg, det, definite)`. 11, 37, 254.

`XY` Choice for "`shading`" in 3D plot commands, where colour depends on $x$ and $y$ coordinates: `shading=XY`. 145.

| | |
|---|---|
| `XYZ` | Choice for "`shading`" in 3D plot commands, where colour depends on all three coordinates: `shading=XYZ`. 145. |
| `Z` | Choice for "`shading`" in 3D plot commands, where colour depends on $z$ coordinate: `shading=Z`. 145, 195. |
| `ZGREYSCALE` | Choice for "`shading`" in 3D plot commands, where darkness depends on $z$ coordinate: `shading=ZGREYSCALE`. 145, 195. |
| `ZHUE` | Choice for "`shading`" in 3D plot commands, where hue depends on $z$ coordinate: `shading=ZHUE`. 145. |

# APPENDIX VI
# Index